Geography of the Physical Environment

The *Geography of the Physical Environment* book series provides a platform for scientific contributions in the field of Physical Geography and its subdisciplines. It publishes a broad portfolio of scientific books covering case studies, theoretical and applied approaches as well as novel developments and techniques in the field. The scope is not limited to a certain spatial scale and can cover local and regional to continental and global facets. Books with strong regional focus should be well illustrated including significant maps and meaningful figures to be potentially used as field guides and standard references for the respective area.

The series appeals to scientists and students in the field of geography as well as regional scientists, landscape planners, policy makers, and everyone interested in wide-ranging aspects of modern Physical Geography. Peer-reviewed research monographs, edited volumes, advance and undergraduate level textbooks, and conference proceedings covering the major topics in Physical Geography are included in the series. Submissions to the Book Series are also invited on the theme 'The Physical Geography of…', with a relevant subtitle of the author's/editor's choice. Please contact the Publisher for further information and to receive a Book Proposal Form.

Balai Chandra Das · Aznarul Islam ·
Biplab Sarkar

The Nadia Rivers

Transition to Transformation

Springer

Balai Chandra Das
Department of Geography
Barasat Government College
Barasat, West Bengal, India

Aznarul Islam
Department of Geography
Aliah University
Kolkata, West Bengal, India

Biplab Sarkar
Department of Geography
Aliah University
Kolkata, West Bengal, India

ISSN 2366-8865 ISSN 2366-8873 (electronic)
Geography of the Physical Environment
ISBN 978-3-031-85970-0 ISBN 978-3-031-85971-7 (eBook)
https://doi.org/10.1007/978-3-031-85971-7

Cover image by Sonja Weber, München

This Springer imprint is published by the registered company Springer Nature Switzerland AG
The registered company address is: Gewerbestrasse 11, 6330 Cham, Switzerland

Confluence of Bhagirathi River (muddy water) and Jalangi River at Swarupganj (2023).
Photo received from District Magistrate's Office, Nadia, West Bengal.

To the enduring spirit of Bengal's rivers, guardians of history and harbingers of change. May their timeless flow inspire us to cherish our past, navigate our present, and safeguard our future!

Foreword by Prof. Hubert Chanson

This monograph *The Nadia Rivers: Transition to Transformation* includes all of the information required to gain a solid physical understanding of the riverine systems in the eastern part of the Indian subcontinent. It combines the geographical context, geological and geomorphological details, and sedimentological insights of the fluvial hydraulic networks. It is complete and self-contained—there is no need to refer to other sources.

Rivers play a major role in shaping the landscapes of our planet. Since Antiquity, rivers have worldwide served as the locus of human development, a major source of freshwater and irrigation water supply for agriculture, as well as an important transportation network. The Nadia Rivers are an obvious illustration. To date, our society has benefited from the numerous services that the rivers offer. River engineers, scientists and geomorphologists play an important role in our society although the technical challenges are enormous, encompassing multiphase flows and interactions with aquatic life and biological forms. The extreme challenges are closely linked to:

- the geometric scale of river systems, from a few square meters for some soil erosion pattern to over 10^6 km^2 for a large river catchment;
- the broad range of relevant time scales, from less than a millisecond for a turbulent burst next to the river bed to a few years for river meandering (Trevethan et al. 2007, Parker et al. 2011);
- the variability of river discharges (IAHS 1984, Herschy 2002);
- the complexity of the governing equations of multiphase flow fluid dynamics and complicated interactions between water, sediments and aquatic life (Chanson 2004); and
- the total dependence of mankind on water.

Over the past two centuries, however, human developments have generated stresses and pressures on the water systems, with threats to their environmental conditions and health. There is an urgent need for scientific resources on our water systems, such as this monograph. The Nadia Rivers are a large trans-boundary river system connecting the Ganga-Padma River to the Bhagirathi-Hooghly River, with waters flowing from Himalayan region through India and Bangladesh. The 21st Century is facing political instabilities centred around freshwater systems and marine environments. A detailed scientific survey of these resources is critical and this book brings together an in-depth understanding of a complex freshwater system around which several millions of people live and rely upon the Nadia Rivers.

All in all, one cannot miss the tremendous enthusiasm the authors have for the Nadia Rivers system. After an introduction and explanatory context, Chaps. 2–4 present in details the three main river systems. The multifaceted investigations encompass a very wide range of parameters and relevant factors at multi levels. The spatial scales cover three levels: 'site', 'reach', and 'catchment'. The temporal scales regroup some important historical background, as well as short-term, medium-term and long-term periods. Such an approach is fascinating and very relevant to the present and future. Each chapter contains a wealth of information, supported by some very nice illustrations and detailed bibliographic references. Chapter 5 develops on the decays of the Nadia Rivers' health and makes a case for restoration of the Nadia Rivers system through integrated river basin management. This is not trivial considering the size of the water systems and the broad range of relevant stakeholders. The final chapter offers an overview of the future challenges and there are no doubt many more.

January 2025

Hubert Chanson
Professor of Civil Engineering
The University of Queensland
Brisbane, Australia

References

Chanson H (2004) Environmental hydraulics of open channel flows. Elsevier-Butterworth-Heinemann, Oxford, UK, 483 pages

Herschy R (2002) The world's maximum observed floods. Flow Meas Instrum 13:231–235

IAHS (1984) World catalogue of maximum observed floods. IAHS—AISH Publication No. 143, Wallingford, UK, 384 pages

Parker G, Shimizu Y, Wilkerson GV, Eke EC, Abad JD, Lauer JW, Paola C, Dietrich WE, Voller VR (2011) A new framework for modeling the migration of meandering rivers. Earth Surf Process Land 36:70–86

Trevethan M, Chanson H, Takeuchi M (2007) Continuous high-frequency turbulence and sediment concentration measurements in an upper estuary. Estuar Coastal Shelf Sci 73(1–2):341–350. 10.1016/j.ecss.2007.01.014

Foreword by Joann Mossa

I appreciate the invitation by Dr. Balai Chandra Das to write a Foreword to the book entitled *The Nadia Rivers: Transition to Transformation*.

This book delves into the historical, geographical, and socio-economic dimensions of the Nadia Rivers, focussing on the distinct but integrated Bhagirathi, Jalangi, and Mathabhanga-Churni rivers, all distributaries of the Ganges and part of its delta plain in eastern India. It explores the evolution of these rivers from the 15th century to present, examining the history and importance of these rivers in literature, navigation, and trade. Topics addressed include the etymology of river names, the portrayal of rivers in epics, changes in forms and hydraulics over time, engineering works, anthropogenic changes, environmental challenges, and the evolving socio-economic structure along the riverbanks.

Our understanding of rivers greatly advances through studies in new areas, where technological advancements, interdisciplinary approaches, and fresh perspectives lead to improved knowledge. The authors and colleagues have been working in this region to fill this gap, starting with one paper in 1978 to fourteen publications in 2024. However, much of this work is at the reach scale, and little prior work examines these rivers at the catchment scale. Thus, this book is a culmination and synthesis of decades of work in this hydrological network.

Whereas some large rivers have centuries of reports and papers that describe their engineering and transformation to support navigation, the authors present a historical synthesis of the early efforts to improve navigability in these rivers. Siltation was a problem at the off takes from the river Ganges for centuries. Regular clearing of the river channel bed from sunken boats, loose timber, and large trees failed to maintain navigation. Many boats were damaged or wrecked while crossing the channels. The inability to establish safe passage meant that cargo-laden ships were often detained for prolonged periods, leading to substantial costs, particularly from demurrage. The British colonial rulers modified the rivers to support navigation and facilitate trade including channelization, embankments, artificial cutoffs, and other infrastructure projects. As elsewhere, such engineering 'improvements' required persistent effort and were problematic. More recently, a pivotal event transforming this system was completion of the Farakka Barrage in 1975. The synthesis of the Nadia rivers fills a void that will likely serve as a foundation for those working here and in nearby areas. The high-quality maps, graphs, and tables of important events or prior works provide background information that will assist others in conducting further work on these systems.

Many factors contribute to this being a dynamic area of interest to scientists and citizens. Tectonic activity, sediment transport, and fluvial processes make this an area of transformation. Flexural subsidence, localized compaction, and faulting control river gradients and sediment accommodation and lead to avulsions, channel migration, and meandering in response to sediment load variations. Low elevations in the delta and its high sediment yield across the delta plain amplify hydrological instability, contributing to avulsions, bifurcations and seasonal variations in flow regimes. Additionally, the shifting courses of these rivers are influenced by subsurface neotectonics and sediment siphoning through deep submarine canyons in the bay redistributes sediment loads within the delta. Further, monsoonal precipitation and anthropogenic interventions, renders the Nadia rivers highly sensitive to changes in sediment supply and hydrodynamic forcing, manifesting in their changing form and high bank erosion rates and channel bar formation across the Bengal delta.

Given the need to fill this geographic hole, I congratulate the authors on their completion of this book about the Nadia Rivers. The region's climate is warming, the rainfall is decreasing, and the rivers are drying. Emerging problems on the upstream end of the system include the growth of undesirable algae and aquatic plants, depletion of fish stocks, and ultimately livelihoods. On the downstream end of this system, threats include relative sea level rise, combining eustatic rise and subsidence, and saltwater intrusion.

Other human impacts include discharge of industrial and wastewater, and diverse land uses including the intensive use of land on riverbeds and adjacent floodplains for agriculture or settlement. The sustainable utilization of available water resources, fostering community engagement in activities such as fish farming, agriculture, aquaculture, duck farming, water tourism, and water entertainment is need in the context of Nadia Rivers. These rivers are necessary for livelihoods and water and food security, but problems with groundwater quantity and quality are emerging. Thus, the book advocates for proactive measures to manage, maintain and improve the condition of these rivers and their water resources sustainably to safeguard human populations and ecosystems.

The transboundary nature of the Mathabhanga-Churni River, crossing in the India-Bangladesh border, adds another complexity and a need for cooperation. While there are substantial socio-economic and environmental opportunities for both nations, the river faces considerable challenges including declining discharge, sedimentation and water pollution that hinder aquatic ecosystem functions.

The authors call for collaborations for data collection and monitoring the entire river system particularly for transboundary rivers, restoring and sustaining natural river functions including native vegetation plantation, riverbank stabilization with grasses, the creation of riparian buffer zones with minimal anthropogenic activity, and large-scale restoration projects with financial support from international collaboration and participation of the local community. The authors hope that the book may grow interest among future stakeholders and researchers for river modelling of varied types, and management and futuristic approach involving climate change, sea level rise and socio-economic changes and ultimately inform hazard management.

January 2025

Joann Mossa Ph.D.
Professor and Associate Chair of Geography
University of Florida
Gainesville, FL, USA

Preface

In the annals of Bengal's history, the Nadia Rivers stand as silent sentinels, bearing witness to the passage of time and the ebb and flow of human civilization. Born of ancient myths and shaped by centuries of interaction, the Bhagirathi, Jalangi, and Mathabhanga-Churni rivers have woven themselves into the very fabric of the region's landscape and identity.

The Nadia Rivers: Transition to Transformation is not merely a scholarly pursuit; it is a journey of rediscovery, a quest to unravel the mysteries enshrined within the waters of these iconic rivers. As we embark on this odyssey, we are guided by a profound sense of purpose—an imperative to fill the void of knowledge that has long shrouded the Nadia Rivers.

Our journey begins with a voyage into the depths of history, where the very name 'Nadia Rivers' resonates with echoes of antiquity. From the epics and legends that enshrine these waterways to the establishment of the Nadia River Division during British colonial rule, we trace the evolution of a landscape shaped by the currents of time. But our exploration does not end with the past; it extends into the present and beyond, delving into the geological, hydrological, and socio-economic dimensions of the Nadia Rivers. Through the lens of interdisciplinary inquiry, we examine the transition and transformation of these once-voluminous rivers into dwindling watercourses, and the impact of human intervention on their fragile ecosystems.

Central to our narrative is the story of rejuvenation and adaptation, epitomized by the Farakka barrage project and its profound implications for the hydrogeomorphic landscape of the region. As we unravel the complexities of this phenomenon, we confront questions of environmental sustainability, socio-economic resilience, and cultural identity.

Yet, amidst the challenges lies a glimmer of hope—a testament to the resilience of nature and the ingenuity of human endeavour. Through our meticulous analysis and critical inquiry, we seek not only to illuminate the past but to chart a course towards a more sustainable future for the Nadia Rivers and the communities they sustain.

This volume is not just for scholars and researchers; it is for anyone with a passion for rivers, deltas, and the intricate tapestry of human-environment interactions. Whether you are a planner, an educator, a student, or simply a curious observer, we invite you to join us on this journey of discovery.

For in the pages that follow, lies not just a chronicle of rivers, but a testament to the enduring spirit of exploration and the boundless possibilities of knowledge. Our endeavour will be fruitful if it sparks dialogue, inspires action, and fosters a deeper appreciation for the rivers that have shaped our past and hold the key to our future. Welcome to *The Nadia Rivers: Transition to Transformation*.

We are very much thankful to the respondents during the field survey and other stakeholders directly and indirectly associated with the data collection process. We kindly acknowledge the active and overwhelming cooperation of our laboratory members and students like Suman Deb Barman, Susmita Ghosh, Mofizul Hoque, and Sadik Mahammad for preparing some maps and diagrams. We are also equally thankful to our family members for their days of support and sacrifice throughout the entire journey. All the concerned persons and institutions from Krishnagar Government College, Barasat Government College, and Aliah University are duly acknowledged. We also show our deep gratitude to all the other persons who promptly replied to our request for study materials and data support.

We are deeply grateful to Hubert Chanson, Professor of Civil Engineering, The University of Queensland, Brisbane, Australia and Joann Mossa, Professor and Associate Chair of Geography, University of Florida, Gainesville, FL, USA for writing insightful 'Foreword's for this volume. We are grateful to Dieter Merkle, Vice President, Dr. Guido Zosimo-Landolfo, Editorial Director/Asset Manager of Springer Nature Switzerland AG, and Dr. Doris Bleier for signing the agreement on this project and for providing the opportunity to use their prestigious pages for manuscripts. We are also grateful to the production team for their meticulous efforts to publish the book as correctly as possible.

We hope the outcomes of this volume will flood the thoughts of scholars, faculties, planners, and stakeholders returning us the apt worth.

Barasat, West Bengal, India — Balai Chandra Das
Kolkata, West Bengal, India — Aznarul Islam
Kolkata, West Bengal, India — Biplab Sarkar
June 2025

Contents

About the Authors

Balai Chandra Das is a distinguished scholar in the field of Geography. He earned his bachelor's degree in 1992 and his master's degree in 1994 from the University of Burdwan, and later received his Ph.D. in Geography from the University of Calcutta. Over the course of his academic career, Dr. Das has authored more than 50 research articles published in a wide range of national and international journals, conference proceedings, and edited volumes. Dr. Das has served as the chief editor for several acclaimed books, including Neo-Thinking on Ganges Brahmaputra Basin Geomorphology, Quaternary Geomorphology in India—Case Studies from the Lower Ganga Basin, Himalayan Neotectonics and Channel Evolution, and Floods in the Ganga–Brahmaputra–Meghna Delta—all published by Springer International Publishing, Switzerland. He is also the lead editor of Anthropogeomorphology of Bhagirathi-Hooghly River System in India, published by CRC Press, Taylor & Francis Group. He has reviewed manuscripts for over seven international earth science journals from publishers such as Springer, Elsevier, and others, and is currently serving on the editorial boards of two international journals. His contributions to the global scientific community include serving as a member of the Scientific Committee for the IWC-2016 and WRAA-2020 conferences held at Sultan Qaboos University, Oman. Dr. Das's current research focuses on the fundamental geomorphology of rivers and lakes, reflecting his deep engagement with the dynamic processes that shape Earth's surface water systems.

Aznarul Islam completed Master of Science in Geography from Kalyani University, West Bengal and M.Phil. and Ph.D. in Geography from the University of Burdwan, West Bengal, India. He is an Assistant Professor in the Department of Geography, Aliah University (AU), Kolkata, India since 2016 and served as the Head (Officiating), Department of Geography, AU from November 2019 to February 2022. Previously he was an Assistant Professor (West Bengal Education Service) at the Department of Geography, Barasat Government College, West Bengal (2014–2016). Dr. Islam served as the Guest Faculty/visiting faculty in a few institutes like Jadavpur University, Kolkata, and Kazi Nazrul University, Asansol. He was awarded the UGC Junior Research Fellowship in Geography in 2012. Besides, he secured All India Rank 018/188 in Earth Sciences in CSIR-UGC NET in 2011. He has already published more than 100 research papers including >70 SCI journal articles. He is the principal author of the book *Riverbank Erosion in the Bengal Delta: An Integrated Perspective* published in 2024 by Springer. He co-authored the book '*Palaeochannels and Water Resource Management A Study in the Western Bengal Delta*' published in 2025 by Springer. He edited another 12 books and contributed more than 30 book chapters in edited volumes and conference proceedings. He has presented papers in more than 35 national and international seminars and conferences. He has successfully guided 4 Ph.D. theses; and 3 theses under preparation. He was awarded one UGC minor research project. He completed one major research project funded by the ICSSR, Government of India in 2019 as the sole Project Director. He received a Department of Science and Technology (Government of India)-Science and Engineering Research Board (DST-SERB) State University Research Excellence (SURE) Award in 2023 as the Principal Investigator. He has been performing the role of editor and reviewer in various journals published by Springer, Elsevier, Taylor and Francis, Wiley, Nature, MDPI, Frontiers etc. He is a life member of the Indian Geographical Foundation (IGF), Kolkata, and the National Association of Geographers, India (NAGI), New Delhi. His principal area of research includes exploring river, water and society through the lens of applied spatial sciences using quantitative techniques and geo-informatics for reducing fluvial and hydrological hazards to enhance community relicense.

Biplab Sarkar holds a Ph.D. in Geography from Aliah University, awarded in 2021 for his thesis "Decaying of River Churni and Its Impact on Ecology, Economy and Society." He completed his postgraduate and undergraduate studies in Geography at the University of Kalyani in 2014 and 2012, respectively. Dr. Sarkar specializes in fluvial geomorphology, with focused expertise in eco-geomorphology of alluvial rivers, floodplain hydrology, and floodplain ecosystem services. His research explores the complex interactions between river dynamics, human interventions, and environmental sustainability, particularly in the Bengal delta. A significant part of his research is devoted to understanding the environmental flow (e-flow) regimes and ecological responses of rivers within the Bengal delta, especially the Bhagirathi-Hooghly, Jalangi, and Mathabhanga-Churni river systems. Dr. Sarkar previously served as a Project Associate-I in a major research project titled "Spatial Mapping and Modelling of Soil and Groundwater Pollution by Nitrate from Agricultural Field in the Lower Ganga Delta of West Bengal," funded by the Department of Science and Technology – Science and Engineering Research Board, State University Research Excellence (DST-SERB SURE), New Delhi, Government of India. Dr. Sarkar has published more than 40 research articles in peer-reviewed SCI-indexed journals and contributed several book chapters. His work reflects a deep commitment to advancing both theoretical understanding and practical applications in environmental geography and river science.

Abbreviations

ASTER	The Advanced Spaceborne Thermal Emission and Reflection Radiometer
BC	Before Christ
BI	Braiding Index
BOD	Biological Oxygen Demand
BRB	Bhagirathi River Basin
C.D. Block	Community Development Block
COD	Chemical Oxygen Demand
CV	Coefficient of Variation
DO	Dissolved Oxygen
FBP	Farakka Barrage Project
FSTPP	Farakka Super Thermal Power Project
GBM	Ganga-Brahmaputra-Meghna
HMR	Hydraulic Mean Radius
HQ	Hazard Quotient
IWAI	Inland Waterway Authority of India
MGNREGA	Mahatma Gandhi National Rural Employment Guarantee Act
MH	Magnesium hazard
MW	Mega Watt
NRD	Nadia River Division
NTPC	National Thermal Power Corporation Ltd.
OPI	Overall Index of Pollution
PWD	Public Works Department
THI	Total Hazard Index
USEPA	United States Environmental Protection Agency
USSL	United States Salinity Laboratory
WQI	Water Quality Index

Scientific Notations

A	Cross-sectional area
A^*	First measure of Channel asymmetry by Knighton
A_1	Second measure of Channel asymmetry by Knighton
A_2	Third measure of Channel asymmetry by Knighton
am	Amplitude
d_{max}	Maximum depth
d'	Expected depth
d	Mean channel depth
D_{50}	Median diameter of sediment
FmI	Meander form index
P	Wetted perimeter
P_1	Poverty gap index
P_2	Poverty severity index
Q	Discharge
Q_m	Maximum discharge
R^2	Co-efficient of determination
r_c	Radius of curvature
R_e	Reynolds number
SmI	Meander shape index
v	Average velocity
w	Channel width
z	Poverty line
θ_a	Arc angle
Ω	Stream power
τ_0	Shear stress
Fr	Froude number

Table of Conversion

Area

1 Hectare	2.4711 Acre
1 Acre	0.4047 Hectare
1 Acre	3.025 Bigha
1 Hectare	7.4751 Bigha
1 Acre	4047 Sq. Metre
1 Acre	100 Satak
1 Acre	43.560 Sq. Ft
1 Bigha	20 Kotha

Weight

1 Maund	37.3242 Kg
1 Quintal	100 Kg
1 Quintal	2.6792 Maund
1 Maund	40 Kg

List of Figures

List of Tables

1 Nadia Rivers: Transition to Transformation Context

Abstract

This chapter delves into the etymology and significance of the term 'Nadia Rivers'. Initially, the meaning of 'Nadia' is explored, followed by a definition of 'Nadia Rivers' in terms of the specific rivers included in this group. The geological and sedimentological characteristics of the region traversed by the Nadia Rivers are then examined. During the era of British colonial rule, the Nadia River Division (NRD) of the Public Works Department (PWD) was entrusted with the maintenance of the Bhagirathi, Jalangi, and Mathabhanga-Churni rivers. The chronological evolution and works carried out by the NRD are thoroughly investigated, providing insights into the historical development of river management practices in the region. Furthermore, this chapter introduces integrated perspectives that serve as the overarching theme of the volume, emphasizing the multidisciplinary approach required to understand the complexities of river systems. The scales of past studies and the importance of studying Nadia Rivers from various perspectives, including historical, geographical, socio-economic, and cultural aspects, are also scrutinized. Understanding these rivers from diverse angles is essential for comprehensively appreciating their multifaceted significance in the broader context of regional development and environmental management.

Keywords

Nadia · Nabadwip · Nadia River Division · Bhagirathi River · Jalangi River · Mathabhanga-Churni River

1.1 Etymology

The word 'Nadia' carries several meanings and has been pronounced severally. Garrett (1910), and Das (2004) used the word synonymous with Nabadwip. 'The town Nadia or Navadvipa (meaning new island) from which the district takes its name, has a very ancient history' (Imperial Gazetteer of India, vol 18, p. 273). Minhāj-us-Sirāj's Tābākāt-i-Nāsiri, circa A.D. 1200, is the earliest source to mention Nadia, spelled as Nudiyā, in relation to Bakhtiyar Khilji's raid on the city (Majumder 1978). The post-Chaitanya Vaishnava literature and popular tradition identify Nadiā as Navadwip or Nabadwip. Chaudhury (2004) reported that Tābākāt-i-Nāsiri mentioned 'Naodiah' and Ain-i-Akbari mentioned 'Nadiamahal'. Long (1846) instead used the word 'Nudiya'. Therefore, the word 'Nadia' is a colloquially distorted form of

B. C. Das et al., *The Nadia Rivers*, Geography of the Physical Environment,
https://doi.org/10.1007/978-3-031-85971-7_1

Nabadwip or Navadvip, a town where Lord Sri Chaitanya was born (Chaudhury 2004). However, since the rules of East India Company, the word connotes a broader sense, the whole district was named Nuddiah, Nediah, Nudiya or Nadia.

During circa A.D. 1200, the word 'Nadiya Rajya' would mean the whole kingdom of Ballāla Sēna. In eighteenth century, during Maharaja Krishnachandra's reign, Nadia was extended from Murshidabad in the north to the Bay of Bengal to the south and Bhagirathi-Hooghly River in the west to Dhulyapur to the east (Majumder 1978).

> রাজ্যের উত্তর সীমা মুরসিদাবাদ।
> পশ্চিমের সীমা গঙ্গা ভাগীরথী খাদ।।
> দক্ষিণের সীমা গঙ্গাসাগরের ধার।
> পূর্ব্বসীমা ধুল্যাপুর বড়গঙ্গা পার।। (Ray 1765)
> [Rājyēra uttara sīmā Murasidābāda
> Paścimēra sīmā Gaṅgā Bhāgīrathī khāda
> Dakṣiṇēra sīmā Gaṅgāsāgarēra dhāra
> Pūrbbasīmā Dhulyāpura baṛaGaṅgā pāra.]

Geographical span of Nadia during undivided Bengal of British colonial rule, as Garrett (1910) mentioned, was from 22°53′00″N to 24°11′00″N and 88°09′00″E to 89°22′00″E.

Therefore, 'Nadia Rivers' seem to be concerned with those rivers flowing through the district of Nadia. The district is traversed by numerous large and small rivers, palaeochannels, beels, and many other water bodies. But when we talk about 'Nadia Rivers,' its sense become narrow. 'Nadia Rivers' is used to mean only three rivers among numerous. During the early phase of British colonial rule, rivers were the main routes for communications. So, the colonial rulers took initiatives to keep some selected rivers navigable throughout the year for the ease of interregional business and revenue collection. In 1819–20, Mr C.K. Robinson was appointed Superintendent and Collector of the river Mathabhanga and out of this appointment a separate division 'Nadia Rivers' of 'Public Works Department' grew. Although the whole district is a network of moribund rivers, Bhagirathi or Baugretty (Colebrooke 1824), Jalangi and Mathabhanga or Matabanga (Colebrooke 1824) are three rivers that have been for over a century and are still distinctively known as the 'Nadia Rivers' (Garrett 1910, p. 6) (Fig. 1.1). Similar has been reported in many other writings (Hunter 1875; Moore et al. 1919; Johnston 1947; Majumder 1978; Basu and Chakraborty 1972). Fergusson (1863) stated that the great group of distributaries of the Ganga (also known as Ganges) is known as 'Kishnaghur Group of Rivers' and consists of three rivers—'Bhagaruttee, Jellinghy, and Matabangah'.

During colonial rule, Jalangi and Mathabhanga rivers were exclusive of the rivers of Nadia because their entire flow paths were within the undivided district of Nadia (Das 2013, 2023a, b, 2024a, b). Bhagirathi is the only river that was (present course also) flowing through Murshidabad and Nadia. After the independence of India in 1947, the upper reach (from offtake from the river Padma to Gede) of the river Mathabhanga has fallen into East Pakistan, and after 1971, it is now the Indo-Bangladesh international river. Therefore, the only river out of these three 'Nadia Rivers', whose entire course flows either through the district of Nadia or along the Nadia-Murshidabad border is Jalangi. Therefore, in the true sense, it may be said that Jalangi is the river of Nadia. Hunter (1877) stated that 'The Jalangi, or Kharia as it is also called, belongs entirely to Nadiya, passing in an exceedingly tortuous course along the northwestern boundary, with Krishnagar (the District Capital) on its bank, and falling into the Bhagirathi opposite Nadiya city'.

However, in this volume, value to the colonial connotation of the term 'Nadia Rivers' is kept intact and Bhagirathi, Jalangi, and Mathabhanga-Churni Rivers are considered as Nadia Rivers (Garrett 1910; Majumder 1978).

1.2 Geology and Sedimentology

The lower Ganga plain, particularly the Bengal delta, is the outcome of fluvial action, shaped by both alluvial and diluvial processes. The foundation for this action was established during Tertiary tectonics linked to the Trans-Eurasian

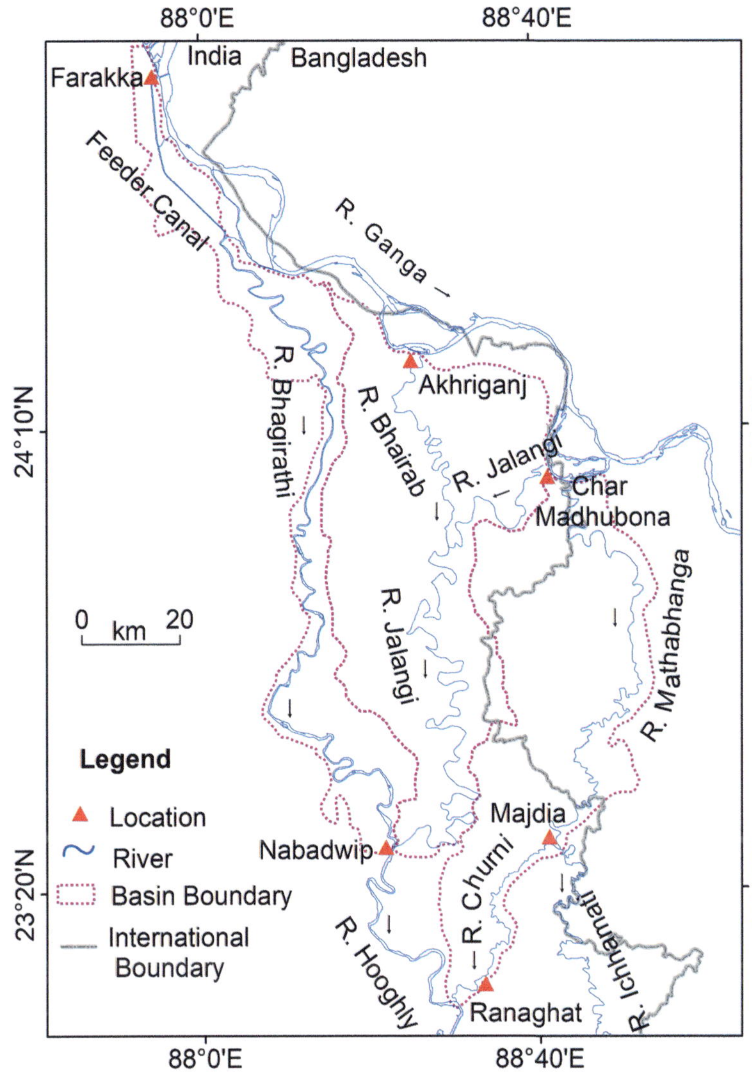

Fig. 1.1 Nadia Rivers: Bhagirathi, Jalangi, and Mathabhanga-Churni

orogeny Singh et al. (1993). The Himalayan orogeny, associated with the Sindhu-Ganga trough and Arakan-Yoma-Andaman-Sumatra arc, played a role in the formation of a depression of the Bengal basin. This geological activity diverted Siwalik Rivers through the Rajmahal-Garo gap and led to substantial sedimentation by the Ganga–Brahmaputra Rivers in the Bengal Basin, resulting in the formation of the Ganga–Brahmaputra-Meghna (GBM) delta (Singh et al. 1993).

Flexural subsidence, in addition to faulting, folding, and localized compaction, facilitates the subaerial section of the GBM delta to capture and retain roughly 30% of the 1.1 billion tons of sediment transported by the rivers (Reitz et al. 2015). Notable research (Auerbach et al. 2015; Reitz et al. 2015; Wilson and Goodbred 2015; Becker et al. 2020; Paszkowski et al. 2021) has demonstrated that the GBM delta is sustained by a collective sediment load of approximately 1 billion tons from the Ganga, Brahmaputra, and

Meghna river systems, with a significant portion of the delta situated within 20–25 m above mean sea level (Fig. 1.2). Nonetheless, the delta is undergoing simultaneous subsidence and sea-level rise.

Early investigations by Broadus et al. (1986) and Milliman et al. (1989) evaluated the impacts of global sea-level rise on delta plains, emphasizing both natural and human-induced processes. A recent study underscores the tectonically active, high sediment-yield nature of the Ganga–Brahmaputra system throughout the Holocene (Nicholls and Goodbred 2004). During the early Holocene climatic optimum (~11–7 ka), sediment discharge from the Ganga–Brahmaputra river system exceeded 200%, maintaining coastal stability despite significant sea-level rise at rates surpassing 1 cm/year (Goodbred and Kuehl 2000). The rapid development of the system, faster than most world deltas, challenges traditional models (Goodbred and Kuehl 2000). This resilience implies that the Ganga–Brahmaputra delta, with its substantial sediment load, may be less vulnerable to expected sea-level rise, particularly under current or wetter climatic conditions.

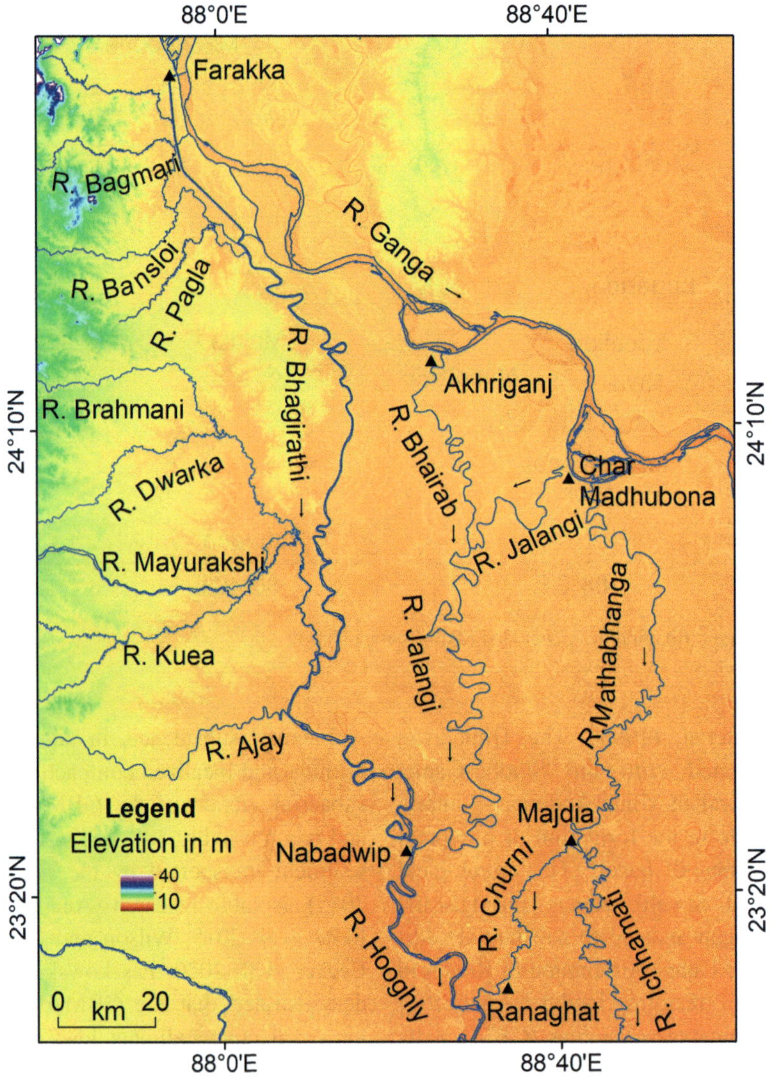

Fig. 1.2 DEM of the Nadia Rivers basins

However, lessons from the Last Glacial Maximum caution that weakened monsoons can influence river discharge and delta response (Goodbred 2003).

Recent research on sediment dispersal in the Ganga–Brahmaputra system suggests a stable pattern over several thousand years (Goodbred and Kuehl 1999). Long-term sediment dispersal is influenced by subsidence, accommodation on the broad shelf, and sediment siphoning via the Swatch of No Ground canyon. The significant sediment load enhances delta stability by preserving land surface elevation and mitigating coastal erosion. Despite receiving minimal sediment input since the Ganga shifted eastward, the western delta remains stable. Further investigation, with improved age control and studies on neotectonic movement, is necessary to better comprehend boundary conditions for this critical period and its implications for flood and sea-level rise scenarios (Coates et al. 1988; Coates 1990). The delta is undergoing multiple mechanisms of subsidence (Goodbred 2003) and/or uplift (Coates et al. 1988; Coates 1990), overlaying changes in delta plain elevation (Milliman et al. 1989).

The Bengal Basin is delineated into six distinct geological regions by rivers like the Ganga, Bhagirathi, Jamuna, Meghna, and others. These regions include: (i) The plains bordering the Himalayas in North Bengal. (ii) The Barind region lies between the Ganga and the Jamuna. (iii) The Rarh region is situated to the west of the Bhagirathi. (iv) The delta area east of the Bhagirathi and the south of the Ganga. v) The Meghna Basin. (vi) Numerous coastal zones comprising river creeks and islands. Nadia Rivers are located within the fourth region, the delta region east of the Bhagirathi and south of the Ganga.

The behavior of the Nadia rivers—Bhagirathi, Jalangi, and Mathabhanga-Churni—is intricately tied to the geology and sedimentology of the Bengal delta, a region shaped by the interplay of tectonic activity, sediment transport, and fluvial dynamics. The Bhagirathi, a former primary distributary of the Ganga, alongside the Jalangi and Mathabhanga-Churni, is embedded in the eastern deltaic plain, where flexural subsidence, localized compaction, and faulting control river gradients and sediment accommodation. These processes lead to pronounced avulsion, channel migration, and meandering in response to sediment load variations. The low elevation of the delta (20–25 m above mean sea level) and its high sediment yield from the GBM rivers further amplify hydrological instability, contributing to frequent bifurcations and seasonal variations in flow regimes. Additionally, the shifting courses of these rivers are influenced by subsurface neotectonic movements and sediment siphoning through features like the Swatch of No Ground canyon (Michels et al. 2003), which redistributes sediment loads within the delta. This dynamic interplay of tectonics and sedimentology, coupled with monsoonal precipitation and anthropogenic interventions, renders the Nadia rivers highly sensitive to changes in sediment supply and hydrodynamic forcing, manifesting in their meandering, braided, and distributary nature across the Bengal delta.

1.3 Nadia River Division

Major Lang (1848) noted that navigability challenges arising from siltation at the offtakes from the river Ganga have persisted since the records began over a century ago, specifically since 1748. This observation implies that monitoring of the rivers Bhagirathi, Jalangi, and Mathabhanga-Churni has been ongoing since the first half of the eighteenth century.

After the battle of Palasi in 1757, the East India Company wielding significant authority in the revenue collection of Bengal, particularly after 1765 when the British East India Company was granted the Diwani (revenue collecting right) of Bengal, sought to facilitate navigation for communication with the northern part of Bengal. This initiative was undertaken to establish a more accessible and convenient route. In the autumn of 1764, Major James Rennell initiated a comprehensive survey of the Ganga River (Lee 1896). During this survey in

1766, Rennell faced the imposing 'Tartarian Mountains' of the Himalayas. The primary objective of Rennell's survey was to identify a navigable waterway connecting Calcutta to the northern regions of Bengal. On January 1, 1767 (Chadha 1991), Lord Robert Clive, who held the position of Governor of Bengal and Bihar, recognized Rennell's expertise and appointed him as the Surveyor-General of Bengal for the East India Company's dominions in Bengal. Initially conceived as a general survey of newly acquired lands, the mapping project's scope expanded significantly under the leadership of Warren Hastings, who assumed the role of Governor-General in 1773 (Rennell 1910). As a recognition of his contributions, James Rennell received the rank of Major of Bengal Engineers on April 5, 1776. Subsequently, he retired from active service in 1777, concluding his notable involvement in the mapping and surveying activities in the region. Rennell (1781, 1788) reported about deteriorated conditions and fluctuating navigability of the Bhagirathi, Jalangi, and Mathabhanga-Churni Rivers.

In an early reference from 1813 (D'Souza 2015), it was documented that the company administration had assigned the responsibility for the 'improvement' of the Mathabanga to the offices of the collector and the local police department. This improvement entailed the regular clearing of the river channel bed from sunken boats, loose timber, and large trees (Sur 1926). The administration believed that such clearances would enhance the safety of river navigation. By 1818, however, the 'obstructions' were deemed so perilous that a significant number of boats were consistently being damaged or wrecked while crossing the channels. The inability to establish safe passage meant that cargo-laden ships were often detained for prolonged periods, leading to substantial costs, particularly from demurrage. Matters had worsened to the point where, in that year, the merchants of Calcutta vociferously petitioned the government to take corrective measures for the 'evil' that was severely affecting commercial interests (Lang 1926).

The British colonial rulers took extensive measures to oversee riverine transport in Bengal, expressing concerns about deteriorating river channels and diminishing navigability. Their persistent efforts were directed at revitalizing the navigability of the Bhagirathi, Jalangi, and Mathabhanga-Churni rivers to facilitate trade with the northern and eastern parts of Bengal, as well as Assam. A significant turning point occurred in 1819–20 when Mr. K. C. Robinson assumed the position of superintendent and collector of the Mathabhanga River (Reaks 1919a, b), marking the formal establishment of the 'Nadia River Division.' Nadia River Division is one of the wings of Public Works Department (PWD) of the then Government of Bengal (Moor et al. 1919). Mr. K. C. Robinson straightened the course of the river Mathabhanga by cutting meander loops. In the year 1821–22, Mr. May succeeded Mr. K. C. Robinson, closed the offtake of the river Pungasi (Pangasi) or Koomar (Kumar) from the river Mathabhanga (Inglis 1909) so that substantial volume is diverted to through the Mathabhanga-Churni River.

In 1835, Nadia Rivers' operations ceased, and the organization was disbanded due to the belief that the outcomes did not justify the expenses. Nevertheless, in June 1837, Mr. May was reappointed as superintendent and observed a decline in river conditions with minimal improvement in the subsequent year. Captain Smyth assumed the role of superintendent in August 1840, and the fluctuating navigability of Nadia Rivers was documented until 1846.

In 1859, Mr. Armstrong, serving as the Superintendent of the Nadia River Division, proposed the excavation of several meander loops of the Mathabhanga River. The following year, in 1860, Mr. Leonard, the Superintending Engineer of the South-Eastern Circle at the time, conducted an inspection of the Nadia Rivers. He expressed the view that there was no established rule or plan to maintain the permanent navigability of the Nadia Rivers.

In 1862, Mr. Franklin Prestage introduced a plan to regulate flow of the Mathabhanga River with locks on the Ganga and Hooghly. Mr. Leonard examined the proposal and highlighted its drawbacks. Bengal Government Letter No. 4558, dated September 23, 1862, stated that due to repeated failures in attempts to keep the Nadia Rivers navigable, the government was interested in the Eastern Bengal Railway project leading to the Ganga at Bhagwangola.

In January 1863, Mr. Bourne suggested a canal connecting the Bhagirathi with the Ganga near the end of the Rajmahal Hills, but the proposal was not accepted. In 1881, Mr. Vertannes advocated the use of dredgers, and Rs. 123,000 was sanctioned (refer to Bengal Government Letter No: 231I-T., dated April 28, 1881) for a hydraulic dredger and two hopper barges for the Nadia Rivers (Inglis 1909).

In 1857, Mr. Thomas Haines Wickes arrived in India and worked under his brother, Mr. Henry Wickes, on the East Indian Railway. During this period, he was involved in the construction of the Jumna Bridge (Tudsbery 1900). Subsequently, after 1872, Mr. Wickes dedicated several years to extensive training works on the Nuddea Rivers, which transformed that division into a lucrative source of revenue for the Government.

In a note dated February 25, 1880, Mr. H. C. Levinge, chief engineer, expressed support for a cross embankment. However, in 1885, Mr. Wickes, serving as the superintending engineer of the Nadia River Division at that time, criticized the Laltakuri embankment for impeding natural spills and causing breaches, leading to severe floods.

As stated by Garrett (1910), regardless of the distance covered, the toll rates for Nadia Rivers, as specified in Act V of 1864, Notification No. 175 dated 30.7.1879, Part-I, page 789, ranged from 4 Annas for boats with personal accommodations to 1 Rupee 8 Annas for pinnaces exceeding 10 oars. The toll rates for Nadia Rivers, under Act V of 1884, were simplified. They range from free for boats with personal accommodations to 1 Rupee for vessels with any cargo (Table 1.1). The charge is based on 100 maunds of measured tonnage. This applies to cargo in flats or steamers surveyed under Act VI of 1884. The tonnage is determined by measurement, whether the vessel is laden or empty. Average annual income, expenditure, and net surplus from Nadia Rivers had a positive budget from 1847-48 to 1900-01 (Table 1.2). After that, net surplus experienced negative values (Fig. 1.3).

Until 1888, the revenue administration of Nadia Rivers was under the control of the Board of Revenue. In 1888, it was transferred to the Public Works Department vide Resolution No. 112, dated March 20, 1888 (Inglis 1909). Inglis said in a note dated 11th September 1909, that dredging through shoals will be filled up again and obliterated each flood season. By a notification No. 4, dated 11.6.1909, and modified Notification No. 6, dated 31.6.1909, the Schedule of tolls of Nadia Rivers was simplified and reduced to half during June to October.

In the fiscal year 1913–14, a committee was established to investigate the current amenities and assess both the present and future needs of the Port of Calcutta. This committee specifically addressed the state of the upper Hooghly, as documented in Sections 139 and 140 of its report. The committee recommended ongoing surveillance and the implementation of necessary measures to maintain the Nadia Rivers in a state that ensures an adequate water supply to the Hooghly River.

In 1914, to implement the suggestion put forth by C. J. Stevenson-Moor, Major Hirst, I.A., the director of Surveys for Bengal and Assam, was assigned the responsibility of investigating the discharge of Nadia Rivers at specific intervals. As part of this initiative, a temporary subdivision called the 'Bhagirathi Observatory' was established under the Nadia River Division. In August 1914, Mr. Cowley fixed nine observation stations on Nadia Rivers as follows: Geriah Jangipur, Berhampur, Jumjumkhali, Databati, Katwa, Satgachia, on Bhagirathi; Punditpur on Jalangi; and Hanskhali on Mathabhanga. Starting from November 6, 1914 (refer to Notification No: 1360 T-R, dated November 2, 1914), and continuing until April 1915, Hirst initiated the first stage of the investigation, with the resulting report being published in 1916. Hirst (1916) suggested 31 investigation stations which was rejected recognizing their impracticality due to the high costs, extensive arrangements, and staffing requirements.

Table 1.1 Showing toll rates revised under Notification No. 175, dated 30th July 1879-vide Calcutta Gazette of 30th July 1879, Part 1, page 789 and Act V of 1884, whatever the distance travelled. Calcutta Gazette, Notification No. 6, dated 31st July 1896. (Source: Garrett 1910)

Schedule of tolls to be levied on the Nadia Rivers under Act V of 1864, whatever the distance travelled. Notification No. 175, dated 30th July 1879-vide Calcutta Gazette of 30th July 1879, Part 1, page 789			
Sl. No	Description	Toll rate	
		Rs	Anna
1	Pinnaces of 10 oars and under that number, each	3	12
2	Pinnaces exceeding 10 oars	6	0
3	Budgerows of 10 oars and under that number	2	4
4	Budgerows exceeding 10 oars	4	8
5	Bholeahs and boats for personal accommodation, cutters not being of the description specially defined, and bulwars, pansways and baggage boats, per oar	0	4
6	Empty boats and boats laden with bricks, tiles or other earthen substances, baked or otherwise, per 100 maunds tonnage	0	2
7	Boats laden with quick-lime, chunam, straw, firewood, gurran-sticks, thatching grass or such like, per 100 maunds tonnage	0	8
8	Boats laden with grain, pulse, seed or vegetables of whatever description and indigo seed, per 100 maunds tonnage	0	12
9	Boats of burthen freighted with timbers and bamboos or with any article not included in the above enumeration, per 100 maunds tonnage	1	0
10	Timbers with chowkars and dowkars, if floated on rafts or otherwise, not being boats, each timber	0	6
11	Unwrought timbers called ghole or floated as above, each	0	3
12	Rafts floating 200 bamboos or fewer, each raft	0	3
13	Rafts floating more than 200 but not more than 400	0	6
14	Rafts floating more than 400 but not more than 1,000	0	12
15	Rafts floating more than 1,000, each raft …	1	8

Schedule of tolls to be levied on the Nadia Rivers under Act V of 1884, whatever the distance travelled. Calcutta Gazette, Notification No. 6, dated 31st July 1896			
Sl. No	Description	Toll rate	
		Rs	Anna
1	Passenger boats, accommodation boats, bholeahs, budgerows, pinnaces, bulwares, pansways and baggage boats, per oar	0	4
2	Empty vessels, except steamers, per 100 maunds measured tonnage… of	0	4
3	Vessels laden with a cargo of any description per 100 maunds of measured tonnage	1	0
4	Cargo carried in flats or steamers and charged on the manifest per 100 maunds of cargo	1	0
5	Steamers that have been surveyed under Act VI of 1884, whose tonnage is determined by measurement, whether laden or empty, per 100 maunds of measured tonnage…	1	0
6	Floats or rafts of roughly squared timbers, each timber	0	6
7	Floats or rafts of unwrought timbers, each timber	0	3
8	Rafts of bamboo, each 100 bamboos	0	2

(continued)

Table 1.1 (continued)

Schedule of tolls to be levied on the Nadia Rivers under Act V of 1884, whatever the distance travelled. Calcutta Gazette, Notification No. 6, dated 31st July 1896			
Sl. No	Description	Toll rate	
		Rs	Anna
9	Reeds, for each hundred bundles of reeds.	0	4
10	All boats whose tollage does not exceed 4 annas will be allowed to pass free	0	0
	N.B-During the flood season (June to October, both inclusive), half the above schedule rates will be charged		

Table 1.2 Income-expenditure and net surplus from Nadia Rivers

Period	Annual average income (Rs)	Annual average expenditure (Rs)	Average annual net surplus (Rs)	
1847–48	–	–	180,250	Garrett (1910)
1861–62 to 1870–71	–	–	104,538	Garrett (1910)
1871–72 to 1880–81	232,938	87,019	145,919	Garrett (1910)
1881–82 to 1890–91	195,632	118,136	77,496	Garrett (1910)
1891–92 to 1900–01	127,479	125,864	1615	Garrett (1910)
1899–1900	–	–	– 2657	Garrett (1910)
1901–02 to 1907–08	76,629	123,689	– 47,060	Garrett (1910)
1906–07	–	–	– 107,804	Garrett (1910)
1907–08	35,229	–	–	Garrett (1910)

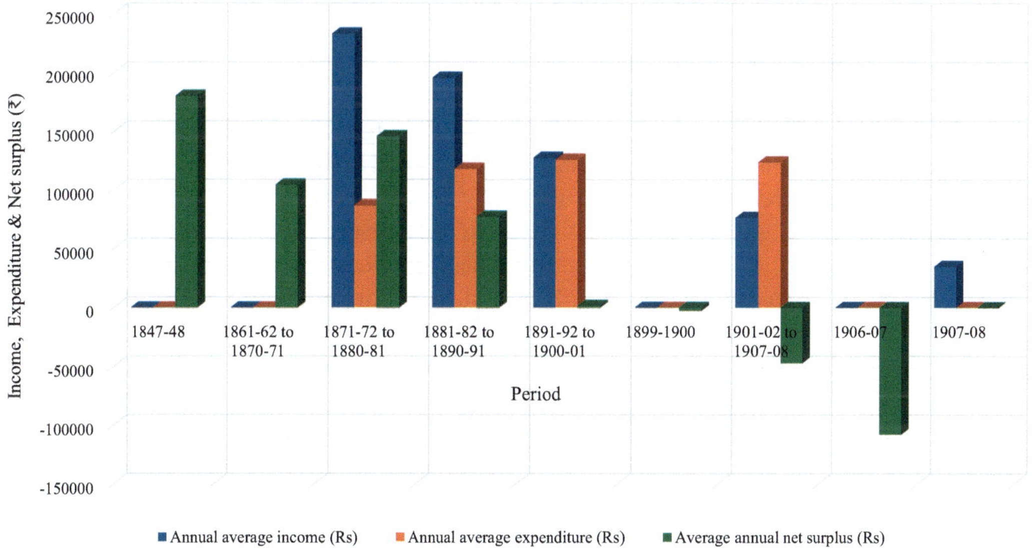

Fig. 1.3 Income-Expenditure and net surplus from Nadia Rivers

Based on Hirst's report, the Government of Bengal passed a resolution (No: 373 T-R, dated October 9, 1916). Subsequently, a committee was established by the Government of Bengal through a communication dated February 17, 1917, with C. J. Stevenson-Moor, Chairman of the Port Commissioners of the Calcutta Port, presiding over it. The committee's objectives were twofold: Firstly, to provide recommendations to the government for the systematic observation of changes in Nadia Rivers. Secondly, to propose actions needed to enhance their navigability and serve as feeders for the lower Hooghly River. The committee's findings, known as the Moor Committee's report, were published in 1919 (Moore et al. 1919).

The committee comprised ten members: C. H. D. Ryder, Manindra Chandra Nandi, Reshee Case Law, H. H. Hayden, J. Campbell, A. R. Murray, C. Addams-Williams, and E. A. Constable. Mr. S. C. Williams and Mr. J. E. Roy did not sign the report as they were out of India at that time. The committee recommended surveys, preparation, and conservation of maps at both large scale (16 inches to 1 mile) and small scale (1 inch to 1 mile). Additionally, they suggested special riverbed surveys and cross-sectional surveys at intervals of 300 feet. The committee also proposed the establishment of five observation stations along the Nadia Rivers (Moore et al. 1919).

A Special Sub-Committee was established within the aforementioned committee with the following members:

Hon'ble Mr. C. J. Stevenson-Moor, C.V.O., I.C.S., serving as President.
Lt.-Col. Ryder, C.I.E. D.S.O., R.E.
Mr. C. Addams-Williams, C.I.E.
Major F. C. Hirst, I.A., Director of Surveys, Bengal.
Mr. S. C. Williams, Vice-chairman of the Commissioner of the Port of Calcutta.
Commander E. A. Constable, R. N., Deputy Conservator of the Port of Calcutta.

At a subsequent date, Mr. James E. Roy of Messrs MacNeill & Co. was added to the committee. The Special Sub-Committee conducted four meetings on May 19, 1917, June 23, 1917, July 31, 1917, and August 17, 1917. Their report, submitted on December 5, 1917, recommended the continuation of existing observation and operational methods until the recommendations of Major F. C. Hirst were fully implemented (Moore et al. 1919).

The Irrigation and Waterways Department was established in 1920, breaking away from the Public Works Department. Its current designation as the Irrigation and Waterways Department was officially adopted in 1946 (Govt. of WB 2024a, b). Annual Report on Major Rivers in the Nadia Rivers Division (1926–37) was published by the Irrigation Department, Government of Bengal. This fact indicates that before 1946, the Irrigation and Waterways Department was named as Irrigation Department. However, it appears from gathered information, that during the period from the 3rd to the 5th decade of the twentieth century, Nadia Rivers Division eventually withered away. In independent India, the Irrigation and Waterways Department, Government of West Bengal, looks after irrigation and waterways related to all the rivers along with Bhagirathi, Jalangi, and Mathabhanga-Churni Rivers.

1.4 Integrated Perspective of Nadia Rivers

The Bhagirathi, Jalangi, and Mathabhanga-Churni rivers, which constitute the Nadia Rivers system, are distinct fluvial entities. Despite their individuality, these rivers display a high degree of interconnectedness. Each river originates as a distributary of the larger Ganga River system, specifically branching off from either the Ganga itself or its distributary, the Padma. These rivers share a common origin, receiving water and sediment inputs from the same primary source. Their combined outflows eventually merge into the Hooghly River. Together, they form a complex hydrological framework known as the Nadia Rivers system. This system plays a vital role in shaping the alluvial deltaic region and managing drainage in the region's comparatively low-energy zones.

While these rivers are distinct, they operate as an integrated fluvial system. The hydrodynamic and sedimentological characteristics of the Bhagirathi-Hooghly segment are profoundly influenced by the Jalangi and Churni rivers, which are integral components of the broader Nadia Rivers system. The catchment processes and basin-level hydrology of these rivers are deeply interlinked. This interdependence highlights the indivisible nature of their functioning within the system. This volume adopts a systemic and integrative perspective, treating the Bhagirathi, Jalangi, and Mathabhanga-Churni rivers not as isolated watercourses but as components of a unified hydrological network. The scope of this research extends beyond the rivers' physical and hydrological characteristics. It explores their intricate relationships with the socio-economic, cultural, historical, and ecological dimensions of the communities residing along their banks. The approach underscores the inherent interdependence of these rivers, revealing that the water resources and environmental health of one river significantly influence the others.

Consequently, planning for the sustainable development of this region necessitates a comprehensive and holistic framework. Fragmented or isolated interventions would fail to address the intricacies of the system. This volume emphasizes the need to view the Nadia Rivers collectively, highlighting their status as an interdependent and dynamic fluvial system. Such a perspective is crucial for devising strategies to manage and mitigate the challenges faced by this network of rivers effectively.

1.5 Scales of Investigation of Past Studies

The study of Nadia Rivers is a multifaceted undertaking, encompassing a range of factors such as the land's topography, water flow dynamics, navigability, flood patterns, bank erosion, and socio-economic and cultural dimensions. Scholars have traditionally delved into this intricate subject by considering specific spatial and temporal scales. In terms of spatial scale, researchers have concentrated on three levels: 'site', 'reach', and 'catchment', often combining these scales to have a comprehensive understanding (Islam and Guchhait 2024). The majority of studies have focused on the 'reach' scale, but there is a noticeable dearth of research covering the entire Nadia Rivers or their catchment areas. Examining the temporal scale, scholars have employed various time frames, including ahistorical, short-term, medium-term, and long-term periods. For studying river bank erosion and lateral channel migration, Lawler (1993) defines short-term scale as a span of a few months to a few years, medium-term covering 1–30 years, and long-term extending to 10,000–20,000 years. However, studies in long-term temporal scales have not extensively explored the Bhagirathi, Jalangi, and Mathabhanga-Churni rivers. The Nadia River System, specifically, has been investigated in short-term and medium-term contexts. However, a comprehensive, ahistorical, and long-term perspective has been somewhat overlooked. A detailed examination of scholars provides insights into the diverse contexts of spatiotemporal dimensions in the study of the Bhagirathi, Jalangi, and Mathabhanga-Churni rivers and tabulated for reference (Table 1.3). It is found that studies related to the small spatial aspects of short-term temporal investigations dominate, overshadowing other perspectives.

The current study introduces a spatial scale division, considering periods as ahistoric (before 1775), historic (1775–1975), and recent (after 1975). The boundary between ahistoric and historic phases is defined by the availability of authentic documents on Nadia Rivers, starting with Rennell's survey (1766–1777) and mapping of Bengal, and documentation. This survey is considered the first authentic document regarding these rivers. Significantly, the boundary between the historic and recent phases is marked by the construction of the Farakka barrage, representing a pivotal anthropogenic intervention in the Nadia River System.

Table 1.3 List of literature covering different spatiotemporal scales and aspects of the rivers and or their basins

Scholars	Year	Spatial Scale			Temporal Scale			Broad aspect of the study
		Site	Reach	Catchment	Short	Medium	Long	
Colebrooke (1803)	1803			*			*	Navigation
Colebrooke (1824)	1824			*			*	Navigation
Fergusson (1863)	1863			*		*		Navigation
Govt. of Bengal (1931)	1931			*	*			
Johnston (1933)	1933		*			*		Navigation
Goodbred and Kuehl (1998)	1998			*			*	Floodplain processes and sedimentation
Paria & Konar (2003)	2003		*			*		Ecology
Sreemani (2003)	2003			*			*	Settlement
Kuehl et al. (2005)	2005			*			*	Sedimentation and delta formation
Rahman et al. (2005)	2005		*			*		Arsenic pollution
Chakrabarty & Das (2006)	2006		*		*			Ecology
Sanyal & Lu (2006)	2006		*			*		Flood
Das & Chakrabarty (2007)	2007		*		*			Ecology
Das & Chakrabarty (2007)	2007		*		*			Ecology
Mukherjee & Fryar (2008)	2008			*		*		Arsenic pollution
Datta et al. (2011)	2011			*		*		Water quality
Kumar et al. (2011)	2011			*		*		Groundwater
Mitra et al. (2012)	2012		*		*			Water quality
Das (2015)	2015	*				*		Degradation & society
Moitra et al. (2015)	2015	*			*			Pollution & health
Mondal & Pal (2015)	2015			*		*		Arsenic pollution
Das & Islam (2016)	2016		*		*			Channel morphology
Guchhait et al. (2016)	2016	*				*		Channel instability
Donselaar et al. (2017)	2017			*		*		Water quality & geomorphology
Islam & Guchhait (2017)	2017			*		*		Society
Pal et al. (2017)	2017	*				*		Channel dynamics
Rudra (2018)	2018			*			*	Evolution
Asian Development Bank (2019)	2019			*		*		
Bag et al. (2019)	2019		*			*		Planform dynamics
Biswas, & Dhara (2019)	2019		*			*		Planform dynamics
Das & Sarkar (2019)	2019			*	*			Land use
Sarkar & Islam (2019)	2019			*		*		Water quality
Das & Bhattacharya (2020)	2020		*				*	Anthropogeomorphology
Das & Das (2020)	2020		*				*	Anthropogeomorphology
Das et al. (2020)	2020			*			*	Anthropogeomorphology
Islam & Guchhait (2020)	2020		*			*		Channel morphology
Sarkar & Islam (2020)	2020		*		*			Water quality
Ashwini et al. (2021)	2021		*			*		Planform dynamics

(continued)

Table 1.3 (continued)

Scholars	Year	Spatial Scale			Temporal Scale			Broad aspect of the study
		Site	Reach	Catchment	Short	Medium	Long	
Das et al. (2021)	2021			*		*		Water quality & geomorphology
Karmokar et al. (2021)	2021			*		*		Paleochannels
Panigrahi & Pattnaik (2021)	2021		*			*		Pollution
Roy et al. (2021)	2021	*						Wetland
Sarkar et al. (2021a)	2021		*		*			Ecology
Das & Bandyopadhyay (2022)	2022			*	*			Flood
Ghosh (2022)	2022	*				*		Bank erosion
Mandal & Bhunia (2022)	2022	*			*			Morphology
Mandal et al. (2022)	2022			*	*			Ecology
Mondal & Bandyopadhyay (2022)	2022			*		*		Morphology
Rudra (2022)	2022			*			*	Channel dynamics
Sarkar & Das (2022)	2022		*		*			Water quality
Sarkar & Islam (2022)	2022		*			*		Water quality
Das (2023b)	2023		*				*	History
Das et al. (2023)	2023		*			*		Flood
De & De (2023)	2023	*				*		Flood
Ghosh & Pal (2023)	2023			*	*			Oxbow lakes
Ghosh & Sahu (2023)	2023		*			*		Social
Ghosh et al. (2023)	2023			*		*		Flood
Mondal et al. (2023)	2023			*		*		Flood
Sarkar et al. (2023)	2023		*			*		Ecology
Das (2024a, b)	2024		*				*	History
Das et al. (2024a)	2024		*			*		Water quality
Das et al. (2024b)	2024	*			*			Bridge construction & channel morphology

1.6 Nadia Rivers from a Bibliometric Perspective

Nadia Rivers (Bhagirathi, Jalangi and Mathabhanga-Churni) bears the marks of the international and transboundary perspectives. Before Indian Independence, India and Bangladesh were a single unit and these rivers were part of the Indian subcontinent. However, these rivers were politically divided between India and Bangladesh in the post-independence period. Thus, these rivers have vivid spatial and temporal scales in the development of the rivers and associated works. To this end, leveraging data from Scopus, four important dimensions were mapped. The first is related to the temporal growth of the works. The works started much earlier on the Bhagirathi River (since 1979) compared to the Jalangi and Mathabhanga-Churni Rivers (Fig. 1.4).

Moreover, the post-2015 spurt in the work is visible for all the rivers. Apart from the temporal dimension, the spatiality of the Nadia Rivers is characterized by the study predominantly by

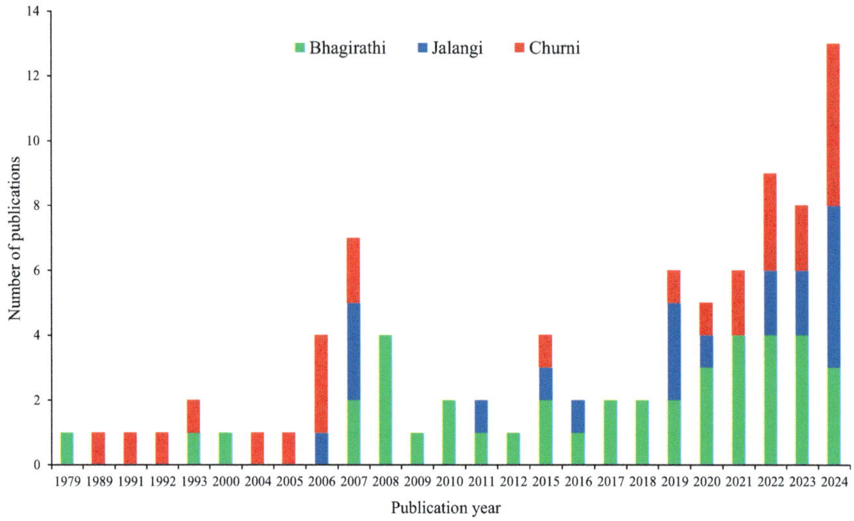

Fig. 1.4 Temporal pattern of study on Nadia Rivers

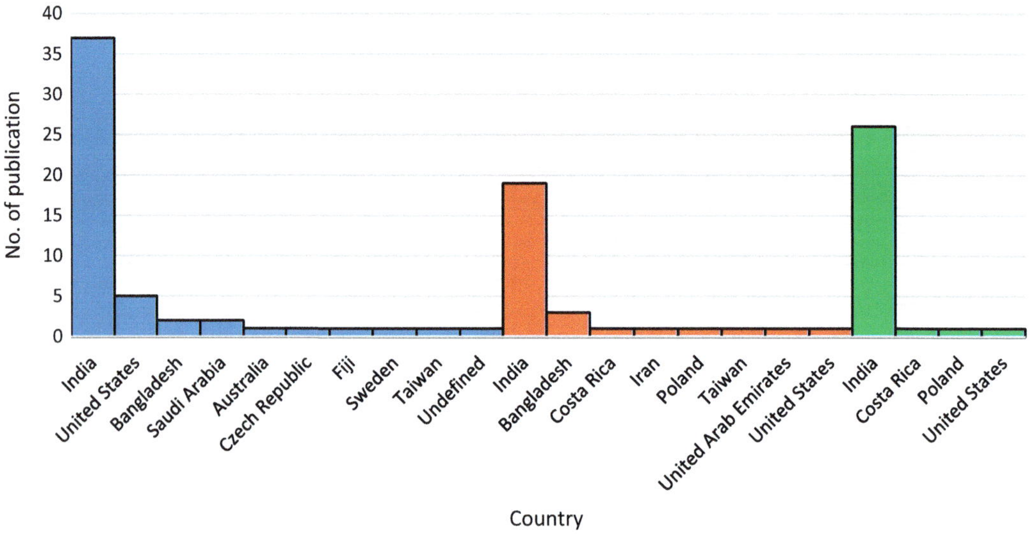

Fig. 1.5 Spatial variation of the Nadia Rivers study (*Note* Blue color bars indicate Bhagirathi, Orange for Jalangi and Green for Churni River)

Indian scholars. However, as these rivers are transboundary, few studies from Bangladesh are also reported (Fig. 1.5). Concerning the work distributions, Islam, A has the maximum works on the Bhagirathi River, Das, BC has the maximum works on the Jalangi River and Sarkar, B has the maximum works on the Churni Rivers (Fig. 1.6). These indicate the contribution of the authors of the present book on the subject is justified. Furthermore, Nadia Rivers is mainly studied from the perspective of the environmental sciences. However, dimensions of the earth sciences and agricultural sciences are also common in the previous scholarships, especially for the Jalangi and Churni Rivers. For the Bhagirathi River, there is an important focus on social

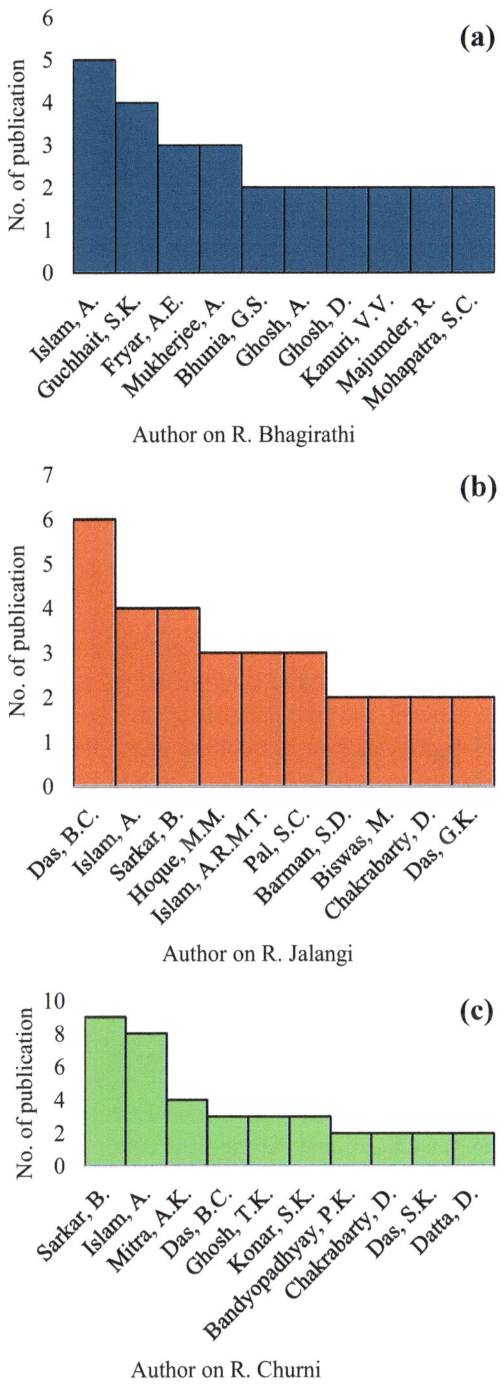

Fig. 1.6 Scholarly Contribution on Nadia Rivers. (a) Bhagirathi, (b) Jalangi and (c) Churni River. The highest number of research articles on the river Bhagirathi has been authored by Islam A (5 articles), on the river Jalangi by Das BC (6 articles), and on the river Churni by Sarkar B (9 articles).

science especially in the context of fluvial hazards like floods and bank erosion (Fig. 1.7). This underscores the multidisciplinary perspectives of the Nadia Rivers.

1.7 Importance of Nadia Rivers

Nadia Rivers, encompassing Bhagirathi, Jalangi, and Mathabhanga-Churni, are integral to the lives of millions, serving diverse roles that underscore their significance in various aspects of society. Nadia Rivers are the lifelines of the region and serve the region in all respect—building and raising the elevation of the delta (Goodbred and Kuehl 2000); offering water for drinking (Hoque et al. 2023; Das and Bhattacharya 2020), irrigation (Gregory 2021), and day-to-day use (Das 2013); providing routes for navigation for intra-regional and interregional transport and communications (Moore et al. 1919; Inglis 1909); etc. for the region since unknown past.

Since getting firman from Mughal Emperor Aurangzeb in 1664 by Englishmen to do business forever in Bengal (Frazer 1897), the foundation of Kolkata in 1690 as the business capital of East India Company (Mukherji 1991), and the inception of colonial revenue rule of the British Government (East India Company) in eastern India in 1765 (Bruce and Dundas 1793; Bowen 1988), Nadia Rivers had been serving as the most important inland trading route till the 1st quarter of the twentieth century. Presently listed under National Waterways (PIB 2018) the river Ganga-Bhagirathi-Hooghly is designated as National Waterway 1 (NW1), and the river Jalangi is labelled as NW47. Except for transport, Nadia Rivers are significant for drinking and irrigation water, fishing, and recreation.

The religious importance of these rivers is evident during the annual pilgrimage from January 14 to 16, 2024, at Ganga Sagar. Ten million devotees, both national and international, congregate at the mouth of Bhagirathi-Hooghly, demonstrating the rivers' sacredness. The remarkable act of pilgrims diving into the

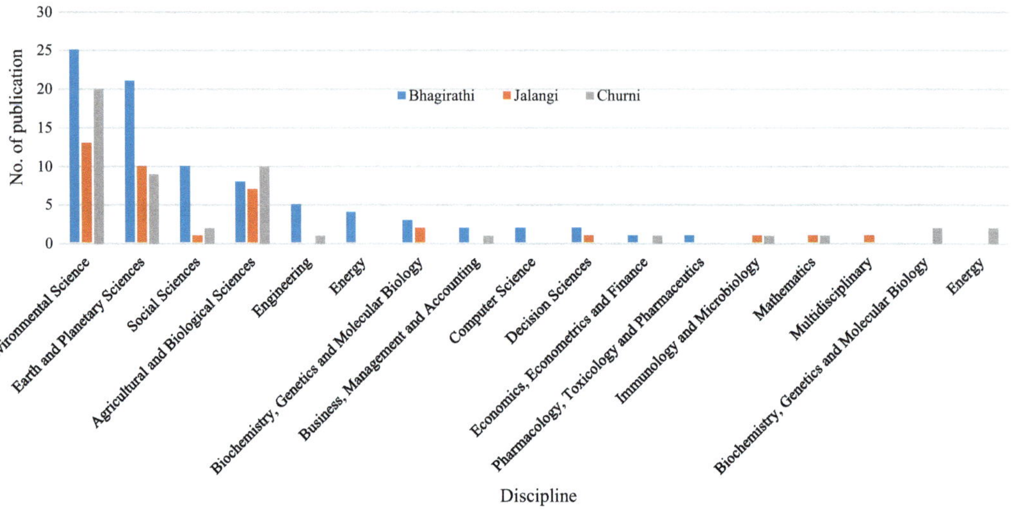

Fig. 1.7 Interdisciplinary nature of the subjects

cold waters at dawn symbolizes the spiritual connection people share with these rivers (The Hindu 2024). Millions of people dive into the Bhagirathi River throughout the year solely for religious purposes.

Economically, the rivers contribute significantly to trade and commerce. Kolkata, situated along Bhagirathi-Hooghly, serves as a crucial port city facilitating millions of dollars in exports and imports and is considered the Infrastructure of Postcolonial Capitalism (Bhaumik 2018; Neilson et al. 2018). The vast hinterland of Kolkata port plays a pivotal role in supporting economic activities (Bandyopadhyay et al. 2009), emphasizing the rivers' role as an economic lifeline for the region. Population density along the riverbanks is a testament to the vital role Nadia Rivers play in sustaining livelihoods. Kolkata, the 3rd largest populated city in India (2018), thrives on the resources provided by these rivers.

Countless individuals depend on Nadia Rivers for essential needs such as drinking water (Kanuri et al. 2020; Sarkar and Das 2022), bathing, and daily activities (Roy et al. 2021b), highlighting the rivers' indispensability to the local population. Agriculturally, the Nadia Rivers are the backbone for millions of farmers. The fertile lands irrigated by these rivers have supported abundant crops, contributing not only to local sustenance but also feeding industries and trade of national and international span. The historical significance of agriculture along these riverbanks is deeply embedded in the economic history of the region (Khasnabis 2008).

Culturally, the rivers are intertwined with a rich heritage. Nabadwip, located on the right bank of Bhagirathi-Hooghly, holds the distinction of being the birthplace of Lord Sri Chaitanya (Das 2023a) and serves as a significant historical center for education. Such is the glory of Nabadwip that it is referred to as the "Oxford of Bengal" and Oxford is often referred to as the "Nabadwip of England," recognizing the rich educational heritage and cultural significance associated with both places. The cultural significance of these rivers is further accentuated by the unique traditions and practices associated with the region. Towns except Nabadwip, like Murshidabad, Jangipur, Azimganj, etc. along Bhagirathi-Hooghly, bear witness to the historical tapestry of the region, adding depth to its cultural heritage (Das et al. 2020). Prominent settlements along the Jalangi River contribute significantly to the importance of Nadia Rivers. These include Krishnagar (a graduated pier of D. L. Roy bridge on the river Jalangi has been used as the gauging staff till 2012), serving as the capital of the Nadia Raj family since the seventeenth century (Ray 1876), and Mayapur, the headquarters of ISKCON (Das 2023a). Other noteworthy locations along the Jalangi River include Bara Andulia, Tehatta, Palashipara, and Karimpur, each playing a crucial role in the historical and cultural tapestry of the region. Similarly, settlements along the Mathabhanga-Churni River further enhance the significance of Nadia Rivers. Majdia stands as a notable location, while Shibnibas, established as the capital of the Nadia Raj family in 1742, adds historical depth to the region. Additionally, Hanskhali (a gauging station on R. Churni was there), Aranghata (home to the Jugal Kishor temple founded by Maharaja Krishna Chandra in 1728), and Ranaghat (the location of Palchaudhury Jamindars residence) contribute to the rich heritage and diverse landscapes along the Mathabhanga-Churni River. These settlements exemplify the historical, cultural, and architectural wealth associated with Nadia Rivers, emphasizing their importance in the broader context of the region.

Industrially, Nadia Rivers provide a conducive environment for the establishment of factories and industries since the sixteenth century (Govt. of WB 2024a, b). Hundreds of large and small enterprises benefit from the strategic locational, transport, and water facilities offered by the rivers, contributing to the industrial growth of the region (Iftikhar-ul-Awwal 1978; Ghosh and Shirodkar 2023). Throughout history, communities that flourished by harnessing the benefits of water resources, river transportation, and elevated levee locations for siting industries along these rivers have played pivotal roles in shaping the industrial landscape and development of the region (Kirk 1956).

The Bhagirathi River, the most important component of the Nadia Rivers, is significant in the context of the Indo-Bangladesh water relations (Rahaman et al. 2019). The river receives its water supply from the Ganga, regulated by the Farakka Barrage, which was executed in 1975. The construction of the Farakka Barrage led to the Indo-Bangladesh Water Treaty of 1977, addressing concerns related to water sharing and distribution between the two countries. This treaty was further modified in 1996 to adapt to changing circumstances and to enhance cooperation in the sustainable utilization of shared water resources (Bhaduri and Barbier 2008). The Nadia Rivers, therefore, not only contribute to regional water security but also serve as a focal point for international agreements and treaties, emphasizing the need for joint efforts in managing and protecting shared water bodies. The cooperative management of these rivers reflects the recognition of the interconnected nature of water resources and the importance of fostering diplomatic ties to address challenges related to water quality, quantity, and equitable distribution.

The Mathabhanga-Churni holds significant importance from an international perspective, primarily due to its status as an Indo-Bangladesh international river (Das et al. 2023c). Mathabhanga serves as a vital watercourse shared by both India and Bangladesh, emphasizing the need for collaborative management and attention to transboundary issues. The river plays a crucial role in fostering bilateral relations between the two countries (Baidya 2022). One of the key challenges facing Mathabhanga-Churni is pollution originating from sources in Bangladesh (Sarkar and Islam 2020). This pollution concern has escalated the importance of the river's conservation and sustainable management, requiring joint efforts between India and Bangladesh to address cross-border environmental issues. Cooperation in addressing pollution challenges is essential to maintaining the health and integrity of the Mathabhanga-Churni River and its ecosystem.

Nadia Rivers emerge as dynamic hydrological sedimentological and morphological entities weaving together the religious, economic, cultural, and historical fabric of the region. Their multifaceted traits and contributions make them indispensable and underscore their pivotal role in the vitality and growth of the surrounding communities. However, previous studies either disintegrated or discrete and partially lacked an integrated and holistic approach to studying Nadia Rivers as a system. Here lies the importance of this present volume weaving together its physical and cultural landscapes in an integrated and holistic approach.

1.8 Transition to Transformation

The Nadia Rivers—Bhagirathi, Jalangi, and Mathabhanga-Churni—serve as poignant examples of waterways in transition, grappling with environmental and anthropogenic pressures that have gradually transformed their ecological and hydrological dynamics. These rivers, once vibrant arteries of the Ganga delta, have been undergoing processes of decay and revival since the eighteenth century, each embodying a unique narrative of transition to transformation.

The Bhagirathi River, a distributary of the Ganga, was historically significant for its role as the lifeline of trade and navigation in the Bengal region. However, the sedimentation and reduced flow from the Ganga rendered it increasingly stagnant, prompting the construction of the Farakka Barrage in 1975. The barrage aimed to divert Ganga waters to revive the Bhagirathi and improve the navigability of the Hooghly River to sustain the Kolkata Port (Sinha 2005). While this intervention initially succeeded in augmenting the Bhagirathi's flow, it also altered the natural hydrological regime of the Nadia Rivers (Sarkar et al. 2024). Downstream, the Jalangi River, which relies on its offtake from the Padma (a major distributary of the Ganga), began to deteriorate rapidly. By the 2020s, the Jalangi had transformed into a near-dead river, clogged with aquatic plants and reduced to a stagnant waterbody (Das 2024a, b; Islam et al. 2024). Its permanent disconnection from the Padma, except for a temporary monsoon link through the Bhairab for a few days per year, highlights the profound ecological imbalance affecting its flow and self-sustaining capacity (Das et al. 2024).

Similarly, the Mathabhanga-Churni system, shared between India and Bangladesh, mirrors the fate of the Jalangi. The Mathabhanga's dwindling flow, coupled with extensive sedimentation, has left the Churni increasingly vulnerable to ecological decline. Although its monsoonal connection with the Padma persists, the steady loss of flow has rendered this system a shadow of its former vitality (Sarkar et al. 2023, 2024).

Natural factors, such as changing monsoonal patterns, and sediment load variations due to climate change, have compounded the issue. Studies indicate that climate change has exacerbated erratic flow regimes, disrupting the seasonal patterns essential for the ecological health of deltaic rivers (Srivastava et al. 2020). Furthermore, unregulated extraction of Ganga waters in upstream regions has reduced the overall discharge reaching the Nadia Rivers, intensifying their desiccation (Das 2013; Sarkar et al. 2021).

The transition to transformation of the Nadia Rivers reflects a complex interplay of human intervention and natural processes. While attempts like the Farakka Barrage were rooted in the vision of revival, they inadvertently triggered ecological challenges downstream. The decline of the Jalangi and Churni highlights the need for a nuanced understanding of riverine systems, recognizing their interconnectedness and vulnerability. The transition to transformation of the Nadia Rivers is a complex interplay of anthropogenic modifications and natural changes. Addressing these challenges requires a comprehensive approach that considers sustainable water management practices, climate adaptation strategies, and regional cooperation to restore and preserve the health of these vital waterways.

1.9 Conclusions

Nadia Rivers encompasses historical, geographical, and cultural significance, reflecting the contributions of Bhagirathi-Hooghly, Jalangi, and Mathabhanga-Churni. The term's historical journey, from association with Nabadwip to colloquial distortion, mirrors the region's evolution. Geographically, Nadia Rivers played a strategic role during British colonial rule, connecting Murshidabad to the Bay of Bengal. The Nadia Rivers Division, established to address navigability challenges, showcases historical efforts to manage these vital waterways.

In terms of geology and sedimentology, Nadia Rivers are integral to the Ganga–Brahmaputra-Meghna delta. Colonial initiatives, such as Major Rennell's survey and the Moor Committee's recommendations in 1915, reveal the then ongoing efforts to maintain navigability. Beyond history, Nadia Rivers are essential to millions, supporting agriculture, trade, industry, and daily life. Culturally and religiously significant, the rivers contribute economically, with Kolkata acting as a crucial port city. Internationally, Mathabhanga's status as an Indo-Bangladesh river and Indo-Bangladesh water treaties related to Farakka Barrage and longevity of Kolkata Port emphasizes collaborative share and conservation efforts. In essence, Nadia Rivers weave together the region's fabric, reflecting human resilience and cooperation. Their multifaceted contributions make them indispensable, symbolizing a saga of human endeavor for the region's well-being and beyond.

In following three chapters (Chapter 2, Chapter 3 & Chapter 4), individual rivers (Bhagirathi, Jalangi, and Mathabhanga-Churni) are illuminated in integrated and holistic approach. These chapters encompass a wide array of topics, starting with etymology and historical context of rivers. They explores the river's significance in epics, history, and recent times. Geographically, it analyzes the river's delineation, size, shape, and changes over time using irrigation waterways reports, SoI toposheets, satellite imageries, and GIS data. The channel and extra-channel geomorphological features, hydrology, ecology, and environmental aspects are examined, highlighting spatiotemporal changes. The riverine society's economy and culture are discussed, including settlement, riverine economy, cultural imprints, and the dynamic interaction between humans and the river across different eras. The chapter concludes by examining planning and management strategies before and after the construction of the Farakka Barrage, encompassing civil, biological, and social engineering approaches. In essence, the chapters provide a concise yet comprehensive overview of the river's diverse dimensions.

Chapter 5, 'Integrated River Basin Management,' presents a holistic framework for managing the Nadia Rivers. It introduces a conceptual approach to an integrated fluvial system and delves into river basin planning and management, covering land, water, biotic resources, and the social system. The chapter addresses land management, including erosion and degradation, water resource management, involving river decay, water quality, and wetland management, and biotic resource management, encompassing afforestation, biodiversity, and aquatic life. It also explores the social system and community behavior, emphasizing responses and management strategies. The chapter culminates with an integrated human-harmony model, providing a comprehensive perspective on river basin management for the Nadia Rivers.

The last Chap. 6, 'Future Challenges and Way Forward,' explores the anticipated obstacles and strategies for managing the Nadia Rivers. It addresses challenges such as climate changes impacting fluvial dynamics, neotectonics jerk affecting river decay, and population pressure influencing fluvial dynamics. The chapter then outlines a strategic 'way forward,' proposing solutions to effectively manage and sustain the Nadia Rivers amidst evolving environmental and demographic factors.

References

Ashwini K, Pathan SA, Singh A (2021) Understanding planform dynamics of the Ganga River in eastern part of India. Spat Inf Res 29:507–518

Asian Development Bank (2019) West Bengal drinking water sector improvement project, sovereign (public) project report, 49107-006. https://www.adb.org/sites/default/files/linked-documents/49107-006-sd-01.pdf. Accessed 7 May 2019

Auerbach LW, Goodbred Jr, SL, Mondal DR, Wilson CA, Ahmed KR, Roy K, Ackerly BA (2015) Flood risk of natural and embanked landscapes on the Ganges–Brahmaputra tidal delta plain. Nature Climate Change 5(2):153–157

Bag R, Mondal I, Bandyopadhyay J (2019) Assessing the oscillation of channel geometry and meander migration cardinality of Bhagirathi River, West Bengal, India. J Geog Sci 29:613–634

Baidya NK (2022) The domestic considerations of Indian Foreign Policy: a case study of Indo-Bangladesh Teesta Water Deal during UPA-II Government. Integr J Res Arts Humanities 2(4):146–154

Bandyopadhyay T, Raghuram G, Sharma Y, Shukla N (2009) Kolkata port trust: a turnaround story. https://doi.org/10.1108/CASE.IIMA.2020.000043

Basu SR, Chakraborty SC (1972) Some considerations on the decay of the Bhagirathi Drainage System. In Bagchi KG (1972). The Bhagirathi-Hooghly Basin. In: Proceedings of the Interdisciplinary Symposium. Calcutta University, pp 59–77

Becker M, Papa F, Karpytchev M, Delebecque C, Krien Y, Khan JU, Shum CK (2020) Water level changes, subsidence, and sea level rise in the Ganges–Brahmaputra–Meghna delta. Proc Nat Acad Sci 117(4):1867–1876

Bhaduri A, Barbier EB (2008) International water transfer and sharing: the case of the Ganges River. Environ Dev Econ 13(1):29–51

Bhaumik S (2018) Kolkata Port: challenges of geopolitics and globalization. In: Neilson B, Rossiter N, Samaddar R (eds) Logistical Asia. Palgrave Macmillan, Singapore. https://doi.org/10.1007/978-981-10-8333-4_6

Biswas M, Dhara P (2019) Evolutionary characteristics of meander cut-off—a hydro-morphological study of the Jalangi River, West Bengal, India. Arab J Geosci 12:1–21

Bowen HV (1988) A question of sovereignty? The Bengal land revenue issue, 1765–67. J Imperial Commonwealth Hist 16(2):155–176

Broadus J, Milliman J, Edwards S, Aubrey DG, Gable F (1986) Rising sea level and damming of rivers: possible effects in Egypt and Bangladesh. Effects Changes Stratosph Ozone Global Climate 4:165–189

Bruce J, Dundas H (1793) Historical view of plans, for the Government of British India, and Regulation of Trade to the East Indies: and outlines of a plan of foreign government, of commercial economy, and of domestic administration, for the Asiatic Interests of Great Britain (No. 15633). J Sewell

Chadha SM (1991) Survey of India through the Ages. Himalayan J 47

Chakrabarty D, Das SK (2006) Fish community structure and ecological degradation in tropical rivers of India. Web Ecol 6(1):27–36

Chaudhury J (2004) Sri Chaitanyadev O Samakalin Nabadwip. Nabadwip Purattatva Parishad. Nabadwip, pp 2, 5–7

Characterising topophilic behaviour in the wake of river decay and pollution through structural equation modelling

Coates DA (1990) The Mymensingh terrace: evidence of Holocene deformation in the delta of the Brahmaputra River, central Bangladesh. Geol Soc Am Abstr Progr 22:310

Coates DA, Whitney JW, Alam AKM (1988) Evidence of neotectonic activity on the Bengal delta, Bangladesh (abstract): Geological Society of America. Abstracts Prog 20:54

Colebrooke RH (1803) On the course of the Ganges through Bengal. Trans Asiatic Soc 7:1–32. https://archive.org/details/asiaticresearche07asia/page/n16/mode/1up?view=theater

Colebrooke RH (1824) Internal navigation in India. Extract from a plan lately submitted to government for permanent communication by water, between the upper and eastern provinces of India and Calcutta; during the dry season. In: The Asiatic Journal and Monthly Register for British India and its dependencies, vol XVII, pp 237–241

Das J (2004) Bangla Bhasar Abhidhan, vol-II. Sahitya Sansad, Kolkata

Das BC (2013) Changes and deterioration of the course of river Jalangi and its impact on the people living on its banks Nadia West Bengal. Ph. D. Thesis. University of Calcutta, pp 47–50

Das BC (2015) Crying with the river: a study on a dying river and her famished fishermen. In: Life and Living Through Newer Spectrum of Geography, vol 3

Das BC (2023a) In Search Of Birthplace Of Sri Chaitanya: Part I. A Study On Hagiographies Up To 17th Century. Studia Universitatis Babeş-Bolyai Geographia, 63–91. https://doi.org/10.24193/subbgeogr.2023.1.05

Das BC (2023b) Cūrṇī ēka curi karā jalēra nadī: Janaśruti āra satyēra sandhānē (In Bengali). Maadhukari.com. https://www.maadhukari.com/apr-2023

Das BC (2024a) Morphological dynamics of mid-channel bars under controlled river regime. Water Resources Management. https://doi.org/10.1007/s11269-024-03901-0

Das BC (2024b) Exploring the past to discern the present: the role of historico-geographical study in differentiating natural river channels from artificial canals. River. https://doi.org/10.1002/rvr2.73

Das BC, Bhattacharya S (2020) The Jalangi: a story of killing of a dying river. In: Anthropogeomorphology of Bhagirathi-Hooghly river system in India. CRC Press, Boca Raton, pp 381–431

Das S, Bandyopadhyay S (2022) The millennium flood of the Upper Ganga Delta, West Bengal, India: a remote sensing based study. Applied geomorphology and contemporary issues. Springer International Publishing, Cham, pp 499–517

Das SK, Chakrabarty D (2007) The use of fish community structure as a measure of ecological degradation: a case study in two tropical rivers of India. Biosystems 90(1):188–196

Das BC, Das D (2020a) The Anjana: a journey from river to canal. In: Das et al. (eds) Anthropogeomorphology of Bhagirathi-Hooghly river system in India. CRC Press. Taylor and Francis Group. https://doi.org/10.1201/9781003032373

Das BC, Das S, Sarkar B (2023a) Floods of Jalangi and Mathabhanga-Churni Rivers, Indo-Bangladesh. In Islam et al (2023). In Floods in the Ganga–Brahmaputra–Meghna Delta. Springer Nature. https://doi.org/10.1007/978-3-031-21086-0

Das BC, Das S, Sarkar B (2023b) Floods of Jalangi and Mathabhanga-Churni rivers, Indo-Bangladesh. Floods in the Ganga–Brahmaputra–Meghna Delta. Springer, Cham, pp 465–489

Das BC, Das S, Sarkar B, Das S, Adhikari R, Saha D et al (2024a) Geospatial assessment of water quality of a dying tropical river and its environmental implications. In: Spatial modeling of environmental pollution and ecological risk. Woodhead Publishing, pp 409–421

Das BC, Ghosh S, Islam A, Roy S (2020) An appraisal to anthropogeomorphology of the Bhagirathi-Hooghly River System: concepts, Ideas and Issues. In Das et al. (2020) eds. Anthropogeomorphology of Bhagirathi-Hooghly River System in India. ISBN (eBook) 9781003032373. Taylor & Francis Group https://doi.org/10.1201/9781003032373

Das BC, Islam A (2016) An enquiry into fitting natural channel shape to geometric shape: a study on River Jalangi, India. In: Neo-Thinking on Ganges-Brahmaputra Basin geomorphology, pp 65–80

Das BC, Islam A, Biswas B (2020) Morphometry as tool to trace out the genealogy of oxbow lake. Environ Earth Sci 79:137. https://doi.org/10.1007/s12665-020-8854-3

Das BC, Islam A, Das S, Adhikary R, Sarkar R (2024b) Bridge Construction Impacts the Adjacent Channel Morphology and Turbidity: a Study on a Dying Distributary of River Padma. Earth Science, Syst Soc 4:10099. https://doi.org/10.3389/esss.2024.10099

Das A, Majumder S, Barman S, Chatterjee D, Mukhopadhyay S, Ghosh P et al (2021) Influence of basin-wide geomorphology on arsenic distribution in Nadia district. Environ Res 192:110314

Das S, Sarkar R (2019) Predicting the land use and land cover change using Markov model: a catchment level analysis of the Bhagirathi-Hugli River. Spat Inf Res 27:439–452

Datta S, Neal AW, Mohajerin TJ, Ocheltree T, Rosenheim BE, White CD, Johannesson KH (2011) Perennial ponds are not an important source of water or dissolved organic matter to groundwaters with high arsenic concentrations in West Bengal, India. Geophys Res Lett 38(20)

De G, De M (2023) Flood frequency analysis with focus on Rivers Mundeswari, Ajoy and Churni. In Emerging Technology and Management Trends in Environment and Sustainability (pp 32–42). Routledge

Donselaar ME, Bhatt AG, Ghosh AK (2017) On the relation between fluvio-deltaic flood basin geomorphology and the wide-spread occurrence of arsenic pollution in shallow aquifers. Sci Total Environ 574:901–913

D'Souza R (2015) Drainage, river erosion, and chaurs: an environmental history of land in colonial Eastern India. In: Nehru Memorial Museum and Library Occasional Paper: History and Society, vol 84, p 22

Fergusson J (1863) On recent changes in the Delta of the Ganges. Q J Geol Soc Lond XIX(1–2):321–354

Frazer RW (1897) British India, vol 46. GP Putnam's Sons, New York; T. Fisher Unwin., London, pp 39–41

Garrett JHE (1910) Bengal District gazetteers. Nadia. Bengal Secretariat Book Depot, Calcutta, pp 1–21

Ghosh D (2022) Identification of prime factors of active river bank erosion in the lower course of Ganga Bhagirathi River: a study. Bull Geogr Phys Geogr Ser 23:71–83

Ghosh S, Hoque MM, Islam A, Barman SD, Mahammad S, Rahman A, Maji NK (2023) Characterizing floods and reviewing flood management strategies for better community resilience in a tropical river basin, India. Nat Hazards 115(2):1799–1832

Ghosh R, Pal S (2023) Delineation of vegetation shaded ox-bow lakes in Ganges flood plain, India. Ecol Inf 74:101954

Ghosh D, Sahu AS (2023) Population shifting and its consequences on women's life: a case study along the river banks of Ganga-Bhagirathi in Jangipur subdivision, West Bengal. Environmental management and sustainability in India: case studies from West Bengal. Springer, Cham, pp 411–424

Ghosh D, Shirodkar AD (2023) Adaptive reuse of abandoned jute mills along the hooghly river in bengal: a sustainable approach. In: Proceedings of the International Conference of Contemporary Affairs in Architecture and Urbanism-ICCAUA, vol 6(1), pp 733–741

Goodbred SL Jr (2003) Response of the Ganges dispersal system to climate change: a source-to-sink view since the last interstade. Sed Geol 162(1–2):83–104

Goodbred SL Jr, Kuehl SA (1998) Floodplain processes in the Bengal Basin and the storage of Ganges-Brahmaputra river sediment: an accretion study using ^{137}Cs and ^{210}Pb geochronology. Sed Geol 121(3–4):239–258

Goodbred Jr, SL, Kuehl SA (1999) Holocene and modern sediment budgets for the Ganges-Brahmaputra river system: evidence for highstand dispersal to flood-plain, shelf, and deep-sea depocenters. Geology 27(6):559–562

Goodbred Jr, SL, Kuehl SA (2000) Enormous Ganges-Brahmaputra sediment discharge during strengthened early Holocene monsoon. Geology 28(12):1083–1086

Govt of Bengal (1931) Annual Report on major rivers in the Nadia Rivers Division. Irrigation Department Bengal Secretariat Press. 1925, 1930, 1929, 1932, 1933, 1931

Govt. Of WB (2024b) Trade & Industry History of West Bengal. http://nriwestbengal.gov.in/nri/common/Industrial_History.aspx#:~:text=The%20first%20%22factory%22%20(that,residency%20was%20separated%20from%20Madras. Retrieved on 30.01.2024

Govt. of WB (2024a) Irrigation and Waterways Department. https://wbiwd.gov.in/index.php/applications/about. Accessed 30 Jan 2024

Govt. of WB (2024b) Department of Industry, Commerce and Enterprises. Government of West Bengal. https://wbindustries.gov.in/index.html. Accessed 30 Jan 2024

Govt. of WB (2024c) Trade and Industry History of West Bengal. http://nriwestbengal.gov.in/nri/common/Industrial_History.aspx#:~:text=The%20first%20%22factory%22%20(that,residency%20was%20separated%20from%20Madras. Accessed 30 Jan 2024

Gregory KJ (2021) Foreword. In: Das et al (eds) Anthropogeomorphology of Bhagirathi-Hooghly River System in India

Guchhait SK, Islam A, Ghosh S, Das BC, Maji NK (2016) Role of hydrological regime and floodplain sediments in channel instability of the Bhagirathi River, Ganga-Brahmaputra Delta, India. Phys Geogr 37(6):476–510

Hirst FC (1916) Report on the Nadia Rivers 1915, The Bengal Secretariat Book Depot, Calcutta, 1916

Hoque MM, Islam A, Islam ARMT, Das BC, Pal, SC, Arabameri A, Khan R (2023) Spatio-temporal assessment of water quality of a tropical decaying river in India for drinking purposes and human health risk characterization. Environ Sci Pollut Res 30(45):101653–101668

Hunter WW (1875) A statistical account of Bengal, vol II. Trubner & Co., London, pp 18–33

Hunter WW (1877) A statistical account of Bengal, vol II. Trubner & Co., London, p 19

Iftikhar-ul-Awwal AZM (1978) The industrial development of Bengal 1900–1939: an examination of the economic features of an underdeveloped area. School of Oriental and African Studies, University of London, London

Inglis WA (1909) The canals and flood banks of Bengal. The Bengal Secretariat Press, Calcutta. Cited from Biswas KR (eds) Rivers of Bengal, vol V(II). West Bengal Dist Gazetteers, Higher Education Department, Government of West Bengal, Kolkata, p 48

Islam A, Guchhait SK (2017) Search for siocal justice for the victims of erosion hazard along the banks of river Bhagirathi by hydraulic control: a case study of West Bengal, India. Environ Dev Sustain 19:433–459

Islam A, Guchhait SK (2024) Riverbank erosion: basic concepts and approaches. In Riverbank erosion in the Bengal delta: An integrated perspective (pp 3–22). Cham: Springer International Publishing

Islam A, Guchhait SK (2020) Characterizing cross-sectional morphology and channel inefficiency of lower Bhagirathi River, India, in post-Farakka barrage condition. Nat Hazards 103(3):3803–3836

Johnston J (1933) Inland navigation on the Gangetic rivers. Thacker, Spink and Co, Calcutta, p 44

Johnston J (1947) Inland navigation on the Gangetic Rivers. Thacker, Spink & Co. (1933) Ltd., Calcutta, p 16

Kanuri VV, Saha R, Raghuvanshi SP, Singh AK, Chakraborty BD, Kumar, VK, Saxena RC (2020) Sewage fluxes and seasonal dynamics of physicochemical characteristics of the Bhagirathi-Hooghly River from the lower stretch of River Ganges, India. Chem Ecol 36(1):30–47

Karmokar S, Halder M, Nandy S (2021) Identification of paleochannels using remote sensing: a study on upper Gangetic Delta in parts of Murshidabad and Nadia Districts, West Bengal. Remote Sens Earth Syst Sci 4(1–2):104–120

Khasnabis R (2008) The Economy of West Bengal. Econ Polit Week 103–115

Kirk W (1956) The cotton and jute industries of India: a study in concentration and dispersal. Scott Geogr Mag 72(1):38–52

Kuehl SA, Allison MA, Goodbred SL, Kudrass H (2005) The Ganges-Brahmaputra delta

Kumar S, Halder S, Singhal DC (2011) Groundwater resources management through flow modeling in lower part of Bhagirathi—Jalangi interfluve, Nadia, West Bengal. J Geol Soc India 78:587–598

Lang (1848) Report to Government, 14 Jul 1848

Lang J (1853) Memorandum on the Ganges and the entrance of the Nuddea Rivers. Bengal secretariate Book Depot Calcutta. p. 109.

Lang J (1926) Report on the Nuddea Rivers and the advantages derived from the measures annually adopted for facilitating their navigation. No. 89, dated Kishnaghur, the 14th July 1848. From Captain John Lang, 36th Regiment, Bengal N.I, Officiating Superintendent, Nuddea Rivers. To the Officiating Superintending Engineer, Lower Provinces, in Nadia Rivers (from 1848–1926), pp 72–89. D'Souza R (2015)

Lawler DM (1993) The measurement of river bank erosion and lateral channel change: a review. Earth Surface Process Landforms 18(9):777–821

Lee S (ed) (1896) Rennell, James. Dictionary of National biography, vol 48. Smith, Elder & Co., London, pp 14–15

Long JR (1846) The banks of the Bhagirathi. Calcutta Rev VI(XII)

Majumder D (1978) West Bengal District Gazetteers Nadia. Govt of W. B, pp 5, 7, 16, 25–34

Mandal AC, Bhunia GS (2022) Spatio-temporal variation of morphological characteristics in Bhagirathi River—case study in Murshidabad District, West Bengal (India). In: Drainage basin dynamics: an introduction to morphology, landscape and modelling, pp 179–191

Mandal MH, Yasmin B, Roy A, Ghosh S, Siddique G (2022) Investigating present status of floodplain wetlands as habitat of water birds and its determinants: an experience from lower part of deltaic West Bengal, India. Wetlands 42(7):70

Milliman JD, Broadus JM, Gable F (1989) Environmental and economic implications of rising sea level and subsiding deltas: the Nile and Bengal examples. Ambio 340–345

Michels KH, Suckow A, Breitzke M, Kudrass HR, Kottke B (2003) Sediment transport in the shelf canyon "Swatch of No Ground" (Bay of Bengal). Deep Sea Res Part II 50(5):1003–1022

Mitra A, Chowdhury R, Banerjee K (2012) Concentrations of some heavy metals in commercially important finfish and shellfish of the River Ganga. Environmental Monit Assess 184:2219–2230

Moitra S, Maity SG, Haldar P, Pandit AK, Sahu S (2015) Trawler fuel exhaust and respiratory impairments: a cross-sectional pilot study among Indian fishermen working in informal sectors. Int J Occup Environ Health 21(3):185–191

Mondal BK, Basu T, Das R, Mahata S, Lee MA (2023) Geospatial assessment of fluvial dynamics and associated flood vulnerability in the Bhagirathi-Hugli Sub-basin in West Bengal, India. Floods in the Ganga–Brahmaputra–Meghna Delta. Springer, Cham, pp 491–527

Mondal D, Pal S (2015) A multi-parametric spatial modeling of vulnerability due to arsenic pollution in Murshidabad district of West Bengal, India. Arab J Geosci 8:8047–8054

Mondal I, Bandyopadhyay J (2022) Morphological landscape mapping of the Bhagirathi Flood Plains in West Bengal, India, using geospatial technology. In: Drainage basin dynamics: an introduction to morphology, landscape and modelling, pp 543–564

Moore et al (1919) Report on the Hooghly River and its head-waters. vol I. West Bengal Secretariat Book Depot, Calcutta. In: Biswas KR (2001) Rivers of Bengal, vol II. Government of West Bengal, Kolkata, pp 27, 49

Mukherji SC (1991) The changing face of Calcutta: an architectural approach, Calcutta: 300. Director of Archaeology, Department of Information and Cultural Affairs, Government of West Bengal

Mukherjee A, Fryar AE (2008) Deeper groundwater chemistry and geochemical modeling of the arsenic affected western Bengal basin, West Bengal, India. Appl Geochem 23(4):863–894

Neilson B, Rossiter N, Samaddar R (eds) (2018) Logistical Asia: the labour of making a world region. Springer, Cham

Nicholls RJ, Goodbred SL (2004, January) Towards integrated assessment of the Ganges-Brahmaputra Delta. In Proceedings of 5th International Conference on Asian Marine Geology, and 1st Annual Meeting of IGCP475 DeltaMAP and APN Mega-Deltas (p 9)

Pal R, Biswas SS, Pramanik MK, Mondal B (2017) Bank vulnerability and avulsion modeling of the Bhagirathi-Hugli river between Ajay and Jalangi confluences in lower Ganga Plain, India. Model Earth Syst Environ 1(3):1–1

Paszkowski A, Goodbred Jr S, Borgomeo E, Khan MSA, Hall JW (2021) Geomorphic change in the Ganges–Brahmaputra–Meghna delta. Nature Rev Earth Environ 2(11):763–780

Panigrahi AK, Pattnaik S (2021) Studies to estimate the pollution load of some South Bengal Rivers and its impacts: a survey. Proc Zool Soc 74:83–90

Paria T, Konar SK (2003) Ecological degradation of some rivers in West Bengal. In: River Pollution in India and Its Management, p 237

PIB (2018) Operational National Waterways in the Country. Government of India

Rahman KS, Islam Z, Navera UK, Ludwig F (2019) A critical review of the Ganges Water Sharing arrangement. Water Policy 21(2):259–276

Rahman MM, Sengupta MK, Ahamed S, Chowdhury UK, Hossain MA, Das B et al (2005) The magnitude of arsenic contamination in groundwater and its health effects to the inhabitants of the Jalangi—one of the 85 arsenic affected blocks in West Bengal, India. Sci Total Environ 338(3):189–200

Reaks HG (1919a) Report on the physical and hydraulic characteristics of the rivers of the delta. In: Moor et al. (eds) Report on the Hooghly River and its headwaters, vol I. The Bengal Secretariat Book Depot, Calcutta, 1919, In: Biswas (2001) Rivers of Bengal, vol II, Govt of West Bengal, pp 87, 107

Reaks HG (1919b) Report on the physical and hydraulic characteristics of the delta. In: Stevenson Moore Committee report on the Hooghly River and its headwaters, vol 1. Bengal Secretariat Book Depot, Bengal

Reitz MD, Pickering JL, Goodbred SL, Paola C, Steckler MS, Seeber L, Akhter SH (2015) Effects of tectonic deformation and sea level on river path selection: Theory and application to the Ganges-Brahmaputra-Meghna River Delta. J Geophys Res Earth Surface 120(4):671–689

Rennell J (1781) Account of the Ganges and Burrampooter Rivers. Philos Trans R Soc Lond LXXI:87–114

Rennell J (1788) Memoir of a map of Hindoostan; or The Mogul Empire: with an introduction, illustrative of the geography and present division of that country: and a map of the countries situated between the head of the Indus, and the Caspian Sea. Printed for the author: and sold by George Nicol ... and W. Richardson, London. urn:oclc:record:1048777295

Rennell J (1910) The journals of Major James Rennell. First surveyor-general of India, written for the information of the governors of Bengal during his surveys of the Ganges and Braghmaputra rivers 1764 to 1767. Baptist Mission Press and Asiatic Society, Calcutta. urn:oclc:record:1047475661

Roy MB, Ghosh A, Kumar A, Roy PK (2021a) Assessing the nature of seasonal meteorological change in people's dependency on wetland: a case study of Bhagirathi-Hooghly floodplain system. Environ Dev Sustain 23(12):17881–17903

Roy MB, Ghosh A, Kumar A, Roy PK (2021b) Study of water productivity model on wetlands: a case study of Bhagirathi-Hooghly sub-catchment basin, Southern West Bengal, India. Environ Monitor Assess 193(10):675

Rudra K (2018) The Bhagirathi-Hugli River system. In: Rivers of the Ganga-Brahmaputra-Meghna Delta: a fluvial account of Bengal, pp 77–93

Rudra K (2022) Changing river courses in Bengal (1780 to 2020). Riverine systems: understanding the hydrological, hydrosocial and hydro-heritage dynamics. Springer, Cham, pp 403–422

Sanyal J, Lu XX (2006) GIS-based flood hazard mapping at different administrative scales: a case study in Gangetic West Bengal, India. Singapore J Trop Geogr 27(2):207–220

Sarkar B, Das BC (2022) A cross-sectional study on the water quality and ecosystem health of the Jalangi and Bhagirathi river and their selected Oxbow-lakes. Fluvial systems in the anthropocene: process, response and modelling. Springer, Cham, pp 353–367

Sarkar B, Islam A, Das BC (2021) Role of declining discharge and water pollution on habitat suitability of fish community in the Mathabhanga-Churni River, India. J Clean Prod 326:129426

Sarkar B, Islam A, Datta D (2023) Characterising topophilic behaviour in the wake of river decay and pollution through structural equation modelling. Environ Dev Sustain 1–32

Sarkar B, Das BC, Islam A, Datta D, Pawlik Ł, Quesada-Román A (2024) Temporal change in channel form and hydraulic behaviour of a tropical river due to natural forcing and anthropogenic interventions. Phys Geogr 45(5):483–517

Sarkar B, Islam A (2019) Assessing the suitability of water for irrigation using major physical parameters and ion chemistry: A study of the Churni River, India. Arabian J Geosc 12(20):1–16

Sarkar B, Islam A (2020) Drivers of water pollution and evaluating its ecological stress with special reference to macrovertebrates (fish community structure): a case of Churni River, India. Environ Monit Assess 192(1):45

Sarkar B, Islam A (2022) Assessing poverty and livelihood vulnerability of the fishing communities in the context of pollution of the Churni River, India. Environ Sci Pollut Res 29(18):26575–26598

Singh KN, Roy BK, Chatterjee B, Das B (1993) Lower Ganga Plain. In Singh RL (eds) India: a regional geography. National Geographical Society of India, Varanasi; UBSPD, New Delhi, pp 252–302

Sinha M (2005) Farakka barrage and its impact on the hydrology and fishery of Hooghly estuary. The Ganges water diversion: environmental effects and implications, pp 103–124

Sreemani S (2003) River and the settlement: the Bhagirathi and Murshidabad; between the 17th and 19th centuries. In: Proceedings of the Indian History Congress, vol 64, pp 741–752. Indian History Congress

Srivastava A, Singhal A, Jha PK (2020) Climate change—implication on water resources in South Asian Countries. In: Resilience, response, and risk in water systems: shifting management and natural forcings paradigms, pp 217–240

Sur SC (1926) Notes on the changes of the heads of the Mathabanga river since 1825 compiled by Mr. S. C. Sur, Executive Engineer, Nadia Rivers Division', in Nadia Rivers (From 1848–1926), p. 35. D'Souza R (2015)

The Hindu (2024, 15th January) Record one crore pilgrims at Ganga. West Bengal Government, Sagar Mela. https://www.thehindu.com/news/cities/kolkata/record-one-crore-pilgrims-at-ganga-sagar-mela-west-bengal-government/article67743135.ece#:~:text=Chief%20Minister%20Mamata%20Banerjee%20in,likely%20to%20surpass%20the%20number%E2%80%9D

Tudsbery JHT (ed) (1900) Minutes of Proceedings of Institution of Civil Engineers; with other selected and abstracted papers, vol CXXXIX. Published by the Institution, London

van Vliet MTH et al (2023) Global river water quality under climate change and hydroclimatic extremes. Nat Rev Earth Environ 1–16. https://doi.org/10.1038/s43017-023-00472-3

Wilson CA, Goodbred Jr, SL (2015) Construction and maintenance of the Ganges-Brahmaputra-Meghna delta: linking process, morphology, and stratigraphy. Ann Rev Mar Sci 7(1):67–88

2 Bhagirathi River: Journey of an Oscillatory River

Abstract

This chapter intends to portray the dynamics of the Bhagirathi River from the spatial and temporal dimensions. The river Bhagirathi is the principal river among the 'Nadia Rivers'. This river is extended from Farakka Barrage outfall at Ahiran to Nabadwip where River Jalangi meets with this one-way non-tidal river. The journey of this river is marked by the oscillatory events. Once a mighty river (National Waterway no. 1) turned into a decaying river due to the eastward titling of the Bengal basin again rejuvenated through the induction of discharge from the Ganga River through the Feeder canal by construction of the Farakka Barrage Project in 1975. This work has attempted a holistic evaluation of the river from the perspectives of the hydrogeomorphic changes, deteriorating water quality, dynamics of the environmental flow and socio-economic transformations in the wake of the fluvial dynamics (e.g. bank erosion, floods, siltation of the river) with an eye to the sustainable management of the river for the development of the local community.

Keywords

Nadia rivers · Farakka barrage project · Controlled hydrology · River decay · Bengal basin · Environmental flow · Water quality · River bank erosion · Floods · Social vulnerability · River restorations

2.1 Etymology and History

2.1.1 Etymology

The name 'Bhāgīrathī' (भागीरथी) holds the connotation of being the daughter of *Bhagīratha* (भगीरथ). The word can be linguistically interpreted to signify "the daughter of Bhagīratha". The etymology of the name can be broken down as follows:

Bhagīratha (भगीरथ): This is a Sanskrit name composed of two elements: Bhagī (भगी): It is derived from the Sanskrit root 'bhaj' which means 'to share' or 'to apportion'. Ratha (रथ): This means 'chariot' or 'vehicle'. Therefore, 'Bhagīratha' can be translated as 'the one who shares or apportions a chariot' or 'chariot of

B. C. Das et al., *The Nadia Rivers*, Geography of the Physical Environment,
https://doi.org/10.1007/978-3-031-85971-7_2

prosperity'. Bhāgīrathī (भागीरथी): The suffix "-rathī" denotes a river, but in this context, it can be understood as 'the daughter of Bhagiratha.' Thus, 'Bhāgīrathī' can be translated as 'the daughter (river) of the king Bhagīratha associated with the chariot of prosperity'. This etymology aligns with the mythological narrative where Bhagīratha's penance and efforts lead to the descent of the Ganges, represented as Bhāgīrathī, as his divine daughter.

2.1.2 River in Epic

As narrated in Ramayana[1] in a bid to assert his sovereignty, King Sagara conducted the '*Ashvamedha*' ritual, allowing a horse to roam freely for a year. However, the success of the ritual was thwarted as Indra stole the horse (Eck 2012). Upon discovering the disappearance of the horse, King Sagara dispatched his sixty-thousand sons to search for it. They eventually located the horse at the ashram of the Sage Kapila. Suspecting the sage of theft, the sons disrupted his deep meditation, enraging sage Kapila. With his ascetic gaze, he incinerated all sixty thousand sons to ashes. In a quest for redemption, King Sagara sent his grandson, Anshuman, to seek guidance from sage Kapila. Sage Kapila revealed that only the heavenly waters of the Ganges could liberate the souls of the sons. Taking up the cause, Bhagīratha, Anshuman's grandson, engaged in ascetic practices and earned the favor of Brahma and Shiva.

देव देवे गते तस्मिन् सो अंगुष्ठ अग्र निपीडिताम् |
कृत्वा वसुमतीम् राम वत्सरम् समुपासत || १-४३-१

deva deve gate tasmin so aṅguṣṭha agra nipīḍitām |
kṛtvā vasumatīm rāma vatsaram samupāsata ||

अथ संवत्सरे पूर्णे सर्व लोक नमस्कृतः |
उमापतिः पशुपती राजानम् इदम् अब्रवीत् || १-४३-२

atha saṃvatsare pūrṇe sarva loka namaskṛtaḥ |
umāpatiḥ paśupatiḥ rājānam idam abravīt ||

प्रीतः ते अहम् नरश्रेष्ठ करिष्यामि तव प्रियम् |
शिरसा धारयिष्यामि शैलराज सुताम् अहम् ||१-४३-३

prītaḥ te aham naraśreṣṭha kariṣyāmi tava priyam |
śirasā dhārayiṣyāmi śailarāja sutām aham ||

ततो हैमवती ज्येष्ठा सर्व लोक नमस्कृता |
तदा सा अति महत् रूपम् कृत्वा वेगम् च दुःसहम् ||१-४३-४

tato haimavatī jyeṣṭhā sarva loka namaskṛtā |
tadā sā ati mahat rūpam kṛtvā vegam ca duḥsaham ||

आकाशात् अपतत् राम शिवे शिव शिरस्य् उत |
अचिन्तयः च सा देवी गंग परम दुर्धरा ||१-४३-५

ākāśāt apatata rāma śive śiva śirasy uta |
acintayaḥ ca sā devī gaṅgā parama durdharā ||

[Meaning: 'When the god of gods, Brahma, departed, Bhagīratha stood on the tip of his big toe, praying to Lord Shiva for mercy for an entire year, while the pressure of his toe bore down on the earth.' Thus, Vishvamitra continued narrating Bhagīratha's efforts to bring Ganga to earth. At the end of the year, Shiva, revered by all worlds, the consort of Uma, and the lord of all creatures, from insects to humans, revealed himself and spoke to the king: 'O best of men, I am pleased with your unwavering dedication and will fulfill your wish. I will bear Ganga, the daughter of the king of mountains, on my head.' Thereafter, Ganga, who is revered by all beings and the elder daughter of Himavan, assumed a fierce and uncontrollable form, and with an unstoppable speed, descended from the sky onto Shiva's sacred head. The mighty goddess Ganga, a river of immense power, even thought to herself, 'I will sweep Shiva away with my torrents and flow into the netherworld.'].

Bhagīratha stood on one foot with strong determination and prayed with raised hands for a whole year, day and night, living on just air. His intense focus caused pressure on the earth. After a year, Shiva, who is respected by everyone, appeared and spoke to the king. 'Great human, I'm pleased with your effort. I'll make your wish come true. I'll hold Ganga, the mountain king's daughter, on my head.' Later, Ganga, an extremely unendurable river, respected by

[1] Valmiki Ramayana. Bala Kanda. 43rd Sarga. Couplets 1–42. https://sanskritdocuments.org/sites/valmikiramayan/vr_index.htm.

everyone and the elder daughter of Himalaya, took on a powerful form and incredible speed. She then fell from the sky onto Shiva's head (Sen 2019). And from there, the mighty river Ganga followed the path in whichever direction King Bhagīratha had advanced.

भगीरथो राजर्षिः दिव्यम् स्यंदनम् आस्थितः ||१-४३-३०
प्रायात् अग्रे महाराजाः तम् गंगा पृष्ठतो अन्वगात् |

bhagīratho rājarṣiḥ divyam syandanam āsthitaḥ ।
prāyāt agre mahārājaḥ tam gaṅgā pṛṣṭhato
anvagāt ॥"

यतो भगीरथो राजा ततो गंगा यशस्विनी || १-४३-३३
जगाम सरिताम् श्रेष्ठा सर्व पाप प्रणाशिनी |

yato bhagīratho rājā tato gaṅgā yaśasvinī ।
jagāma saritām śreṣṭhā sarva pāpa praṇāśinī ॥

[Meaning: In this manner, the illustrious and royal sage Bhageeratha, seated in a divine chariot, steadily moved forward, with Ganga faithfully following behind him. Wherever King Bhagīratha advanced, the magnificent River Ganga, the foremost among all rivers and the purifier of sins, flowed in his wake, following him in every direction.]

গঙ্গাকে কহিয়া এই বাক্য জগৎপতি।
শঙ্খ দিয়া কহিলেন ভগীরথ প্রতি।।
আগে আগে যাহ তুমি শঙ্খ বাজাইয়া।
পশ্চাতে যাবেন গঙ্গা তোমাকে দেখিয়া।।
(ভট্টাচার্য 1940).

Gaṅgākē kahiẏā ē'i bākya jagaṯpati.
Śaṅkha diẏā kahilēna bhagīratha prati;-
āgē āgē yāha tumi śaṅkha bājā'iẏa.
Paścātē yābēna gaṅgā tōmākē dēkhiẏa.

[Meaning: The Lord of the Universe spoke to Ganga, then handed a conch to Bhagīratha, instructing him to move forward while sounding the conch, and the river Ganga would follow him.]

When Ganga descended from Shiva's locks to the earth and followed the chariot of King Bhagīratha, her waters caused havoc in Jahnu's fields and disturbed his penance. In anger, the sage Jahnu drank up all of Ganga's waters as a punishment. The devas intervened, pleading with Jahnu to release Ganga for Bhagiratha's mission to liberate ancestors' souls. Jahnu, relenting, released Ganga from his ear. As a result, the Ganges River is also called *Jahnavi*, signifying 'daughter of Jahnu'. Ultimately, the Ganga (Bhagirathi River) flowed into the Bay of Bengal at Sagar Island, cleansing the sins of King Sagar's sixty thousand sons with the holy waters of the Ganges, and granting them heavenly salvation.

अथ तत् भस्मनाम् राशिम् गंग सलिलम् उत्तमम् |
प्लावयत् पूत पाप्मानः स्वर्गम् प्रप्ता रघु उत्तम ||१-४३-४१

atha tat bhasmanām rāśim gaṅga salilam uttamam ।
plāvayat pūta pāpmānaḥ svargam prāptā raghu uttama ॥

[Meaning: Oh, Rama, the best of Raghu's dynasty, then Ganga inundated that mound of ashes of Sagara's six-thousand sons, by which those souls obtained heaven, while the sins of souls are cleansed with the water of Ganga. Thus Vishvamitra continued his narration.]

The River Ganges (Ganga) undergoes a bifurcation near Farakka, with its right branch named Bhagirathi extending up to Nabadwip. Beyond Nabadwip, this watercourse is referred to as Hooghly. Despite the transition, the Bhagirathi-Hooghly segment retains its sacred significance for the general populace and is commonly referred to as Ganga. On the other hand, the left branch takes the name Padma and lacks any sacred significance. The Valmiki Ramayana does not provide information on the existence and origin of the Padma River. According to the rendition in the Ramayana by Krittibas, a 15th-century Bengali poet, as king Bhagīratha and river Ganga reached Gour in the present-day Maldah district, sage Padma blew his conch towards the east, causing the Ganga to flow in that direction and become the river Padmavati (Padma). Subsequently, Bhagirath discovered the easterly course of the Ganga, sounded his conch, and proceeded southward to Ganga Sagar. Thus, the mythological origin of the river Padma is narrated in this context.

পদ্ম নামে এক মুনি পূর্ব্বমুখে যায়।
ভগীরথ বলি গঙ্গা পশ্চাৎ গোড়ায়।।
যোড়হাত করিয়া বলেন ভগীরথ।
পূর্ব্বদিক যাইতে আমার নাহি পথ।।

পদ্ম মুনি লয়ে গেল নাম পদ্মাবতী।
ভগীরথ সঙ্গেতে চলিল ভাগীরথী॥
শাপবাণী সুরধুনী দিলেন পদ্মারে।
মুক্তিপদ যেন নাহি হয় তব নীরে॥ (ভট্টাচার্য 1940; Banglalibrary 2025).

padma nāme ek muni pūrvamukhe yāy |
bhagīrath bali gaṅgā paścāte goṛāy ||
joṛahāt kariyā balilen bhagīrath |
pūrvabadik yāite āmār nāhi path ||
padmamuni loye gel nām padmāvatī |
bhagīrath saṅgete calil bhāgīrathī ||
śāpabāṇī suradhunī dilen padmāre |
muktipad yena nāhi hoy tab nīre ||

[Meaning: This passage describes an event where the river Ganga, under Bhageeratha's guidance, was momentarily diverted by a sage named Padma, and the resulting river, Padmavati, received a curse that its waters would not grant liberation (moksha) like the Ganga.]

Mentioning Mahabharata as the source of the narrative, similar story with minute distortion regarding the origin of the river Padma was narrated by Willcocks (1930). '… as Bhagīrath followed by Ganga descended the Ganges Valley near the head of the Delta, Bhagīrath took rest to eat his meal and Ganga, hearing the sound of Padmavati's shell, thought it was Bhagirath's and followed her in her eastern course down the Padma. It was then that, Bhagirath sounded his shell and Ganga recognized her mistake. She retraced her steps and went southwards following the course of Bhagirathi-Hooghly'.

2.2 Geography and Location

2.2.1 Bhagirathi River System

The Ganga–Brahmaputra–Meghna delta includes the Bhagirathi River Basin (BRB). This river, a distributary of the Ganga, travels approximately 230 km across the state of West Bengal until joining the Jalangi River in Nabadwip. As a result, the BRB extends from 23°23′8″ N to 24°49′35″ N and 87°54′15″ E to 88°22′2″ E. The BRB is the Ganga basin's lower region, as opposed to an ideal drainage basin, which is defined by zones for production, transfer, and deposition. This makes it difficult to draw an explicit border around the basin. In its 2016 annual flood report, the Irrigation and Waterways Department defined the Bhagirathi-Hooghly drainage region as 5452 km^2, of which the Bhagirathi River accounts for 1687.45 km^2 (Fig. 2.1). In order to determine the effects of the anthropogenic interventions, a buffer with a width of 1.381 km on either side of the river has been drawn for the current study. The buffer width is justified by taking into account the historical, current, and projected channel oscillation, or safety factor (Ness 2004).

While the Hooghly River runs both ways depending on the tides, the Bhagirathi River only flows in one direction and is not tidal. The tributaries of the Bhagirathi River influence its fluvial dynamics in part. The Jalangi River is the only tributary on the left bank, while prominent tributaries on the right bank are the Bansloi, Babla, and Ajay (Fig. 2.1). The Bhagirathi River was conveniently divided into three sections by the Kolkata Port Trust to facilitate a thorough hydrogeomorphic analysis: the head reach (upstream) from the feeder canal outfall at Ahiran to Nasipur, the middle reach from Nasipur to Sitahati, and the tail reach from Sitahati to Nabadwip (Guchhait et al. 2016). In a similar manner, the head reach consists of three sectors (1–3), the middle reach consists of four sectors (4–7), and the tail reach consists of an additional four sectors (8–11), for a total of 11 sectors or the river's sector. The Feeder Canal in the head reach, Berhampore in the middle reach, and Katwa and Nabadwip in the tail reach are the four main gauge stations that currently monitor the Bhagirathi River (Fig. 2.1). The BRB is distinct from the standpoint of anthropogeomorphology, meaning that human activity has significantly changed the basin, primarily through the Farakka barrage (Fig. 2.2). Additionally, the BRB's sediment flux and hydrologic regime have been significantly regulated by the dams and barrages in its tributary basins, such as the Sikatia barrage over the Ajay River, the Massanjore dam and the Tilpara barrage over the Mayaukashi River.

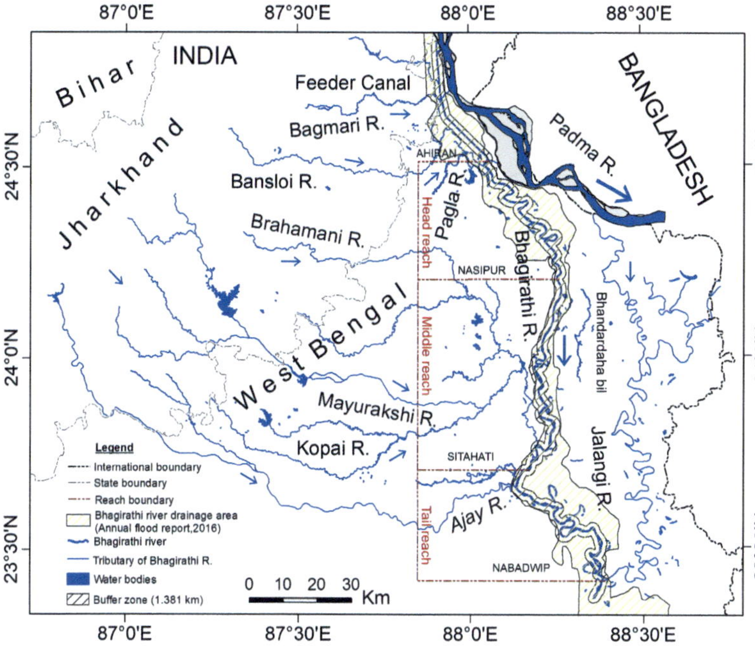

Fig. 2.1 Bhagirathi River system with Bhagirathi drainage area, reach delineation and buffer

2.2.2 Transformation of Bhagirathi Over Time and Space

Once the river Ganges used to flow through the river Bhairab which is expressed in the *Śrīrām Pācālī* by Krittibas Ojha, a 15th-century poet of Bengal.

> একবার গেল গঙ্গা ভৈরব-বাহিনী।
> আরবার ফিরিলেন সাগর-গামিনী।। (ভট্টাচার্য, আশুতোষ 1940, p-26).

> Ēkabāra gēla gaṅgā bhairababāhinī.
> Ārabāra phirilēna sāgaragāminī.

> [Meaning: The Ganga River first flows through the Bhairab River, and later, she returns on her path towards the sea.]

Until the twelfth century, Bhairab was the main river of Bengal carrying merchandise of the province into the Bay of Bengal (Mukherjee 1938).

The evolution of the Ganga River is marked with myth and disagreement. Given the paucity of scientific data on the Ganga and its restricted scope within a broad historical framework, it is challenging to accurately anticipate the river's origin and evolution in a spatiotemporal sequence. Based on earlier research, a number of academics have tried to unravel the dynamics in the context of the Ganga evolution's puzzle in various historical periods. Examining earlier research both clarifies and complicates the Ganga's evolutionary history. Thus, there are differing opinions about where the Bhagirathi River in West Bengal originated. Regarding earlier records, it is a well-established fact that the Bhagirathi is the oldest and was the first to achieve supremacy (Umitsu 1987; Bagchi 1944; Majumder 1942; Mukherjee 1938; Reaks 1919; Sherwill 1858; Oldham 1870). According to Rudra (2010), there are three distinct lines of evidence that indicate Bhagirathi's antiquity: (a) textual, (b) travelogue, and (c) deltaic posture.

Textual evidence consists of descriptions and interpretations of Bhagirathi found in religious books and epics such as the Ramayana, Mahabharata, Baiyu Purana, and Matsya Purana. According to some texts, Ganga is

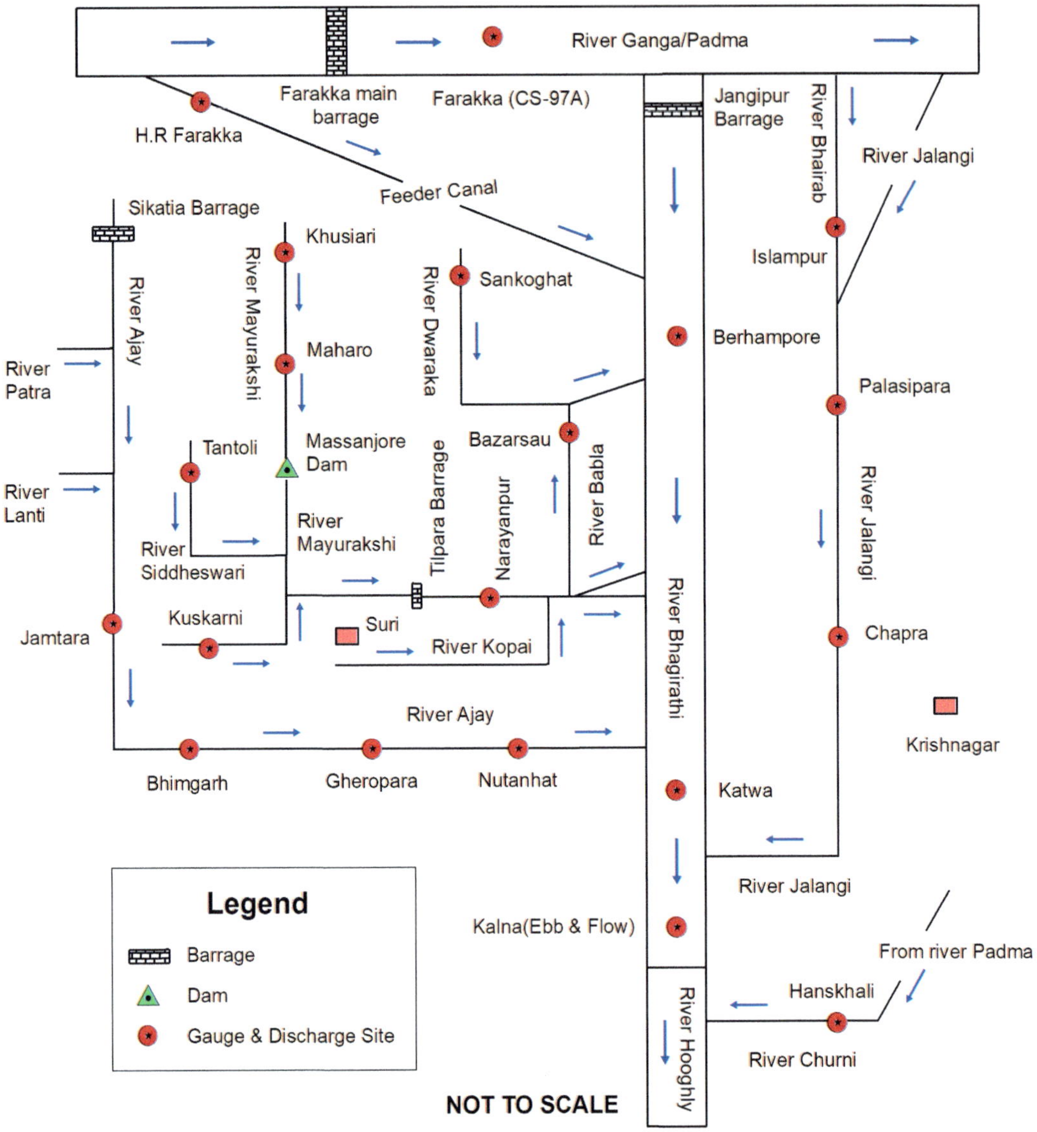

Fig. 2.2 Bhagirathi River system using flow diagram (*Source* CWC, 2015–16)

Bhagirathi. Western rather than eastern parts of the delta saw the development of early settlements. In the Purana interpretation, Tamralipta, Karnasubarna, Chandraketugarh, and others bear witness to this. Ganga's ancient channel was called Bhagirathi. These communities were the product of an earlier, river-centric civilization that flourished in the delta's western region. However, their descriptions are not based on scientific analysis, but rather on subjective feeling and life experiences. Because of this, these linguistic proofs seem questionable when compared to historical components (Rudra 2010). Furthermore, the river Bhagirathi and port Tamralipta are mentioned in detail in the travelogues of foreign diplomats like as Megasthenes (300 BC), Ptolemy (150 BC), Fa-Hien (300 AD), and Hsuan Tsang (639 AD). There was continuous trade from Tamralipta to South East Asia and Rome between 300 BC and 700 AD. This account carries the stamp of Bhagirathi's antiquity. However, trade along waterways is dependent on a variety of circumstances, including the area's socio-economic development and hinterland, in addition to the river's navigability. In addition, the historical period and the geological time scale diverge greatly. Moreover, the form and orientation of the Bengal delta is sometimes invoked as proof that Bhagirathi is older than Padma. The delta's western portion extends farther southward than its eastern portion. The higher southerly expansion shows evidence of the early delta building process as the delta moves southward. This validates the idea that the ancient Ganga passed through Bhagirathi.

The aforementioned proofs support Bhagirathi's ancient status. Most prior scientists, including Umitsu (1987), Bagchi (1944), Majumder (1942), Mukherjee (1938), Reaks (1919), Sherwill (1858), Oldham (1870) endorsed the concept of Bhagirathi's antiquity. They held that the original and natural Ganges path was Bhagirathi. On the question of when Padma became supreme at Bhagirathi's expense, two different scientific groups disagree with one another. While the second group believes that Padma's supremacy is relatively recent, the first group believes that Padma's power dates back centuries. One of the leading figures of the first group, Bagchi (1944), claimed that despite the river Bhagirathi's superior holiness and earlier origins, it was unimportant even in 300 BC. Padma was born before 300 BC, and the Ganga had been the main flow of the river continuously from 300 BC to the present. The oldest course and the first to gain popularity was Bhagirathi. However, it started to silt up some 2000 years ago. It was Padma who eventually gained notoriety, presumably at Bhagirathi's expense. This suggests that the delta was progressively expanded eastward and southward when it was still in its embryonic state close to the apex. One of the second group's proponents, Mukherjee (1938), believes that Padma assumed its domination no earlier than the fifteenth century. This unsettling occurrence was thought to be caused by the Bengal basin tilting eastward in the sixteenth century. As a result, the Padma was regarded as a secondary Ganga channel and the Bhagirathi as the main one until the fifteenth century. These opposing philosophical stances convey reasons, but none provides a thorough comprehension of the evolutions; rather, they each propose opposing points of view. Willcocks proposed the opposite of Bhagirathi's antiquity (1930). He believed that Bhagirathi and the other south Bengal channels were man-made irrigation canals dug by the Hindu kings of the time to lessen extreme drought. "Bhagirathi as the older path of Ganga" is a more sentimental remark than a logical geographic one, according to Rudra (2010). Since the beginning, Padma's dominance has been backed by Rudra (2010), who also offered three more arguments. Firstly, the creation of the Para delta and the building of canals follow the logic that Ajay, Mayurakshi, Damodar, Rupnarayan, and Haldi created their tiny deltas along the western bank of the present-day Bhagirathi in the past. Para deltas are the name given to these little deltas. The estuary was now filled with these rivers. Because these rain-fed rivers were running low on fresh water during the dry spell, the estuary turned saltwater. As a result, a canal named Bhagirath was built to remove salt from the surroundings. The second argument relates to the asymmetry of channel width and belt breadth and asserts that the width of the rivers Bhagirathi and Padma in the Murshidabad district are 300 m and two kilometers, respectively. There is no proof of the Bhagirathi flood plain that contradicts the massive Ganges discharge that flows via Bhagirathi. The third is related to the flood impulse and suggests that the Ganga-Padma river currently receives a discharge of 18–27 lac cusec (Rudra 2010). During peak hours, excess water spills the bank. According to Rudra (2010), Bhagirathi most likely originated from a previous high-magnitude flood.

Upon detailed examination of the prior arguments and counterarguments concerning Bhagirathi's evolution, it can be stated that, at this point, none of the explanations have a solid foundation in geology or scientific logic because river study in this area is still in its early stages. Since 1980, a number of researchers—Allison et al. (2003), Goodbred Jr and Kuehl (1998,

2000a, b)—have used contemporary dating methods to conduct thorough geological exploration. Through quantitative analysis, they hope to solve some mysteries and advance scientific understanding of the Ganga–Brahmaputra delta's evolutionary history. The stages of delta evolution, lobe migration (the Ganga delta moving eastward), the effect of climatic change on delta evolution, and the stratigraphic succession of the delta have all been scientifically demonstrated. The historical reconstruction of the channel's evolution remains unfinished despite those efforts. The discussion above makes clear that there are still many unanswered questions regarding the birth and development of the Bhagirathi River. Unquestionably, however, eastward tilting caused by Neo-tectonic movements is not a recent phenomenon. Since the Eocene, it has been happening at the Indo-Burmese plate border. However, the most recent instance of neo-tectonic movement occurred in the sixteenth and eighteenth centuries (Sen 2010), which hastened the relative decline of Bhagirathi and reinforced Padma's dominance due to the Bengal basin's eastward tilt. Therefore, the Farakka Barrage was built in 1975 to protect Calcutta Port and supply saline-free water to the urban and industrial setup in and around Kolkata. Therefore, human interventions have played a major role in the geomorphic history of the Bhagirathi river (?) since the construction of the Farakka barrage, or the geomorphic evolution of the Bhagirathi canal (?) since its beginnings. The primary anthropogenic signals over the river and their hydrogeomorphic effects would be methodically addressed in this investigation.

2.3 Geomorphology and Landforms

2.3.1 Sedimentology

The sediment loads carried by the Bhagirathi River are regularly measured by the Kolkata Port Trust in the post-monsoon season. The survey during 2006–08 indicates that median particle size (D50) varies from 0.25 to 0.3 mm from Ahiran (source) to confluence (Nabadwip) (Fig. 2.3a). It is found that where river Ajay has joined with Bhagirathi at Katwa, the D50 of the bed materials has increased because the Ajay and Mayuakshi system contributes coarse-grained peninsular sediments (Guchhait et al. 2016). Similarly, surface sediment concentrations below Katwa are also higher (0.3–0.45 mg/l) (Fig. 2.3b), although with a fluctuating trend because of local topography and human interventions. The bed sediment concentration is relatively low (0.1 to 0.2 mg/l) for the entire river (Fig. 2.3c). Here, below Katwa, the bed sediment concentration is recorded as low. It is interesting to note that surface and bed sediment concentrations are poorly correlated. A feeble positive correlation ($R^2 = 0.08$) with a logarithmic curve indicates the randomness of the relationship at the various starches of the river.

2.3.2 Plan View Morphology-Meandering and Braiding

Using Landsat satellite imagery, planform changes between 1973 and 2019 are analyzed. Channel oscillation, stream meandering, sinuosity, and braiding have all been observed throughout this observation period; these features are mostly the result of hydrologic changes that occurred after 1975.

2.3.2.1. Channel oscillation and stream avulsion

Over the past few centuries, the path of the Bhagirathi-Hooghly River has experienced alterations (Chatterjee 1989). Accounts from multiple scholars, including Rudra (2010), Bhattacharya (2000), Ray (1999), Bandyopadhyay (1996), Mukherjee (1938), Hirst (1915), Sherwill (1858), Colebrooke (1801) and Rennell (1788), have depicted the river Ganga's propensity for frequent shifts. Thus, the Bhagirathi-Hooghly system or the Ganga's channel changing has garnered interest for ages.

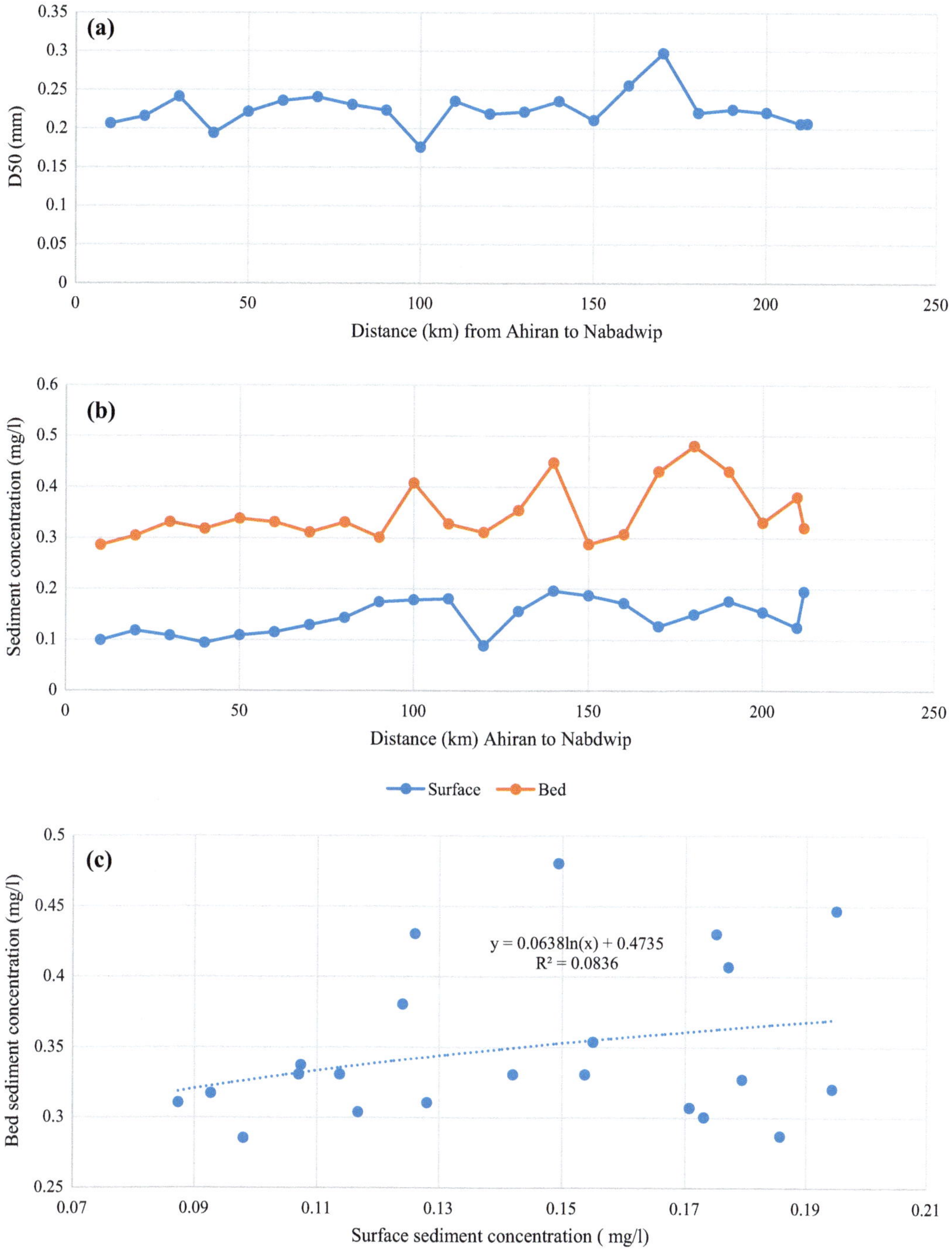

Fig. 2.3 **a** D50 of the bed material (2006–08), **b** surface and bed sediment concentrations (2006–08), **c** relationship between surface and bed sediment concentrations

Around 1911, a loop at Nadia (Nabadwip) was severed and cut through, and in 1913, the river abandoned a winding channel spanning 14 miles east of Agradwip. Instead, it forged a new path through a passage created by Patuli Khal, effectively shortening its course by 9 miles

(Reaks 1919, p. 57). In 1913, Harubhanga loop of 8.5 miles at Kaliganj was abandoned. In 1914 a natural cut-off took place at the Dadpur loop of 4.5 miles.

Recent decades have seen a surge in the Bhagirathi's channel modifications, particularly with the completion of the FBP in 1975. As soon as the barrage was built, the Bhagirathi's whole course became unstable (Figs. 2.4, 2.5 and 2.6). The lower middle and lower stretch continue to oscillate at a significant rate, despite the upper stretch having somewhat adjusted to the morphogenic factors. This is because of the Ajay-Mayurakshi River's contribution during the monsoon season as well as the combined effects of the FBP during the lean months. There was a noticeable amount of channel instability and meander advancement in 1973 and 2019.

Three locations close to Beldanga, Agradwip, and Nabadwip were treated through windows A, B, and C, respectively, in order to trace the evolution of the ox-bow lakes (Fig. 2.7). The chute

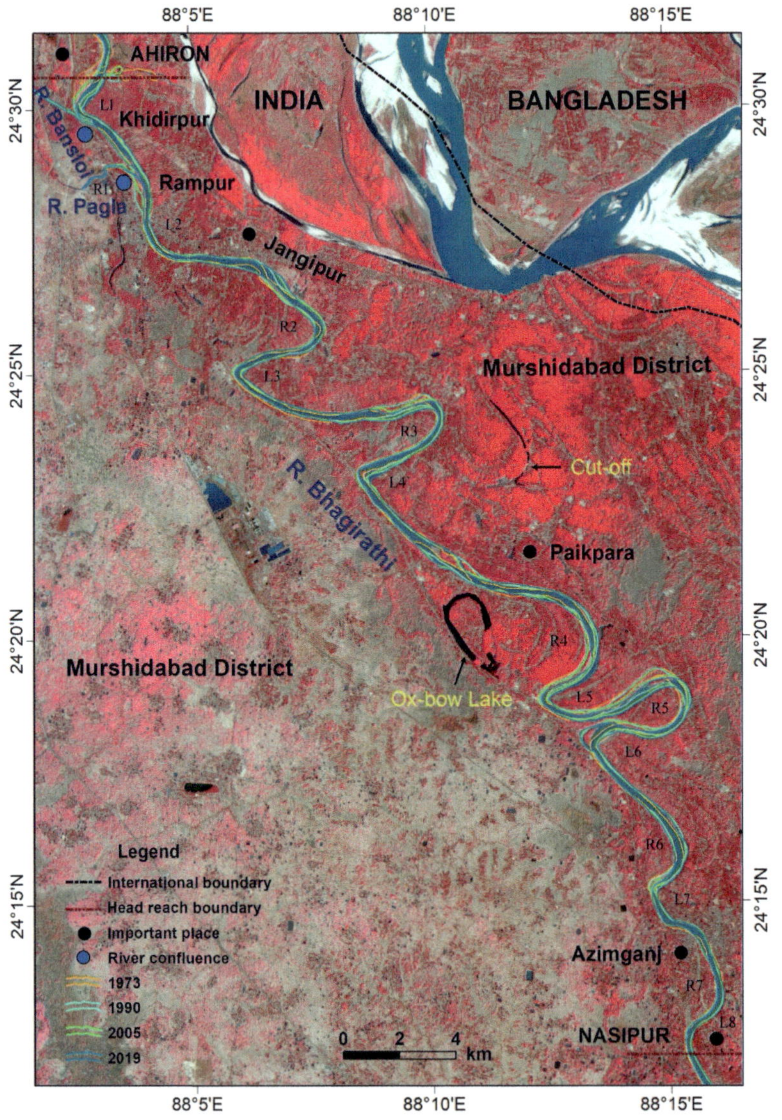

Fig. 2.4 Bhagirathi River oscillation in the head reach from 1973 to 2019 (Islam et al. 2020)

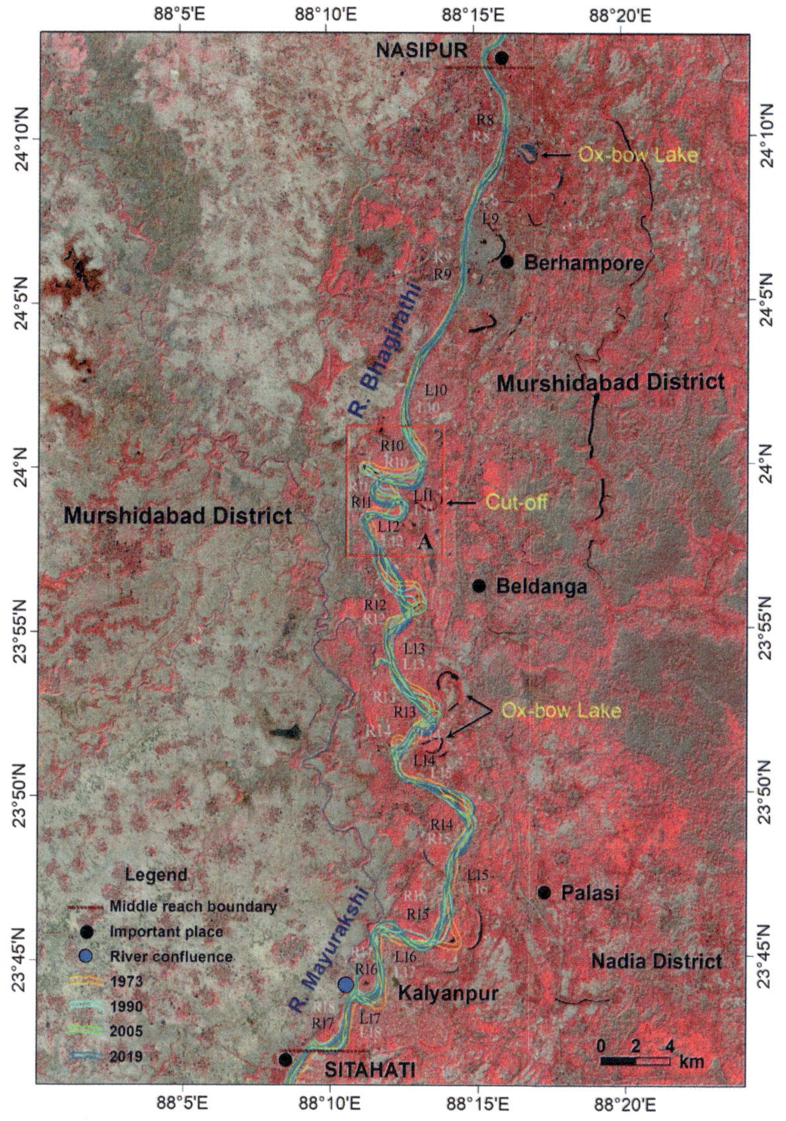

Fig. 2.5 Bhagirathi River oscillation in the middle reach from 1973 to 2019 (Islam et al. 2020)

cut-off known as the Mayapur-Sankhapur cut-off happened in the channel's lower reach in 1989. This was an unforeseen event. The neck distance has changed since Rennell's time in 1788, going from 5.62 to 2.20 km in 1975, right before the Farakka Barrage Project (FBP) began to operate. It did, however, undergo an abrupt flood impulse following the FBP's deployment, which resulted in the formation of this channel cut-off during the 1989 major monsoon influx (Fig. 2.8). Accordingly, the flood impulse of 1989 was the real cause of this cut-off (Basu et al. 2005). This cut-off consequently resulted in a 9 km reduction in channel length and a more straighter channel (Islam and Guchhait 2017a).

In 1994, there was another cut-off near the Bishnupur Char-Chakundi area, resulting in a reduction of the channel length by 11.85 km (Fig. 2.8). Subsequently, in 2008, another separation occurred in the same vicinity, named Char Chakundi cut-off-II, shortening the channel by 3.6 km. The recurrence of such

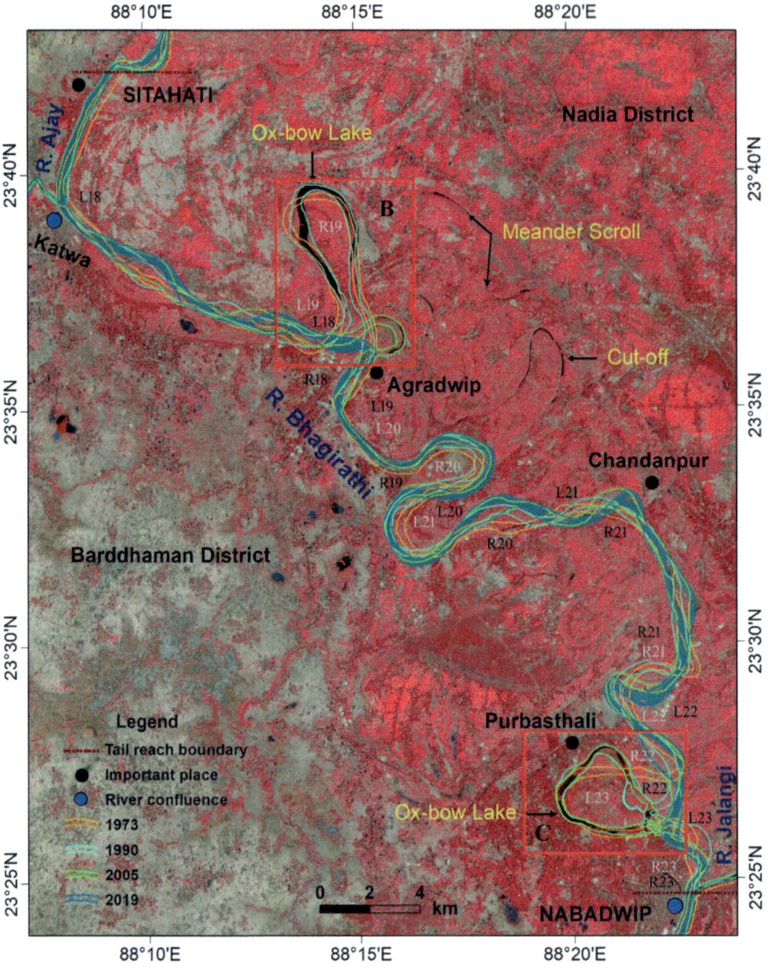

Fig. 2.6 Bhagirathi River oscillation in the tail reaches from 1973 to 2019 (Islam et al. 2020)

separations in this region highlights the ever-changing nature of the channel, influenced by hydrological and geological factors, as elaborated in the following section. It's crucial to observe the specific spatial and temporal patterns of these separations. Spatially, three out of four of them are concentrated in the lower stretch of the Bhagirathi (Katwa-Nabadwip). Temporally, all four separations happened after the construction of the Farakka Barrage Project (FBP), influenced by both the project's discharge and flood events. Overlaying channels reveals that in the recent decade (2004–2014), the channel has been oscillating, albeit not as rapidly as immediately after the operation of FBP (Islam and Guchhait 2017a, b).

Consequently, The Bhagirathi River is now unstable as a result of the Farakka Barrage Project's (FBP) introduction of a novel hydrological regime, characterized by a new dimension of attempting to accommodate flood impulses through cut-offs. Beyond the FBP's impact on channel instability and meandering in a broader context, it's essential to explore the role of lithofacies in explaining the pattern of acute swinging and meandering observed to the left-flood plain of the Bhagirathi (Islam 2016).

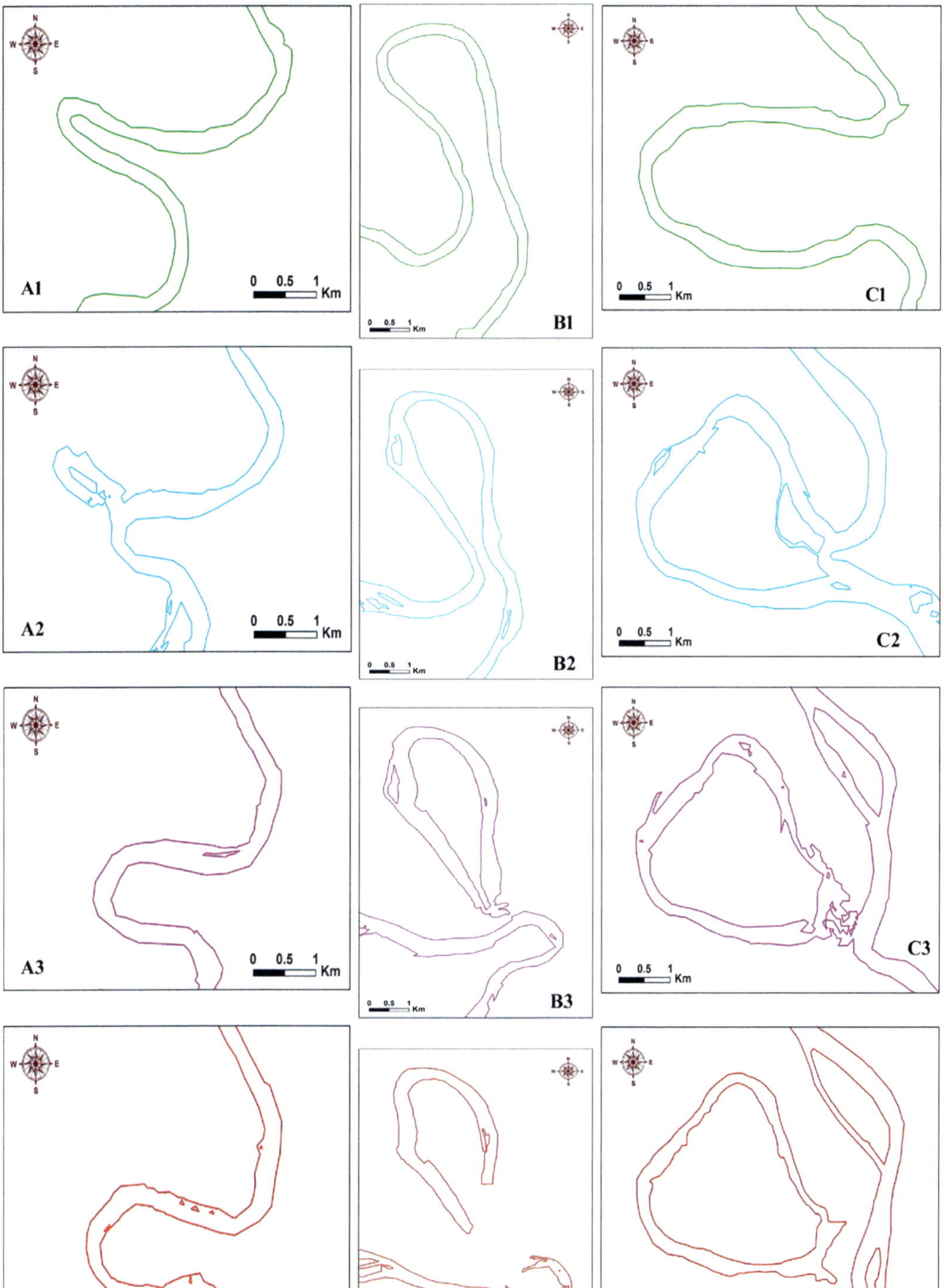

Fig. 2.7 Evolution of Bhagirathi river cut-off for 1973 (A1, B1, C1), 1990 (A2, B2, C2), 2005 (A3, B3, C3) and 2019 (A4, B4, C4) (Islam et al. 2020)

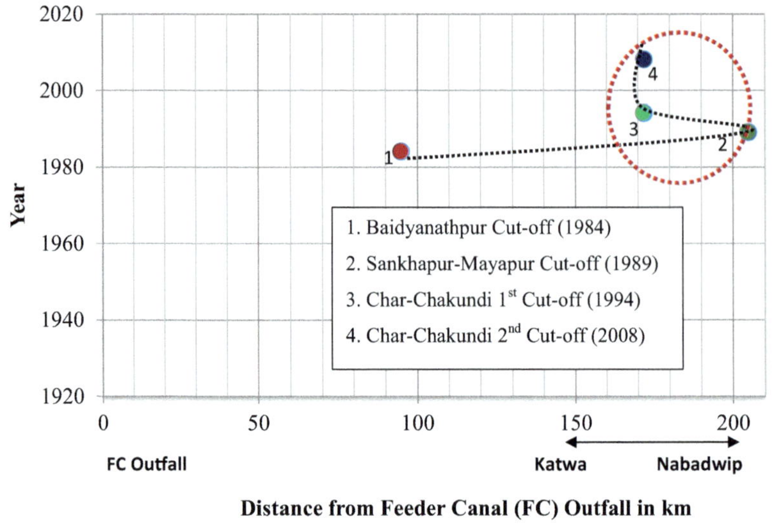

Fig. 2.8 Spatial–temporal specificity in channel cut-offs (Derived from topographic maps and satellite imagery)

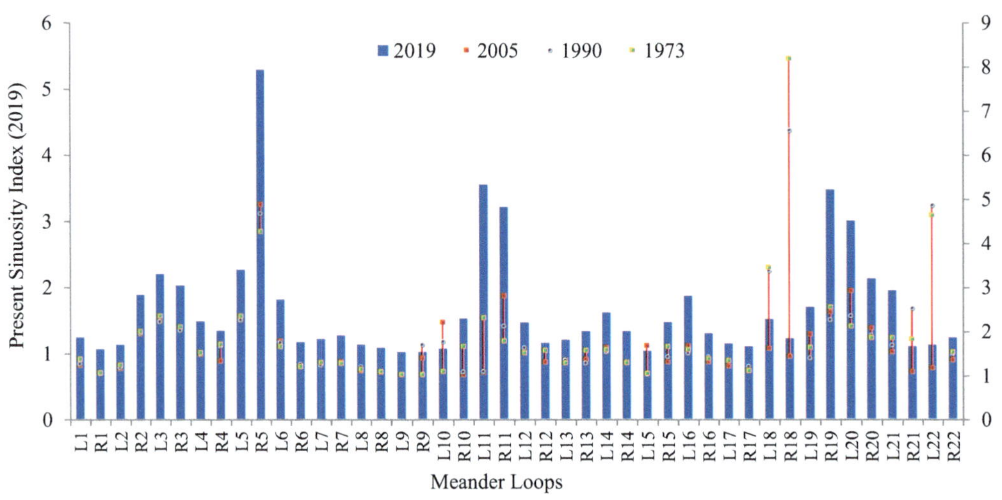

Fig. 2.9 Loop-wise sinuosity index of the Bhagirathi River

2.3.2.2. Channel sinuosity and channel length

For the years 1973, 1990, 2005, and 2019, a channel sinuosity index (SI) of 44 meander loops (22 on each bank) with a SI>1.05 was calculated for the river Bhagirathi. According to Islam and Guchhait (2017a), the current state of sinuosity shows that the tail reach has greater sinuosity than the head and middle reaches. Furthermore, regular variations in the SI in the river's tail reach from one measurement year to the next suggest that the channel's evolution is dynamic. Most of the loops in the tail reach in our analysis have clearly shown this trend (Fig. 2.9).

For instance, the mean SI at the tail reach was recorded as 2.96 in 1973, dropped to 1.76, and then increased to 1.85 once again. Furthermore, in 1973, the SI's CV was relatively high (70%) (Table 2.2). On the other hand, the head reach's SI ranges from 1.76 to 1.77, indicating a relative stability over time. Nonetheless, there is a spatial variance in the loop-wise SI, as evidenced by the head reach's comparatively greater CV value (around 50%) across all measurement years. Similarly, the SI shifting about 1.45 in different years indicates that the center reach is relatively more stable than the tail reach. Moreover, the lower CV of almost 20% in 1973, 1990, and 2005 illustrates how uniform this reach is in loop properties. On the other hand, the loop characteristics in the last year show regional variation (Table 2.1).

The channel length is directly impacted by changes in the SI. Channel shortening is caused by a meander's rapid evolution and eventual cut-off, which shortens the channel. Reach-wise channel length has been measured in this attempt for the following years: 1973, 1990, 2005, and 2019 (Fig. 2.10). The outcomes demonstrate that the channel shortening clearly has a spatial component. From 1973 to 2019, the channel length did not change, indicating that the head reach is nearly steady. Nonetheless, the tail reach depicts a significant decrease in the channel length after post-Farakka, while the intermediate reach displays a little shift. As a result, the total channel length has fallen off over time (1973: 234 km, 1990: 223 km, 2005: 216 km, and 2019: 205 km). This is a result of the growing hydraulic impulse under the FBP's new hydraulic regime.

Table 2.1 Dynamism in channel sinuosity from 1973 to 2019

	1973	1990	2005	2019
Head reach	1.77 (45.79)	1.77 (50.73)	1.76 (54.43)	1.77 (59.66)
Middle reach	1.42 (23.05)	1.39 (20.68)	1.45 (30.26)	1.50 (46.77)
Tail reach	2.96 (69.98)	2.84 (58.71)	1.76 (33.08)	1.85 (44.30)

Source Calculated by authors 2019 (Values in parentheses denote the coefficient of variation of the sinuosity index)

Table 2.2 Comparison of water share (in Cusecs) under the 1977* and 1996** treaties

Period	Bangladesh			India			Remarks
	1977	1996	Incr. /Decr	1977	1996	Incr. /Decr	
1–10, Jan	58,500	67,516	9016	40,000	40,000	0	* Flow calculations were based on the 75% dependable flow from 1948 to 1973, prior to the operation of the Farakka Barrage
11–20, Jan	51,250	57,673	6423	38,500	40,000	1500	
21–31, Jan	47,500	50,154	2654	35,000	40,000	5000	
1–10, Feb	46,250	46,323	73	33,000	40,000	7000	
11–20, Feb	42,500	42,859	– 359	31,500	40,000	8500	** Flow calculations were based on the average flow (50% dependable) from 1948 to 1988, encompassing a period of 25 years (1948 to 1973) before the Farakka operation and 15 years (1974 to 1988) after the operation, as per mutual agreement. The period from 1988 to 1996, consisting of 8 years of unilateral withdrawal, was not taken into account
21–28/29, Feb	39,250	39,106	– 144	30,750	40,000	9250	
1–10, Mar	38,500	35,000	– 3500	26,750	39,419	12,669	
11–20, Mar	38,000	35,000	– 3000	25,500	33,931	8431	
21–31, Mar	36,000	29,688	– 6312	25,000	35,000	10,000	
1–10, Apr	35,000	35,000	0	24,000	28,180	4180	
11–20, Apr	34,750	27,633	– 7117	20,750	35,000	14,250	
21–30, Apr	34,500	35,000	500	20,500	25,992	5492	
1–10, May	35,000	32,351	– 2649	21,500	35,000	13,500	
11–20, May	35,250	35,000	– 250	24,000	38,590	14,590	
21–31, May	38,750	41,854	3104	26,750	40,000	13,250	

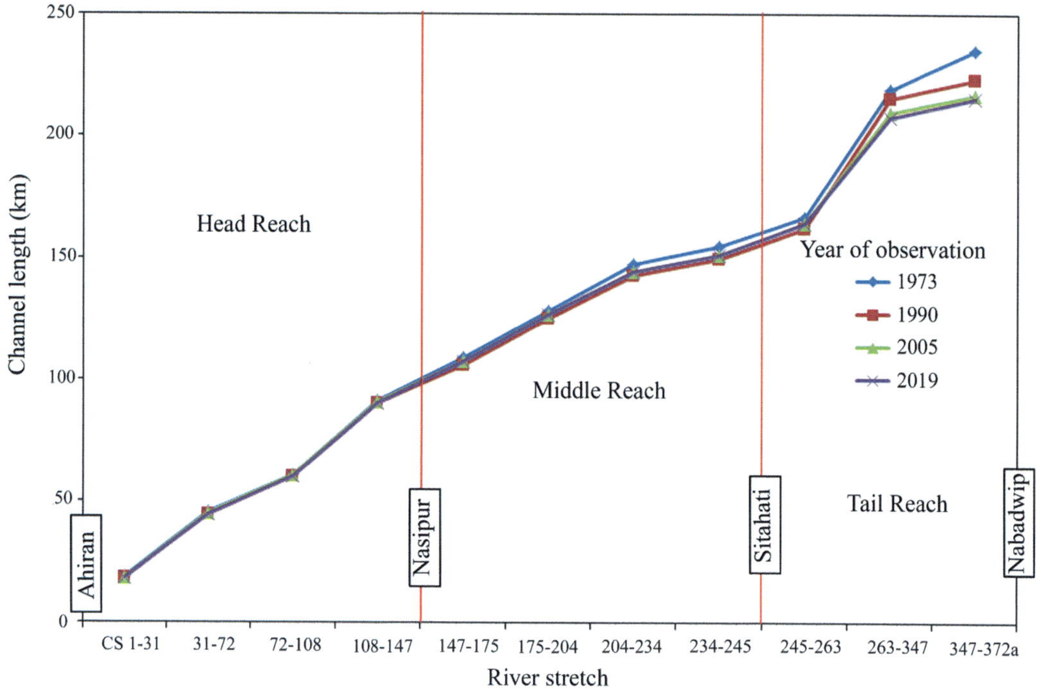

Fig. 2.10 Channel shortening of the Bhagirathi River in its different stretches from 1973 to 2019 (Islam et al. 2020)

2.3.2.3. Channel braiding and bar growth

Channel braiding suggests a link between energy and load. When a channel can no longer support its weight, bed deposition occurs. Over time, this leads to the development of bars and islands, which eventually create a braided channel. In this instance, the braiding index (BI) for the years 1990, 2005, and 2019 has been calculated using Brice's (1964) methodology. All three reaches show an increase in BI from their 1990s baseline in the recent past (2019). For instance, in 2019, tail reach reached a BI of approximately 0.4, up from approximately 0.3 in 2005 (Fig. 2.11). But in the year 2005, the BI fell in the head and middle reaches. Mid-channel bars are present at the moment (head reach: 3, middle reach: 7, and tail reach: 9). The development of the six bars from 1990 to 2019 has been charted using those bars. While some of the bars have a tendency towards decrease, most have an increasing area (Fig. 2.11). Additionally, a process that operates in the Bhagirathi, particularly in the middle and tail sections, converts the mid-channel bars into side-attached bars.

Six bars have been identified from the tail reach and lower middle reach of the river: (a) Sarpakhia bar, located 71.12 km from the feeder canal outfall, emerged in 1994 with an initial area of 0.036 m^2, which expanded to approximately 0.091 m^2 by 2006. Presently (2019), it has grown to about 0.261 m^2 (Fig. 2.12a). (b) Uday Chandrapur bar, situated 223.96 km from the feeder canal outfall, began in 2006 with an area of around 0.261 m^2, and by 2019, it had increased to 0.4432 m^2 (Fig. 2.12b). (c) Mangonpara bar, positioned 160.33 km from the feeder canal outfall, originated in 1994 with an area of about 0.3 m^2. By 2019, it has expanded to 0.6 m^2 (Fig. 2.12c). (d) Palash Purnima bar, located 233.83 km from the feeder canal outfall, commenced in 1994 with an area of approximately 0.271 m^2. By 2019, it has grown to 0.958 m^2 (Fig. 2.21d). (e) Rajnagar bar, situated below the Ajay-Bhagirathi confluence at Katwa, is positioned 192 km from the feeder canal outfall. This mid-channel bar emerged in 1994 with a small area of 0.354 m^2, which gradually increased to 0.771 m^2 in 2006 and currently

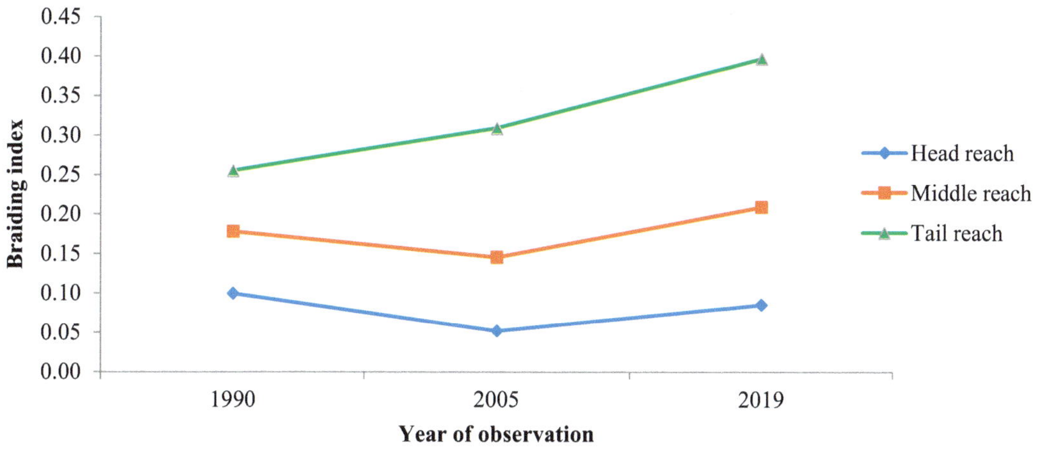

Fig. 2.11 Nature of bar growth in different reaches of the Bhagirathi River between 1990 and 2019 (Islam et al. 2020)

stands at around 1.057 m^2. (f) Kashthasali bar, near Nabadwip and 244.17 km from the feeder canal outfall, has also displayed a progressive increase in area over time (1994: 0.385 m^2; 2006: 0.765 m^2; 2019: 1.092 m^2).

2.3.3 Cross Profiles Morphology

Hydraulic changes are quickly followed by changes in channel shape. The post-Farakka stage has seen a dramatic shift in channel shape as a result of the FBP's hydrologic alternation. Four key variables—channel width, cross-sectional area, hydraulic mean radius, and cubic capacity—have been used to identify the morphological alterations. The entire 11-section river, as measured by the Kolkata Port Trust, from cross-section 1 to cross-section 372a, has been the subject of the analysis (Fig. 2.13).

2.3.3.1. **Channel width**

Channel width serves as a particularly sensitive indicator of hydrological modification (Islam 2011). Measurements have been taken at the datum level, revealing that among the 11 stretches of the river, 10 exhibit a positive expansion in channel width (Fig. 2.14). The most significant increase, approximately 36%, is observed in the stretch spanning from CS 347 to CS 372, whereas the stretch from CS 234 to 245 shows a negative change in width growth index, around – 5%, over the measurement period. Nevertheless, overall, a rising trend is evident for the river during post-Farakka. This augmentation is primarily attributed to the increasing hydraulic impulse from the Farakka Barrage Project during in lean months (January to May).

2.3.3.2. Cross-sectional area

The average cross-sectional area has increased by 23% from a Pre-Farakka minimum of roughly 723 m^2 to a height of 941 m^2. Variations in reach are present in this developing tendency. The areal increase reaches its greatest in the higher reach, where the proximity effect maximizes the influence of the hydraulic impulse. For instance, stretch 1 has observed a rise of almost 46% over the observation period (Fig. 2.15). There has also been a notable growth in the middle reach, ranging from 10 to 25%. The area has increased less (~3–24%) in the tail reach, where the impact of the major tributaries, such as Ajay and Mayurakshi, is noticeable. This minimum increase (~3%) in the CS region is shown in Stretch 8, which runs from the CS 234 to the CS 245. The hydrologic input from the Farakka may have a distance decay effect, which could explain this. Here, Ajay Mayurakshi has excellent control over the lower stretch during the monsoon season, but

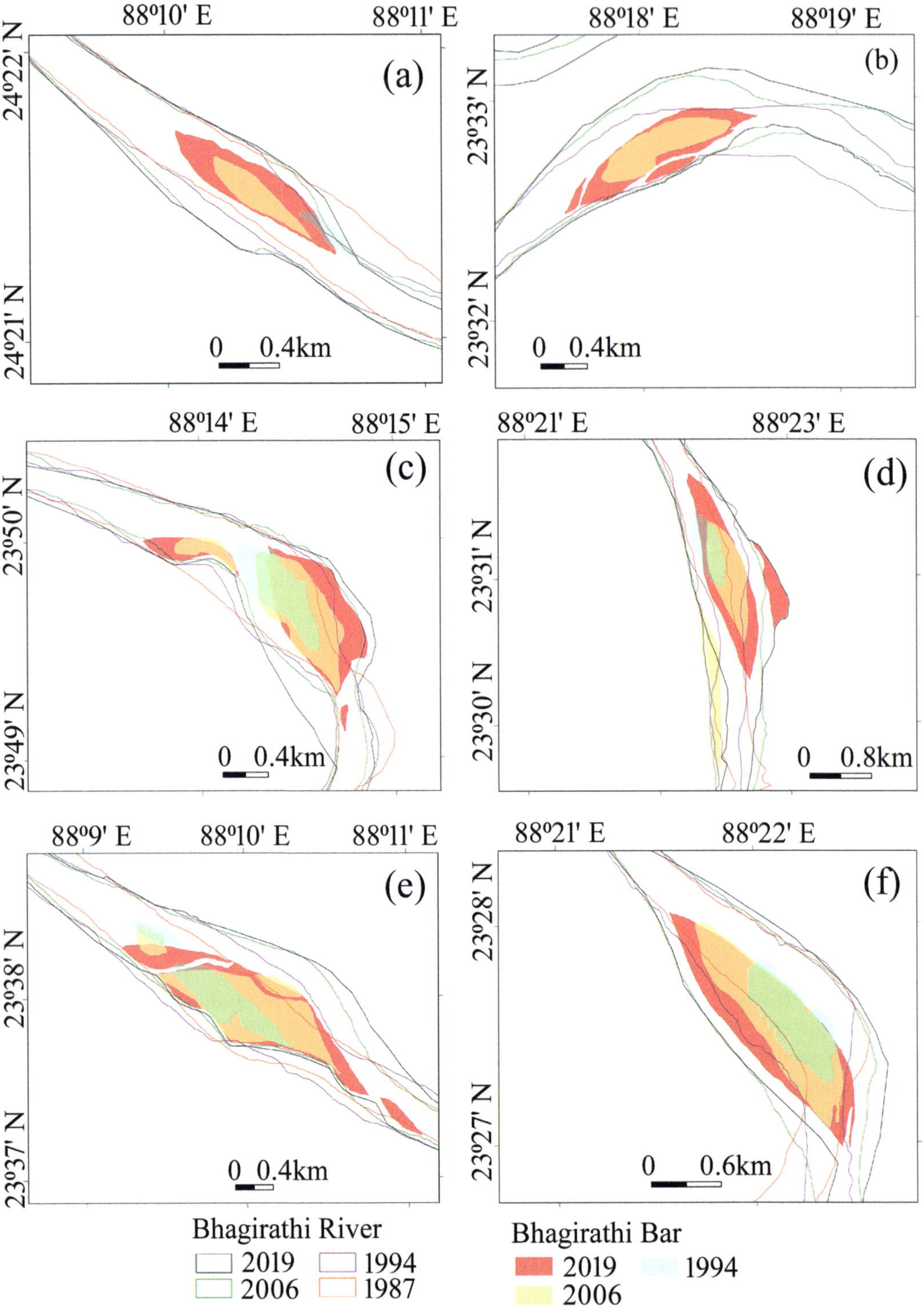

Fig. 2.12 Windows showing the evolution of bar growth amid the Bhagirathi River, **a** Sarpakhia bar, **b** Uday Chandrapur bar, **c** Mangonpara bar, **d** Palash Purnima bar, **e** Rajnagar bar, **f** Kashthasali bar

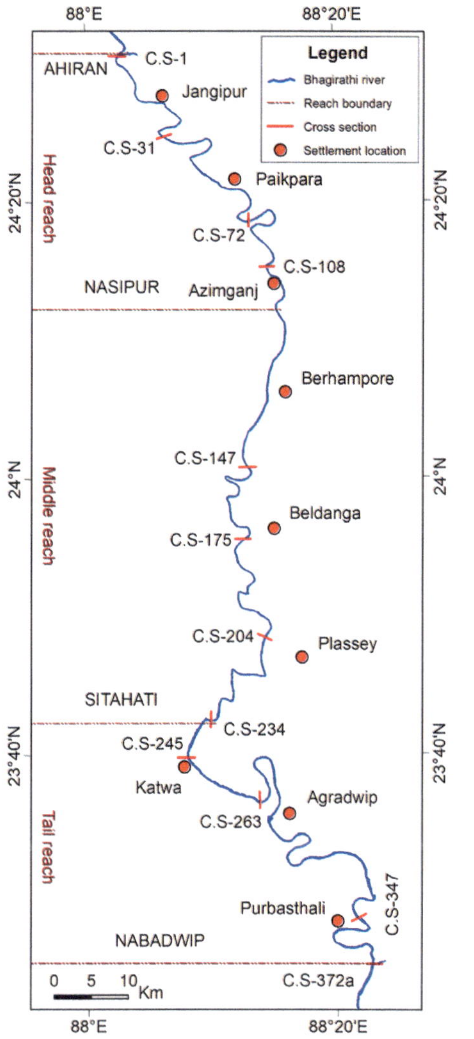

Fig. 2.13 Cross-sectional locations of the Bhagirathi River from Ahiran to Nabadwip considered in the present study

the FBP only has control over the channel during the lean months.

2.3.3.3. **Hydraulic mean radius**

When compared to the 1975 episode, the hydraulic mean radius (HMR) has grown in the last ten years. The increase's pattern corresponds with the CS area's trend. The stretch average growth for the observation period is almost 8%. But from the feeder canal outfall at Ahiran to the Jalangi Bhagirathi confluence at Nabadwip, there is a noticeable spatial pattern, i.e., falling (Fig. 2.16). There has been a 10–37% increase in head reach. The intermediate reach, on the other hand, shows a wide range, from – 4.7% to 9.6%. Similar to this, the tail reach likewise exhibits the most variable outcomes, with a range of – 7.4% to 15%. In comparison to the upper levels, it suggests that channel efficiency has dropped in the middle and lower reaches.

2.3.3.4. **Cubic capacity**

Cubic capacity denotes the channel's ability to support loads and has been measured at zero level. During pre-Farakka period, there was a notable decrease in flow during the lean season, particularly in the upper portion of the river (Fig. 2.17). As a result, the hydrologic input through FBP enhanced a high flow during the lean months, which in turn raised the head reach's cubic capacity (Islam and Guchhait 2018). According to the findings, head reach saw highest increases (41–51%) while a slow decline was noted. Between 21 to 49% is the range of the middle reach. In a similar vein, certain regions in the tail reach have also shown a decline in cubic capacity. Stretch 11 (CS 347–372a), for instance, exhibits a decline of roughly – 57%. This results from deposition in the river's lower reaches and bed scouring in its upper portion.

2.3.4 Long Profile Morphology

Bank erosion along the Bhagirathi River was more noticeable during the monsoon season in the pre-Farakka era, but it dramatically diminished during the lean months as a result of lower water flow. Nonetheless, erosion has become a year-round occurrence in the post-Farakka era, even if the total volume of bank erosion has significantly diminished (Islam and Guchhait 2015, 2019). Following the start of the Farakka Barrage Project, the river saw an abrupt rejuvenation in the lean months as a result of increased water intake from the barrage, causing a spike in erosion in the years 1976–1977 and 1977–1978

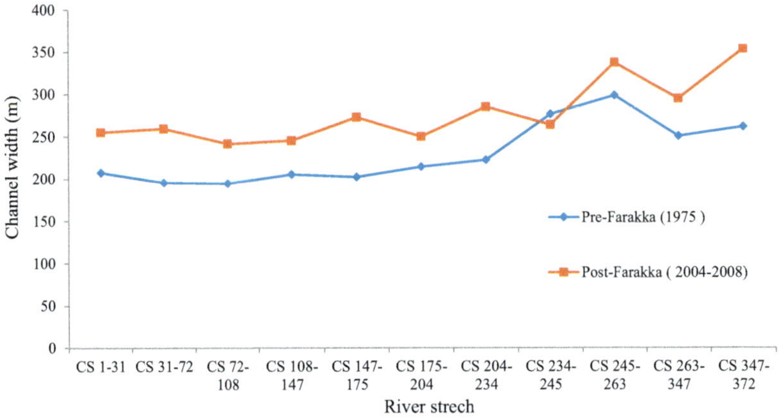

Fig. 2.14 River width dynamics in the pre and post-Farakka barrage period (Islam et al. 2020)

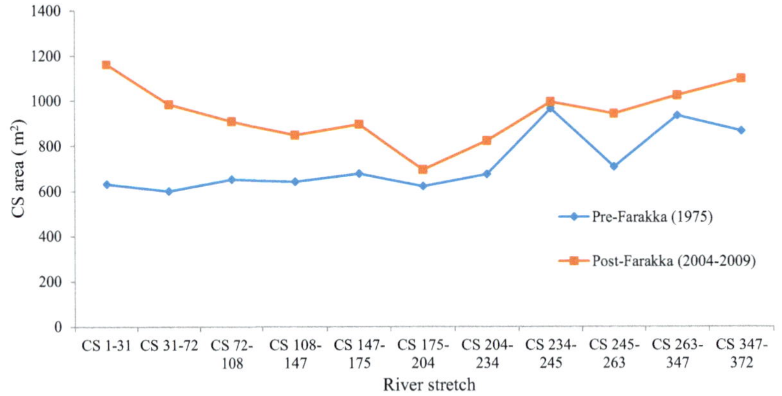

Fig. 2.15 River cross-sectional area dynamics in the pre and post-Farakka barrage period (Islam et al. 2020)

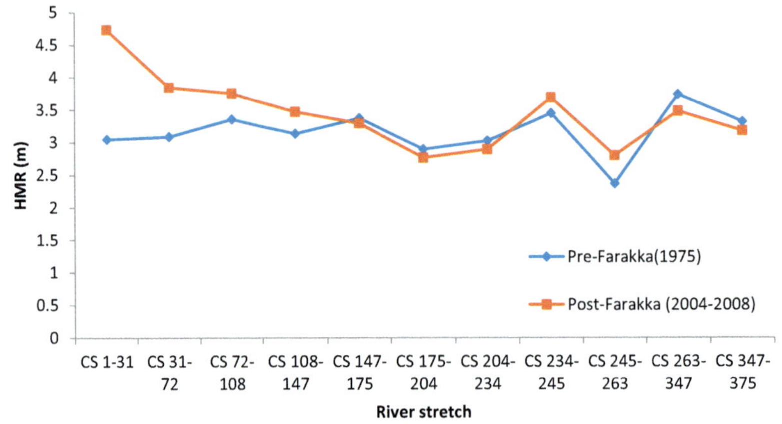

Fig. 2.16 River hydraulic mean radius dynamics in the pre and post-Farakka barrage period (Islam et al. 2020)

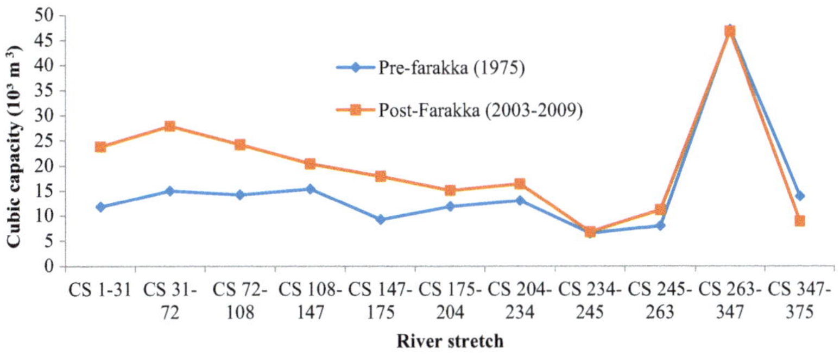

Fig. 2.17 River cubic capacity dynamics in the pre and post-Farakka barrage period (Islam et al. 2020)

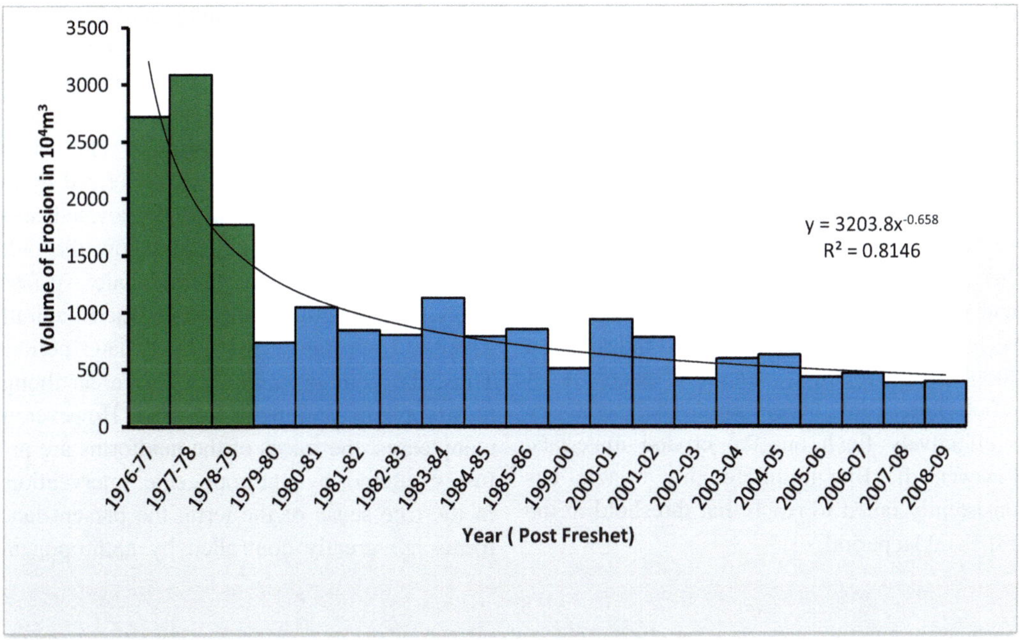

Fig. 2.18 River bank erosion after commissioning of Farakka Barrage Project in 1975 (Computed from Parua 1992 and annual reports on river Bhagirathi of KoPT: 2003–04 to 2008–09)

(Fig. 2.18). According to Parau (1992), the Bhagirathi River experienced bank erosion totaling 2720×10^4 m^3 and 3086×10^4 m^3, respectively, during these years.

The upper Bhagirathi River (from the Bhagirathi offtake to Nasipur, 13.5 km north of Berhampore) stabilized and adjusted to the flow after this initial period (1976–1978). The lower reaches (from Sitahati to Nabadwip) and middle (from Nasipur to Sitahati, 5.6 km north of Katwa) have not yet adapted to morphogenic factors, though. The lower Bhagirathi river still experiences severe bank erosion (Islam 2011; Islam et al. 2012). Between the Ajay-Bhagirathi confluence at Katwa and the Jalangi-Bhagirathi confluence at Nabadwip is the main erosion-prone area in the lower reach of the Bhagirathi. The degree of recent erosion in various reaches has fluctuated according on how unpredictable the monsoon season's flood output is.

Despite fluctuations, the tail reach consistently experiences the highest annual erosion, except for the year 2007. Bank erosion at the tail reach often surpasses 50% of the total erosion, indicating its severity. This segment covers almost one-third of the total river length. The nature of erosion in the tail reach, though fluctuating, shows a higher rate than the other two reaches (Islam and Guchhait 2015).

Moreover, there's noteworthy bed scouring and sediment deposition occurring in the aftermath of the Farakka project. Alongside hydraulic impulses, both the upper and middle sections of the river have undergone some bed scouring, alongside bank erosion. The degree of bank erosion exceeded that of bed scouring due to the very gradual slope of the channel (ranging from 3° to 5°). In the post-Farakka era, the overall slope of the channel ($R^2=0.65$) is lower than that of the pre-Farakka era ($R^2=0.74$). This shift is a result of material erosion from the upper stretches, subsequently deposited onto the riverbed of the lower stretches (Figs. 2.19, 2.20a, b). This phenomenon is primarily due to insufficient discharge to transport the eroded materials (Islam and Guchhait 2020). Bhattacharyya (2000) asserted that a minimum discharge of 40,000 cusecs of water would be necessary to effectively flush out the eroded materials. However, the Bhagirathi-Hooghly system has consistently failed to reach that threshold in the post-Farakka period.

2.3.5 Typical Landforms

The BRB exhibits several landforms such as meander scar, oxbow lake, floodplains, natural levees, etc. Being located in the lower part of a river basin, embodies mostly flood-related forms, as per the Geological Survey of India, diversified geomorphic forms are observed in the BRB. The common landforms include abandoned channels, back swamps, channel bars and islands, older deltaic plains, active floodplains, meander scar and oxbow lakes, pediplain, point bar, and water bodies (Fig. 2.21). The most conscious landform is the meander and oxbow lakes that prove the dynamicity of the river over space and time. Moreover, different types of flood plains from active to older indicate the spatiality in the geomorphic evolution. The annual flood-prone areas are often characterized by the active plain while the relatively elevated areas of the BRB with lesser floods show the older floodplains. Even the (para)deltaic surfaces imply the proximity of the seas to the Bhagirathi River and sea transgression in the later periods. Thus, the geomorphology of this area, though mostly plain, is quite diversified. However, in many cases, the traces of the landforms are profusely altered by anthropogenic interventions. In the true sense of the term, the present landforms are greatly controlled by anthropogenic

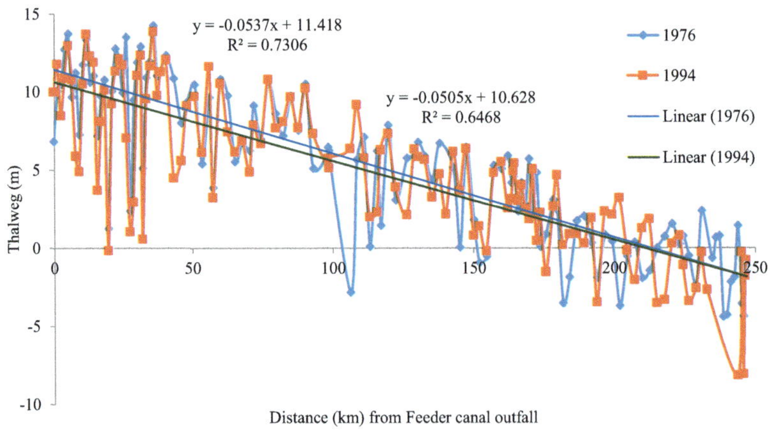

Fig. 2.19 Channel gradient dynamics of the Bhagirathi River in the pre and post-Farakka period

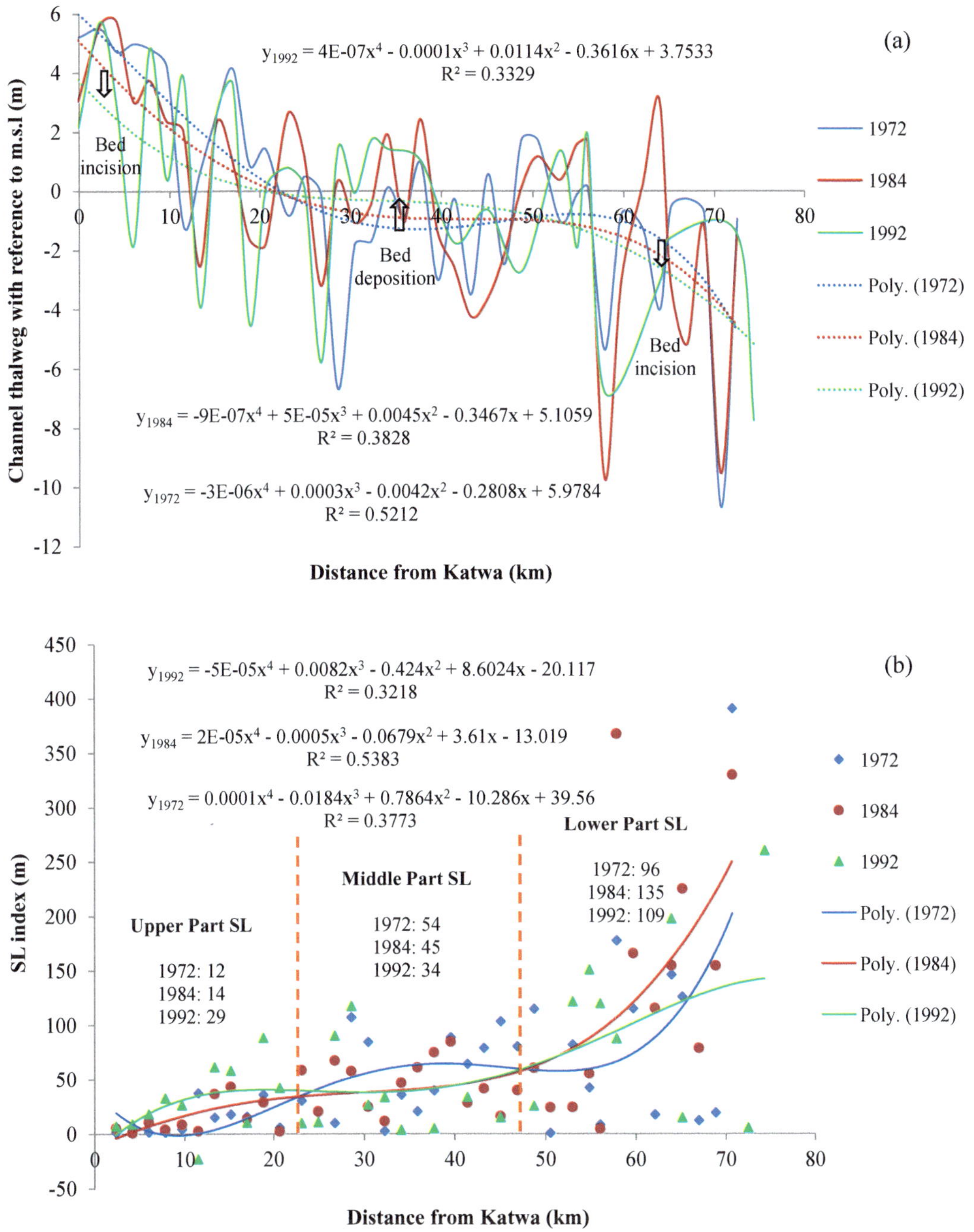

Fig. 2.20 Alterations in the channel bed slope of the Lower Bhagirathi River from Katwa to Nabadwip **a** Long profile, **b** SL index (*Source* Islam and Guchhait 2020)

activities and hence the anthropic landforms are also developed in the basin. The Szabo classification of human-made landform is quite appropriate here. Urbanogenic, montanogenic, and agrogenic landforms are found in this area. Therefore, landform transformation from purely natural to dominantly anthropic is noted future investigations.

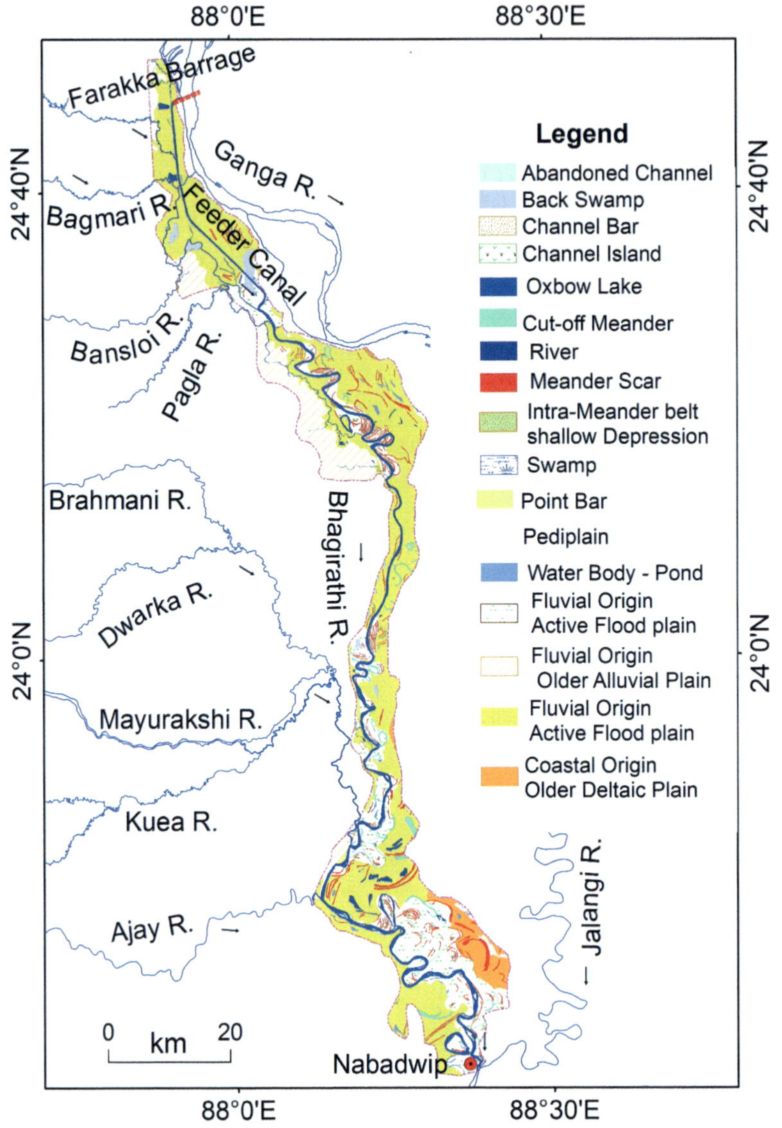

Fig. 2.21 Typical landforms of the Bhagirathi River basin

2.4 Hydrology and Water Resources

2.4.1 Tributaries, Palaeo-Channels and Spill Channels

The Bhagirathi River is influenced by the Himalayan and peninsular river regimes. On the right of the Bhagirathi river, several peninsular rivers (Pagla, Bansloi, Ajay, and Mayurakshi) join with the Bhagirathi river. These peninsular rivers are mostly rainfed and contain water during the monsoon season. Thus, a strong seasonality in the river regime is observed. These rivers contribute huge coarse-grained sediment coupled with discharge to the Bhagirathi system in the monsoon season thereby altering the fluvial dynamics of the Bhagirathi River. The sedimentological composition of the peninsular river is mostly cratonic and resistant to erosion which induces a lesser westward shift of the Bhagirathi River (Guchhait et al. 2016). The left bank tributaries are limited—(1) Jalangi and (2) Churni. These tributaries are Himalayan rivers. Once these were perennial but are now being disconnected from the source and hence maintain flow only during the freshet (monsoon) period. These rivers contribute fine sediments (silt/clay) to the Bhagirathi-Hooghly River making provisions for the easy swing of the Bhgairathi River on the east (Fig. 2.22).

Apart from the tributaries, the Bhagirathi River exhibits many palaeo-channels and spill channels in its basins. The palaeo-channels indicate the trajectory of the river evolution in different periods and eventual channel cut-off and concealment under sediment. The presence of the spill channels indicates frequent floods with some major flood events. During the floods, the spill channels are used to pass the excess flood water during the monsoon periods (Fig. 2.23).

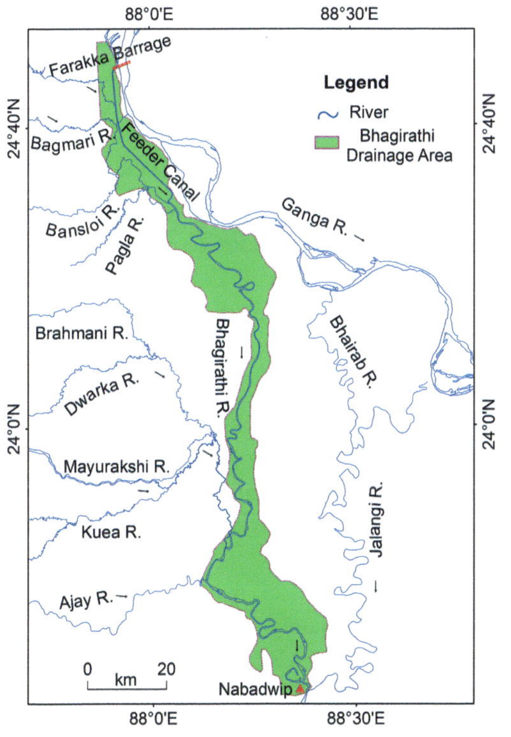

Fig. 2.22 Bhagirathi River system with its major tributaries

The BRB also exhibits the relative abundance of inland waterbodies like lakes, ponds and other forms of wetlands. Thus, the freshwater bodies are abundant for various uses. However, the rampant growth of urban settlements, industries like brick kilns and intensive agricultural practices with irrigation exploit the waterbodies at a greater pace than in the portfolio of the freshwater system in recent times. These anthropogenic interventions are tale-tell in the wake of the rapid population growth and huge demand for freshwater.

2.4.2 Controlled Hydrology and River Regime

The Farakka Barrage Project (FBP), which spans the West Bengal districts of Murshidabad and Malda, is an important engineering achievement started by the Indian government. Its main goals were to restore the Kolkata Port by transferring 40,000 Cusecs of water from the Ganga River into the Bhagirathi-Hooghly River system via a feeder canal, and to give the people of Kolkata access to fresh water and a saline-free environment (Rudra 2011; Islam and Guchhait 2017b).

The Farakka Barrage (2245 m long), the Jangipur Barrage (213 m long), and the 38-km Feeder Canal are the three primary components of the project, which was started in 1961 and finished in 1975 (Fig. 2.24). 112 gates make up the complex construction; there are 4 fish lock gates, 108 main gates, and 11 head regulator gates. The purpose of the gates is to redirect roughly 40,000 cusecs of discharge into the Feeder Canal (Parua 2009). By controlling the flow from the feeder canal outfall at Ahiran to the Bhagirathi, the Jangipur Barrage keeps the water from being diverted back into the Padma through the Bhagirathi.

Additionally, the road and rail bridges (Fig. 2.25a, b) spanning the Ganga River at Farakka establish a vital link between North Bengal and South Bengal. Furthermore, the Feeder Canal serves as a crucial component of National Waterway Number 1, connecting

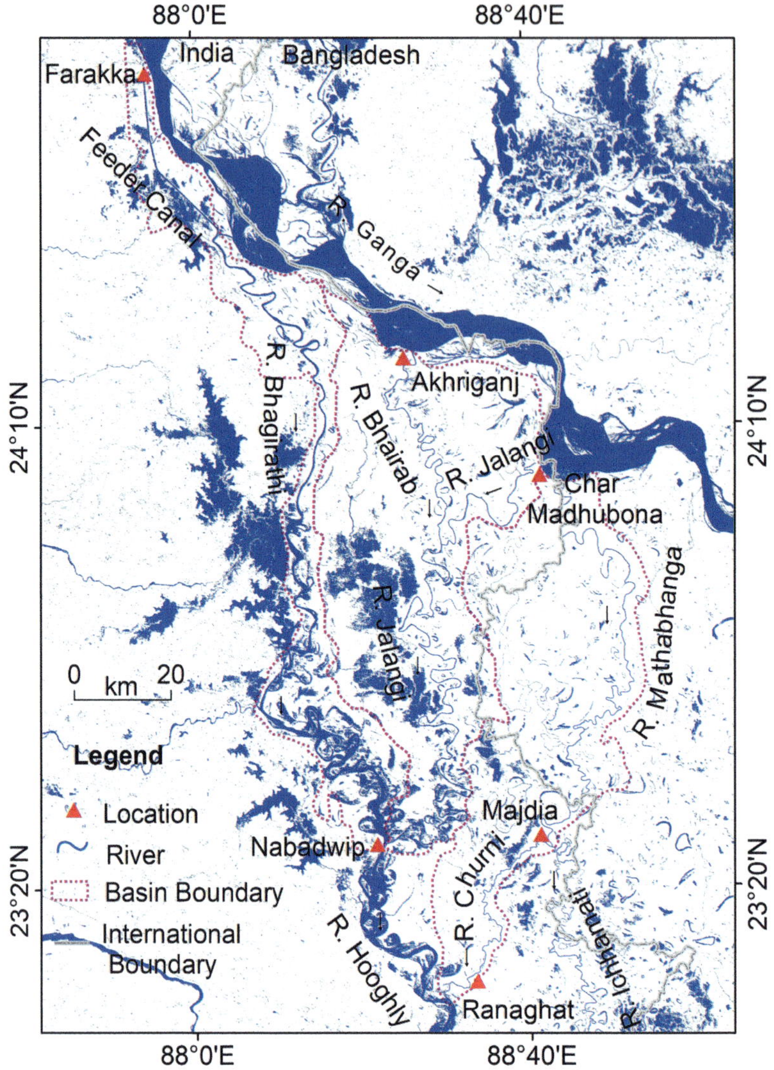

Fig. 2.23 Paleochannel and spill channel of the Bhagirathi river

Haldia to Allahabad inland. Moreover, it provides water supply to the 2100 MW Farakka Super Thermal Power Project (FSTPP) operated by NTPC Ltd. at Farakka (Parua 2009).

The Farakka Barrage Project (FBP) has significant implications for the channel morphology and hydraulic behavior of the Ganges system, both upstream and downstream of the barrage (Parua 2009; Mirza 2004). Along the course of the Ganges (Malda) and Bhagirathi (Murshidabad & Nadia), inundation and river bank erosion are very common hazards linked with the Farakka Barrage Project (Rudra 2011; Islam 2016). The human-induced change in flow pattern barrage operations since 1975 has effectively regulated the fluvial regime of the Lower Ganges system.

In the pre-Farakka period, discharge variability relied mainly on fluctuations in the monsoon climate. River discharge exhibited seasonal variability and was less frequent during that time. However, following the construction and commissioning of the Farakka Barrage Project (FBP) in 1975, aimed at maintaining the

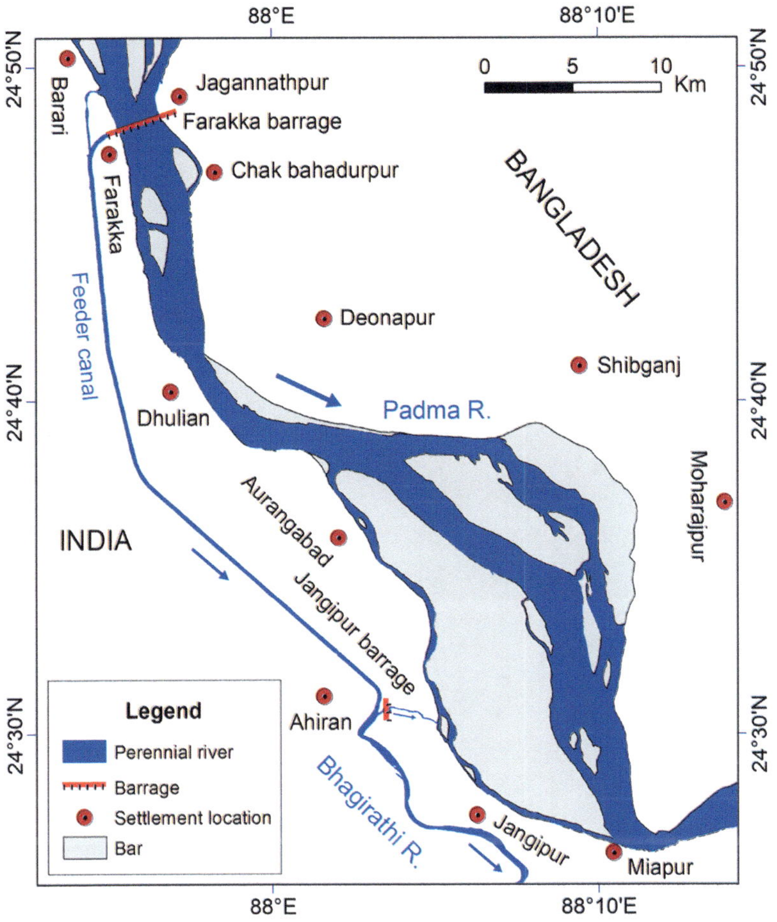

Fig. 2.24 Plan of Farakka Barrage Project with Farakk barrage, feeder canal and Jangipur barrage (*Source* Islam and Guchhait 2024a)

navigability of the Bhagirathi River, particularly during the lean months by flushing out sediments deposited in the Hooghly River, a flow of 40,000 cusecs of water was induced through a feeder canal. This measure was intended to preserve the Kolkata Port and provide saline-free water to the residents of Kolkata (Parua 2009). As a result, the river's flow patterns have been artificially managed, guided by arithmetic hydrological principles. The erection of the Farakka Barrage on the Ganga triggered an immediate water dispute between India and Bangladesh. This disagreement led to the formulation of water-sharing agreements in 1977 and 1996, outlining fluctuating discharge regulations for both nations over ten-day intervals during the dry months from January to May (Table 2.2).

Table 2.1 provides a clear depiction of the river regime's frequent fluctuations, particularly noticeable from March to May, in alignment with the stipulations outlined in the 1996 water-sharing treaty. Discharge demonstrates increased variability in the months that are lean, specifically from March to May, and comparatively lower variability in the remaining months of the year. The discharge hydrograph at the gauge stations at Berhampore and the Feeder Canal (Farakka) shows this pattern (Fig. 2.26a, b). At the gauge stations at Berhampore, Katwa, and the Feeder Canal

Fig. 2.25 Photographs **a** and **b** showing Rail cum Road Bridge across the river Ganga (*Source* Todaytimesnews.com)

(Farakka), there is a unique peak and valley pattern on the discharge hydrograph throughout the treaty period (January to May). On the other hand, the hydrograph shows oscillations at Katwa but appears regular at the Feeder Canal and Berhampore during the usual period (June to December) (Islam and Guchhait 2017a). A smoother curve and a pattern of peak and valley structure are revealed by the ten-day moving average.

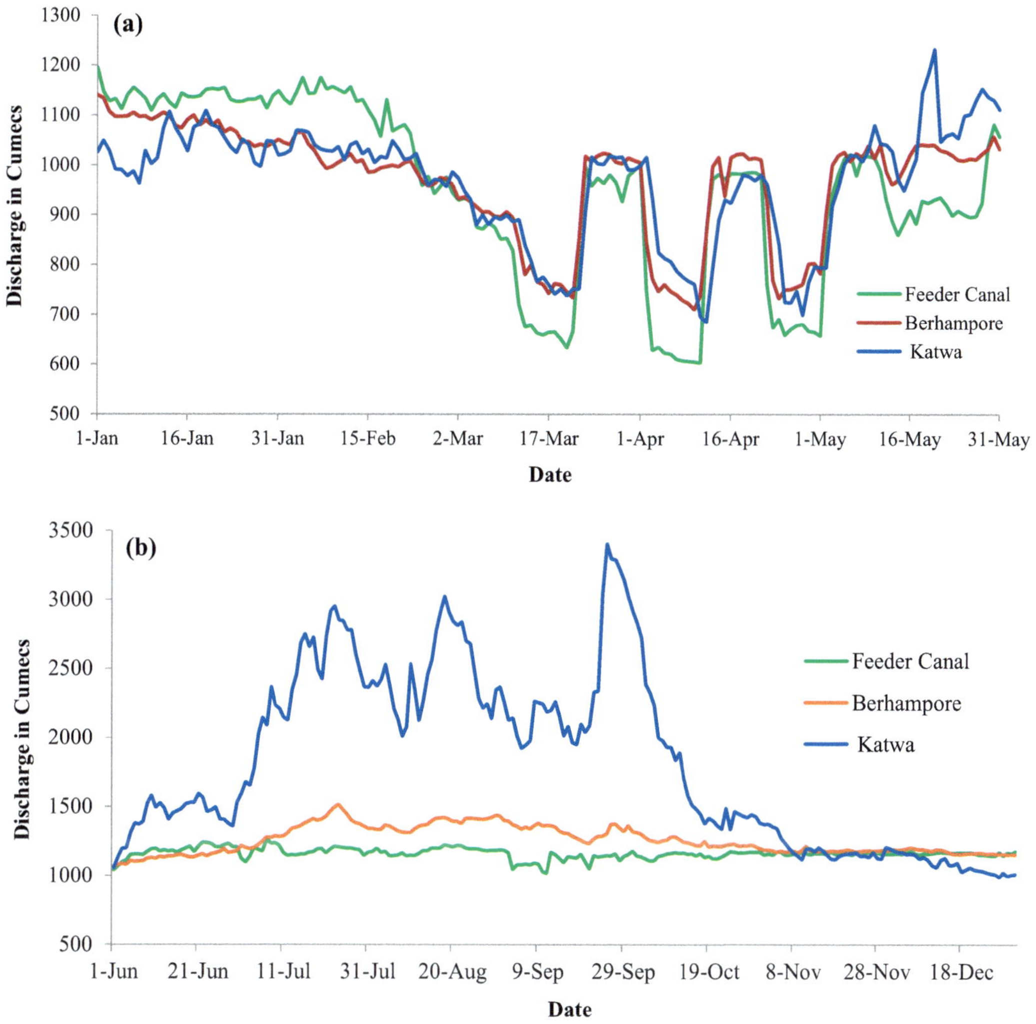

Fig. 2.26 Discharge hydrograph of the river Bhagirathi. **a** Treaty period (January to May), **b** Normal period (June to December) (*Source* Islam and Guchhait 2024a)

2.5 Ecology and Environment

2.5.1 Ecological and Drinking Water Quality

The ecological and environmental parameters are quite oscillatory. When the river was active, the flow was maintained throughout the year. However, before the installation of the FBP, the flow was negligible in the lean months (January to May). After the operationalization of the FBP, the flow was regained and aquatic ecosystems transformed from lentic (standstill) to lotic type (running). This oscillation in the flow behavior is also recorded in the distribution and proliferation of the aquatic flora and fauna. After the implementation of the Ganga Action Plan and Namage Gange, the river's health was substantially improved due to the lessening of river pollution and the increase in the aquatic biota. At present, the water quality is moderate to good with spatial and seasonal variations.

Ecological water quality is assessed in terms of BOD and DO. The higher BOD and lower

DO indicate poor water quality. The pristine water samples have a value < 1 mg\l, moderately polluted water samples 2–8 mg\l, and treated municipal sewage 20 mg\l. In the context of the Bhagirathi River, most water samples belong to moderately polluted water while few samples maintain pristine water quality. DO levels > 5 mg/l are good for fish and levels 3-5 mg/l fish growth is unsupported while DO levels < 1 mg/l is considered hypoxic supporting no life at all (Table 2.3). In the context of the Bhagirathi River, DO concentrations lie above levels > 5 mg/l supporting the growth of the aquatic organisms. The West Bengal Pollution Control Board data indicates that the upper part of the Bhagirathi (Farakka) portrays better water quality during the monsoon season compared to the pre and post-monsoon seasons while the middle (Gorabazar, Berhampore) and lower (Nabadwip) show better quality in the pre and post-monsoon seasons compared to the monsoon seasons. It indicates that the lower and middle parts receive more effluents from agriculture, urban and industrial areas of the BRB during the monsoon season.

Regarding the drinking water quality measures of bacterial count are considered. As per the World Health Organization, a zero count of *E. coli* per 100 ml of water is considered safe for drinking while a count of 1–10 MPN/100 ml is considered low risk and 11–100 MPN/100 ml medium risk. The count of *E. coli*, *Fecal coliform*, and *Fecal streptococci* shows that river water is not at all fit for drinking. Farkka has the lower risk while Gorazbar shows the maximum risk (Table 2.3). Moreover, seasonal variations depict that pre-monsoon has the maximum count of coliform compared to the monsoon and post-monsoon seasons.

2.5.2 Irrigation Suitability

This section examines several hazards such as sodicity and alkalinity permeability problems, to assess the suitability of Bhagirathi river water for irrigation. SAR is a measure of the sodium hazards, which has an impact on crops. High sodium concentrations in irrigation water have an impact on the permeability of the soil. The river SAR values vary from 0.38 to 1.12. Based on the SAR analyses, irrigation water is categorized as excellent (SAR < 6), dubious (6 < SAR < 9), and inappropriate (SAR > 9) (Table 2.4). All of the samples fall into excellent categories and are suitable for irrigation based on this classification. Furthermore, using the US Salinity Laboratory (USSL) diagram with SAR plotted against the EC, the suitability of river water for irrigation has been established (Fig. 2.27). The graphic showed that all of the samples fell into S1, or low sodium threat, meaning that there is no sodium hazard at any of the sample sites. The good water quality for irrigation is also supported by the values of SSP and %Na. Due to the high water absorption between magnesium and clay particles, a high concentration of magnesium in irrigation water causes soil alkalinity and reduces soil infiltration capacity. When the MH value is < 50, it is appropriate to use the water for irrigation. All the samples except monsoon at Farakka indicate river water free from magnesium hazard.

2.6 River and Society

2.6.1 River-Based Economy and Society

River-human interactions in the BRB are quite fascinating. The river has been beneficial for agricultural economics because of the presence of highly fertile soils in the BRB. About 70% of people are engaged as farmers. The cropping intensity has increased over time in this. For example, Nadia district has now the cropping intensity of 230% (2008) from 190% (1990). Once the Kolkata port was on the verge of extinction due to sedimentation and lower water supply through the Bhagirathi-Hooghly River. In a classical work 'Changing face of Bengal,' RK Mukherjee has beautifully sketched that scenario. However, after the installation of the FBP, the port industrial economy is still operating. Otherwise, the Kolkata port-based

Table 2.3 Hydrochemical properties of the Bhagirathi River

Parameter	Unit of measurement	Pre-monsoon			Monsoon			Post-monsoon		
		Farakka (22 March 2023)	Gorabazar (23 March 2023)	Nabadwip (22 March 2023)	Farakka (25 July 2023)	Gorabazar (24 July 2023)	Nabadwip (19 July 2023)	Farakka (28 Nov 2023)	Gora-bazar (29 Nov 2023)	Nabadwip (23 November 2023)
Ammonia-N	mg/l	BDL	0.15	0.16	0.10	0.15	0.12	0.12	0.15	0.10
BOD	mg/l	2.75	1.90	1.70	2.40	2.60	2.80	2.25	2.20	2.10
Conductivity	µs/cm	405.40	382.50	377.40	436.30	422.18	468.80	239.60	246.20	298.20
Dissolved O_2 (DO)	mg/l	7.60	7.20	7.90	6.50	7.00	6.70	8.00	8.10	7.30
E. coli	MPN/100 ml	170	780	330	45	780	490	130	450	230
Fecal coliform	MPN/100 ml	490	2200	790	310	3500	1400	1700	2400	1400
Fecal streptococci	MPN/100 ml	78	270	130	23	490	170	33	330	130
Nitrate–N	mg/l	0.60	0.11	0.16	0.94	0.55	0.12	0.34	0.32	0.88
pH	Unit	7.62	7.86	7.98	7.24	7.10	7.79	7.43	7.64	7.43
Temperature (Water)	°C	28	28	28	34	32.5	32	25	26	28
Total coliform	MPN/100 ml	700	7900	2400	940	13,000	2800	3500	13,000	5400
Boron	mg/l	BDL	BDL	BDL	BDL	BDL	BDL	BDL	BDL	BDL
Calcium	mg/l	38.48	46.49	29.66	26.72	28.29	23.58	33.67	51.30	27.25
Chloride	mg/l	16.99	14.99	17.61	8.00	10.00	9.00	14.00	12.00	14.00
COD	mg/l	11.76	7.84	8.82	10.00	11.00	13.00	10.78	10.78	8.00
Fluoride	mg/l	0.19	0.19	0.23	0.16	0.20	0.19	0.25	0.22	0.19
Magnesium	mg/l	13.61	11.18	17	7.15	4.29	8.58	18.47	13.61	10.69
Nitrite N	mg/l	BDL	BDL	BDL	0.01	0.01	0.12	BDL	BDL	BDL
Phenolpthalein Alkanity	mg/l	Nil	Nil	Nil	Nil	Nil	Nil	Nil	Nil	11.72
Phosphate-P	mg/l	0.19	0.13	0.21	0.15	0.13	0.07	0.15	0.08	0.20
Potassium	mg/l	5.60	5.40	5.00	4.50	4.10	4.60	4.60	5.10	4.90
Sodium	mg/l	15.50	15.50	15.20	13.40	13.60	14.40	13.30	13.30	13.10

(continued)

Table 2.3 (continued)

Parameter	Unit of measurement	Pre-monsoon			Monsoon			Post-monsoon		
		Farakka (22 March 2023)	Gorabazar (23 March 2023)	Nabadwip (22 March 2023)	Farakka (25 July 2023)	Gorabazar (24 July 2023)	Nabadwip (19 July 2023)	Farakka (28 Nov 2023)	Gora-bazar (29 Nov 2023)	Nabadwip (23 November 2023)
Sulphate	mg/l	24.79	22.63	22.37	18.82	17.41	24.68	25.13	24.99	22.08
Total alkalinity	mg/l	150.00	157.41	152.26	85.32	85.32	87.30	128.90	130.86	138.67
Total dissolved solids (TDS)	mg/l	234.00	210.00	208.00	228.00	224.00	244.00	126.00	142.00	154.00
Total fixed solids (TFS)	mg/l	270.00	204.00	210.00	206.00	198.00	196.00	106.00	94.00	136.00
Total Hardness as $CaCO_3$	mg/l	152.00	162.00	144.00	96.07	88.24	94.12	160.00	184.00	112.00
Total suspended solids (TSS)	mg/l	94.00	48.00	58.00	58.00	26.00	32.00	28.00	30.00	44.00
Turbidity	NTU	88.14	58.91	46.28	44.12	22.13	27.74	16.90	19.80	37.90
Human activities		Fishing	Bathing and washing, Navi-gation, Fishing	Bathing and washing, Fis-hing	Navigation, Fishing	Navigation, Fishing	Navigation, Fishing	Barrage area, Fishing	Navi-gation, Fishing Cattle farming	Navigation
Color and inten-sity		Muddy	Muddy	Muddy	Muddy	Muddy	Muddy	Slight turbid	Slighty turbid	Muddy
Odour		Odour free	Odour free	Odour free	Odour free	Odour free	Odour free	Odour free	Odour free	Odour free
Approximate depth (bottom/table)		3.00 m	3.00 m	3.00 m	3.00 m	3.00 m	3.00 m	1.50 m	1.50 m	3.00 m

BDL below detectable level

Table 2.4 Spatial variation of the irrigation suitability indices in different reaches of the Bhagirathi River

Seasons	Sites	SAR	% Na	SSP	KI	MH	RSC	RSBC	RI	PI
Pre-monsoon	Farakka (22 March 2023)	0.54	21.09	18.06	0.22	37	2.40	0.54	0.09	57.85
	Gorabazar (23 March 2023)	0.53	19.99	17.17	0.21	28	2.48	0.26	0.07	56.12
	Nabadwip (22 March 2023)	0.55	21.39	18.56	0.23	48	2.64	1.01	0.09	60.75
Monsoon	Farakka (25 July 2023)	0.50	20.15	17.40	0.21	51	0.34	0.06	0.07	50.97
	Gorabazar (24 July 2023)	0.63	28.21	25.02	0.33	20	1.33	– 0.02	0.09	71.87
	Nabadwip (19 July 2023)	0.64	28.20	24.84	0.33	37	1.28	0.25	0.08	69.08
Post-monsoon	Farakka (28 Nov 2023)	0.46	17.77	15.21	0.18	47	1.47	0.43	0.08	51.85
	Gorabazar (29 Nov 2023)	0.43	16.08	13.52	0.16	30	1.06	– 0.42	0.07	46.34
	Nabadwip (23 November 2023)	0.54	23.58	20.18	0.25	39	2.79	0.91	0.08	70.45

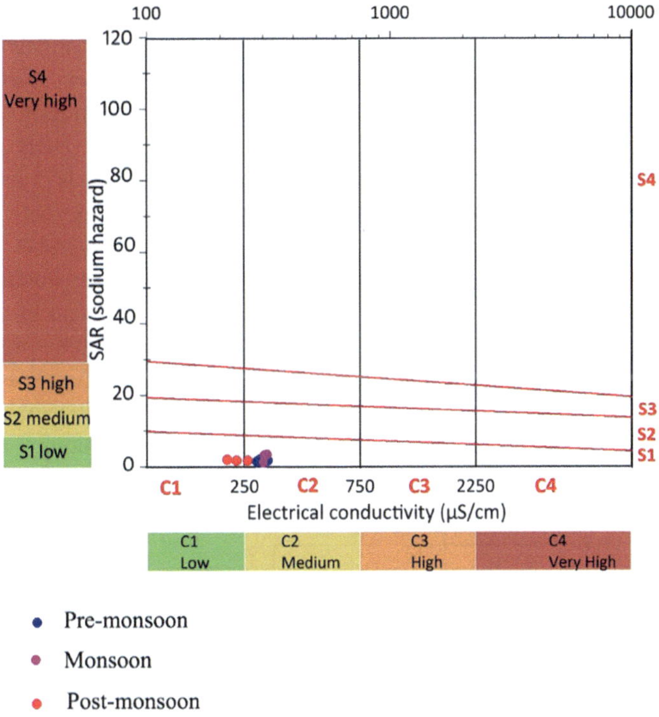

Fig. 2.27 USSL diagram showing the irrigation suitability of the Bhagirathi River

economy could have been derailed in the absence of the FBP. The industrial agglomerations (paper, jute, Special Economic Zone, petrochemical, etc.) along the banks of the Bhagirathi-Hooghly River also settled. Though the jute industry has faced a great decline after the partition of Bengal, other industrial activities are still depending on the flow of the river.

The riverbank is the place of the recreations and aesthetics. Beatification along the river also attracts tourism based on the scenic view of the river. Apart from this beneficial aspect, the river has a colossal history of few fluvial hazards like flood, river bank erosion, and channel oscillations.

2.6.2 Fluvial Hazards

2.6.2.1 Floods

The Farakka barrage, tributaries of the Bhagirathi River and southwest monsoonal regimes exacerbate the BRB flood. The BRB has experienced two different kinds of flooding: (1) monsoonal floods and (2) inundations during lean periods. The feeder canal of the Farakka barrage and the tributaries of the Bhagirathi River receive an enormous volume of floodwater during the monsoon or rainy season. A significant portion of this basin floods during the monsoonal flood, resulting in the loss of lives and property. The majority of BRB's floods, according to historical data, happened in September. Between 1900 and 2020, there will be an increase in the frequency of floods. Flood frequencies are 5, 5, 8, and 11 between 1900 and 1930, 1931 to 1960, 1961 to 1990, and 1991 to 2020, respectively (Fig. 2.28a). The Bhagirathi River exhibited an extremely high volume of flood discharge. Agradweep recorded a significant magnitude (> 40,000 $m^3 s^{-1}$) flood flow, according to Das (2013). The mean annual peak flood discharge was 1985 $m^3 s^{-1}$, with a minimum of 1189 $m^3 s^{-1}$ and a maximum of 3359 $m^3 s^{-1}$ recorded. From the initial construction of the Farakka barrage reservoir in 1972 until 1983, there was relatively little distortion of the peak annual discharge curve; nevertheless, from 1984 onward, the trend of the annual peak flood flow increased (Fig. 2.28b).

Based on the average flood extent area of the BRB between 1984 and 2020, significant areas that are prone to flooding were identified in the Rarh and Bagri regions; however, there were differences in their spatial distribution. Due to the presence of large wetlands in and around the northern BRB's Kandi low-lying area, the rarh region has a larger flood extent area than the Bagri zone. Nevertheless, in the southern region of the BRB, the main oxbow lakes and cut-offs, as well as the eastern bank of the Bhagirathi River, were the places with the highest concentration of flood extent areas (Fig. 2.29a). The BRB was shown to have highly seasonal floods in terms of the seasonality of flood occurrence between 1984 and 2020. From July to September, which is monsoon season, is when most floods occur. As a result, floodwaters are present in the BRB's low-lying sections for one to three months. This rarh region's area had the highest seasonal concentration of water because of the drainage congestion in the BRB's Hijal sections (Fig. 2.29b).

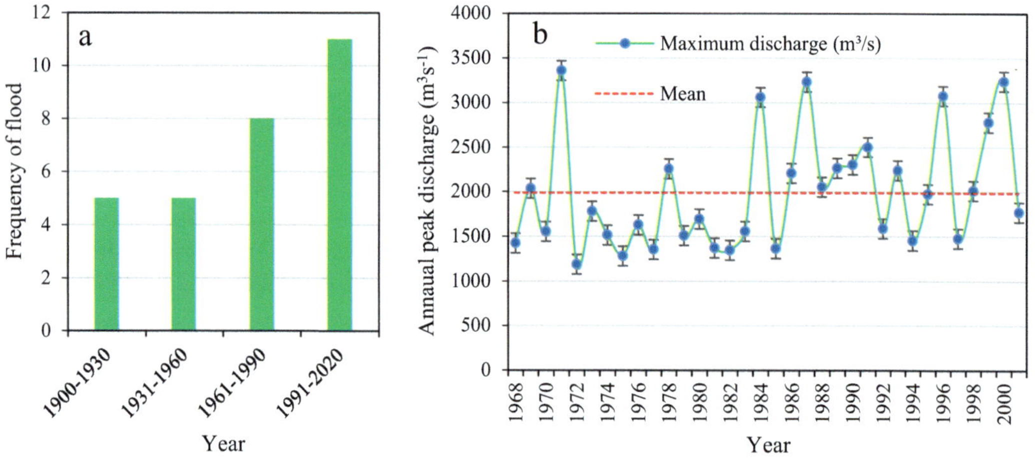

Fig. 2.28 Flood in the Bhagirathi River Basin. **a** Flood frequency (1900–2020). **b** Annual flood discharge of Bhagirathi River at Agradweep station, Burdwan (Das 2013; Ghosh et al. 2023)

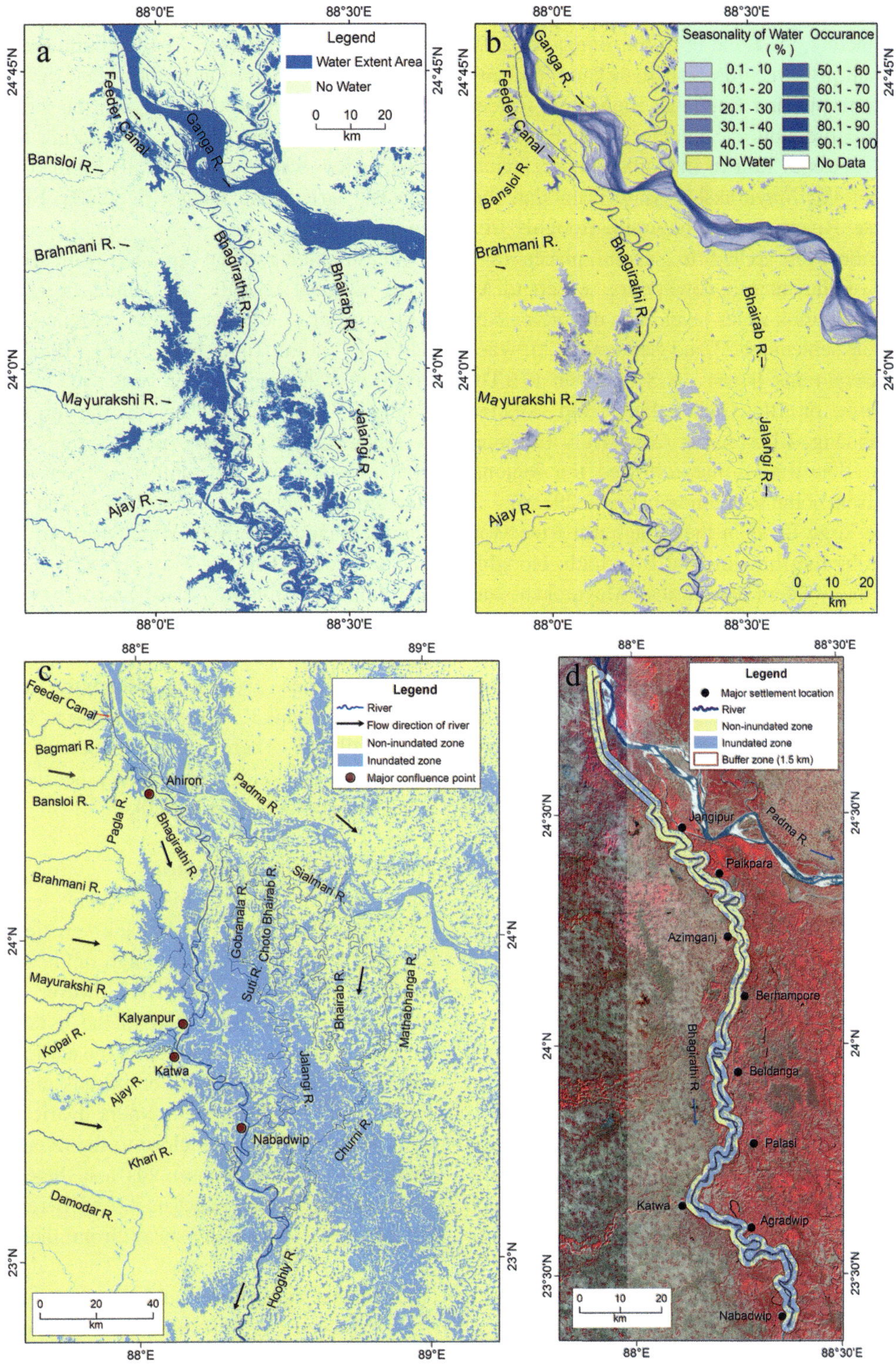

Fig. 2.29 Flood inundation along the Bhagirathi River, **a** Water extent area (1984–2020), **b** Seasonality of occurrence of water (1984–2020), **c** Flood during the September 26, 2000, in and around the Bhagirathi River, **d** Flood inundation zones within the buffers (1.5 km) for the Bhagirathi River (*Source* Ghosh et al. 2023)

When it came to intensity, size, damages, and other factors, the flood in South Bengal in 2000 broke all prior records. The Jharkhand Plateau's Chhotanagpur and Rajmahal mountains formed a deep depression in September, which resulted in an excess of rain in the sub-Gangetic catchment of the Mayurakshi, Ajay, Damodar, and Barakar Rivers. The Bhagirathi-Hooghly rivers' tributaries spilled in several places as a result of the intense downpours, which raised the Bhagirathi's water level. The deluge peaked between September 23 and September 25, when the heavy rains began on September 17. The interfluve of Bhagirathi-Jalangi and Jalangi-Mathabhanga-Churni was extensively damaged by the floodwaters that followed the regional slope from south to southeast (Fig. 2.29c, d).

The large flood in the Bhagirathi River was caused by the flow into the Bhagirathi-Hooghly from the western tributaries, Mayurakshi and Ajay. In addition, the situation became worse as a result of dam and barrage releases of water. The flood of 2000 flooded almost the entire Nadia district, with the exception of the northeast and southwest parts, and the southern portion of the Murshidabad district. Except for three CD blocks in the Murshidabad District (Raninagar-I, Bhagwangola-I, and Domkal Blocks) and one CD block in the Nadia District (Haringhata Block), 27 CD blocks in the two districts—11 in Murshidabad and 16 in Nadia—were drowned.

Since monsoon rains are only concentrated during the monsoon months (June to October), major rivers in the Ganga–Brahmaputra-Meghna Delta do not frequently flood during the lean season (January to May). However, because of the water released from the dams and barrages during the lean month, several areas have experienced minor flooding. The BRB observed that, throughout the treaty period (January to May), controlled hydrology from the Farakka Barrage Project caused inundation along both sides of the Bhagirathi River, albeit with a negligible area-wise growth. For example, flooding was discovered in several areas next to the river in the Murshidabad District, particularly from Beldanga to Palasy. As Bhagirathi received a more equitable portion of the water, the water level rose and, in accordance with the terms of the 1997 Indo-Bangladesh water-sharing agreement, river water was laterally spread into adjacent agricultural lands (Fig. 2.30a, b) on a ten-day basis.

Measurements taken in 2020 with Google Earth revealed that the area under water was approximately 0.8 km^2 near Melani and 0.14 km^2 on the right bank of the Bhagirathi River at Chadpur. There was an area of around 0.3 km^2 under water on the left, close to Jalkar Dadpur. Moreover, flooding was a regular occurrence in another section of the Murshidabad District, which stretched from Char Ramnagar (south) to Mirzapur (north). The crop calendar's January–May period is particularly important for the Kharif and Rabi season crops in this area. But the land is susceptible to agricultural practices because of the regular and recurrent floods. This lush, frequently flooded region has essentially been turned into a "belt of no agriculture". There is no induction of large-scale socio-economic vulnerability because the area extension is only pocket-like.

2.6.2.2 Riverbank Erosion

The temporal pattern of riverbank erosion is quite fascinating. The bank erosion of Bhgaritahi River was quite high immediately after the installation of the FBP. However, in the later phases, it was adjusted to the hydrogeomorphic variables as discussed in the earlier section. Here, impact of bank erosion is depicted with field survey data from Nadia district. The study found that villages close to the river bank exhibit that 60% of households lost their land. The distance decay effect is found here. The villages (e.g., Sujanpur) away from the river bank show little or no land loss. The study also reported that agricultural land loss is the maximum followed by the settled areas. For example, 92% of households in Ganjadanga lost agricultural land while 33% of households in

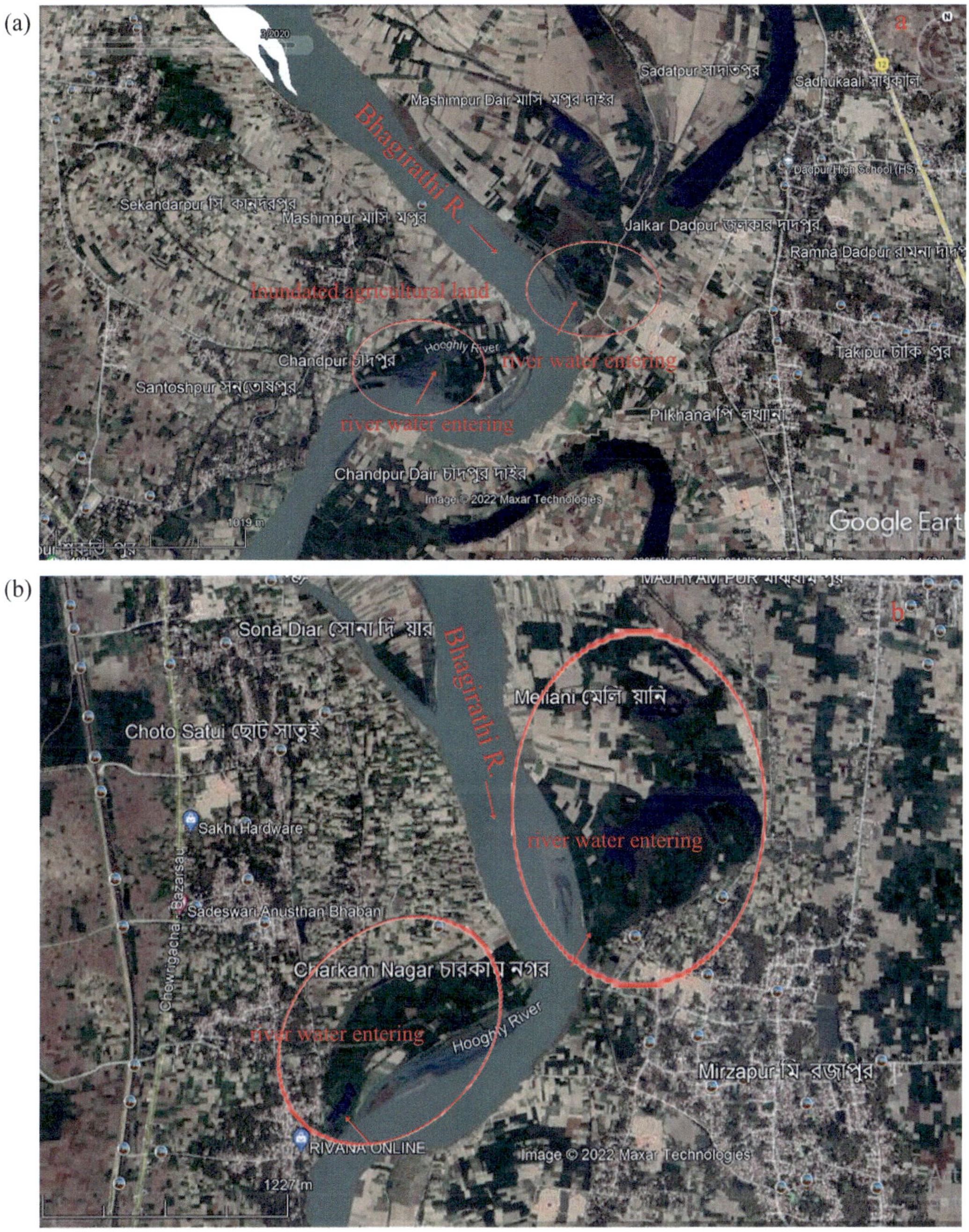

Fig. 2.30 During the treaty period (January to May), there was a belt of frequent small-scale inundations in two locations: **a** near Chadpur and **b** Melani. This was evidently caused by the increased pressure of river water entering the agricultural fields (based on Google Earth Images, 26 March 2020)

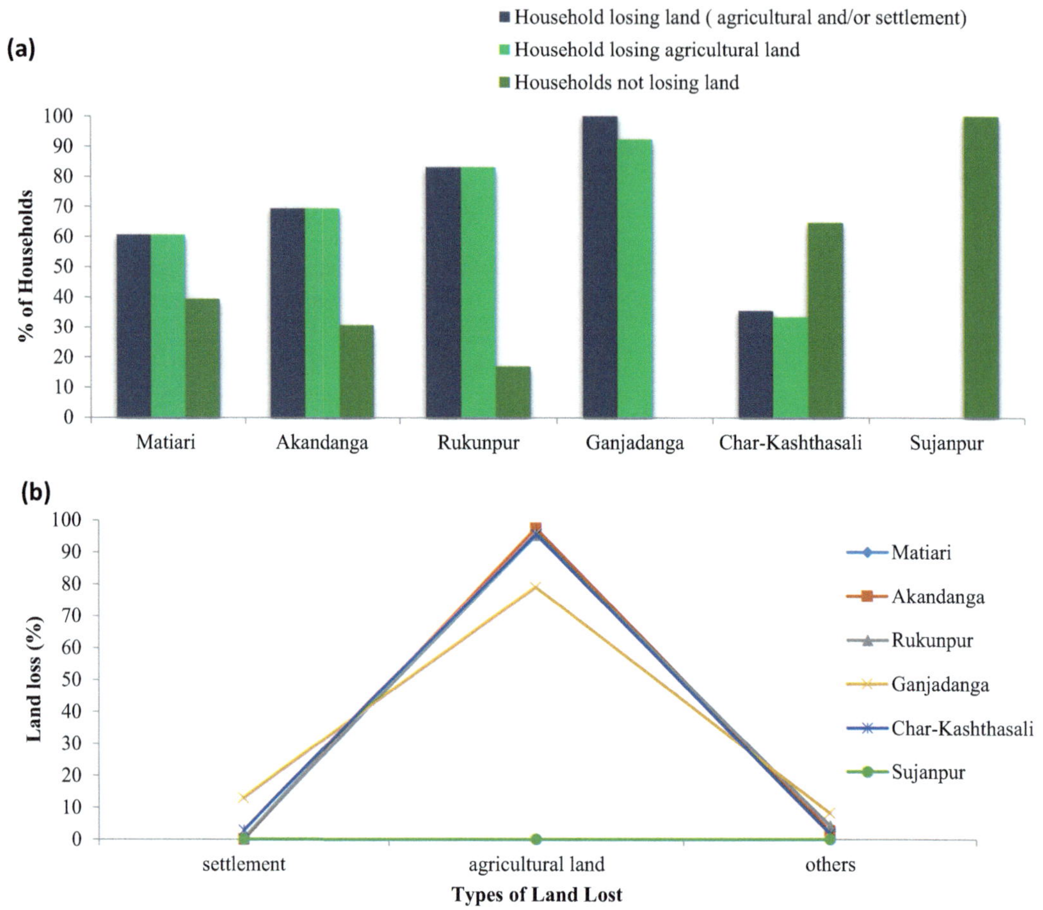

Fig. 2.31 Bhagirathi River study villages exhibit two distinct dynamics related to land loss: firstly, the nature of the land loss (as determined by the Field Survey, 2012–13; sample size: Matiari 597; Akandanga 127; Rukunpur 183; Ganjadanga 78; Char–Kashthasali 144; Sujanpur 57); and secondly, the types of land loss in the study area (as determined by the Field Survey, 2012–13; sample size: Matiari, 362; Akandanga 88; Rukunpur 152; Ganjadanga 78; Char–Kashthasali 51; Sujanpur 5) (*Source* Islam and Guchhait 2024b)

Char-Kashthasali lost land (Fig. 2.31a, b) (Islam and Guchhait 2017a, b).

Other types of land loss are pastures, roads, orchards, and playgrounds, etc. As Ganjadanga has shifted from west to east, different types of land loss are also maximum. Therefore, agricultural economy has become devastated by the bank erosion that attract special attentions. The study portrays that farmers of the Matiari and Rukunpur villages are severely affected by bank erosion induced agricultural land loss, and crop damages (Fig. 2.32a–d). The similar land loss scenario in respect of amount of the land loss and the proportion of the land loss are also documented in the village level survey (Fig. 2.33a, b). However, lower land loss is reported for Char-Kashthasali and Sujanpur with no land loss at all. It is observed that loss of settlement area is severe for Matiari, Rukunpur, and Ganjadanga. These villages have reflected that 10–40% people have become homeless for 15–30 days and spent their anxious nights

Fig. 2.32 Loss of properties by bank erosion, **a** Agricultural land in Matiari, **b** Agricultural land in Rukunpur, **c** Banana orchard in Rukunpur, **d** Loss of crops by bank erosion induced flood in Akandanga (*Source* Field survey, 2012–13)

under trees or under open sky (Fig. 2.34a). Similarly, those three villages show that 5–30% of the people have become homeless 1–4 times (Fig. 2.34b). Ganjadanga is the most victim in this respect followed by Rukunpur and Matiari.

2.6.3 Human-Induced Channel and Basin Scale Modifications

There has been a paradigm shift in the interactions. The field investigation indicates that around the 1990s, there was a good bond between humans and rivers for various reasons—agriculture, fishing, cults, and other socio-political issues. Nowadays, that bond is diminishing and present-day humans are distancing from the river. Therefore, the BRB is vehemently altered by human interventions in various ways—changes in the land use and land cover, mushrooming of brick fields, etc. Waterbodies and vegetation cover are now replaced with agriculture and built-up areas (Fig. 2.35).

2.6.3.1 Expansion of the Settlements and Urbanization and Related Channel Changes

Because of the hydrological and agroecological benefits, West Bengal's delta region is heavily populated. The population of the BRB is growing as a result of both natural and migration-induced expansion. The population of the basin's main urban centers has significantly increased throughout the past ten years (2001–2011). According to the District

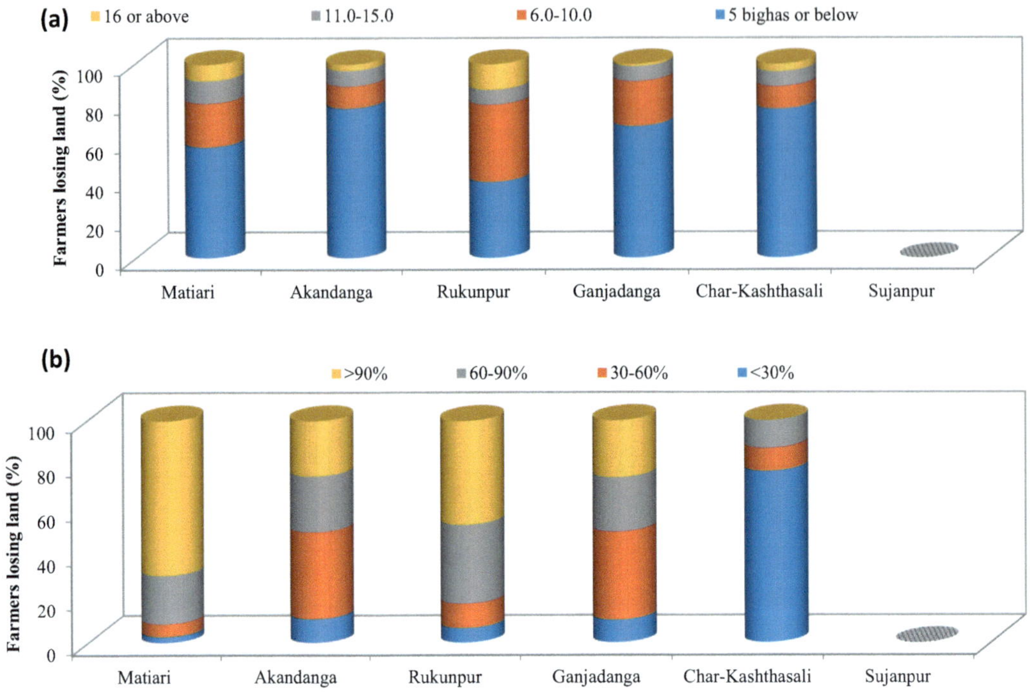

Fig. 2.33 Dynamics of agricultural land loss, **a** Amount of agricultural land loss, **b** Percentage of agricultural land loss in the study area (*Data Source* Field Survey, 2012–13; Sample size: Matiari, 362; Akandanga 88; Rukunpur 152; Ganjadanga 72; Char-Kashthasali 48; Sujanpur 0) (*Source* Islam and Guchhait 2024b)

Census Handbook (2001, 2011), Berhampore Municipality, for instance, saw a consistent rise in population throughout the various census years (1951: 55,613, 1961: 62,317, 1971: 92,889, 1981: 94,907, 1991: 115,144, 2001: 160,143, 2011: 195,223). Similarly, in 2011, the population of Katwa municipality increased from 71,573 to 81,615. There has been a noticeable shift in the spatial urbanization process due to urban expansion and sprawl. The settlement area in the buffer zone (the total buffer area 1367.05 km^2) has grown from 310.5 km^2 to 317.1 km^2 during the course of the decade (2005–06 to 2015–16) (Fig. 2.36a, b). Stated otherwise, there has been an increase in the settlement area from approximately 22.71–23.19%.

In general, when the impervious area of an urban setting grows, run-off increases and lag time decreases. This causes a shift in the river's sediment flow, which exacerbates the river's biological and physical deterioration as well as the expansion of its channel (Chin 2006). It is discovered that urbanization stabilizes channel movement. In comparison to other stretches, the channel oscillation is smaller along the major metropolitan areas in 1973 and 2019. The Bhagirathi's left bank, which is next to Berhampore city, is more stable than its right bank, which is home to a few scattered rural villages. In this stretch, the left bank's yearly bank line shifting rate is roughly 0.39 m, while the right bank's is 0.87 m. (Fig. 2.37a). This

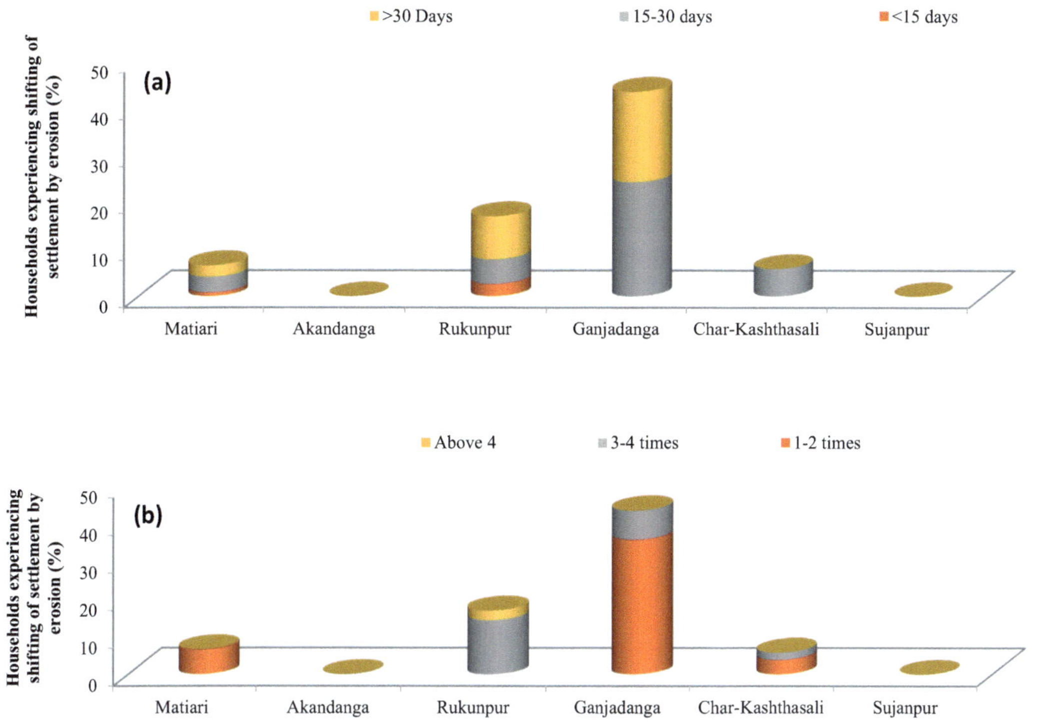

Fig. 2.34 Riverbank erosion, relocation of settlement and homelessness in the study villages, **a** Shifting of settlement by bank erosion, **b** Homelessness induced by bank erosion (*Data Source* Field Survey, 2012–13; Sample size: Matiari 362; Akandanga 88; Rukunpur 152; Ganjadanga 78; Char-Kashthasali 51; Sujanpur 57) (*Source* Islam and Guchhait 2024b)

results from forced stabilization and the concretization of the bank (Fig. 2.37b). On the left bank, the concrete bank extends 5.94 km, whereas there is no comparable civil structure on the right. Furthermore, a comparable stabilization has been noted on the river's right bank at Katwa Township (Fig. 2.38). Bank line shifting occurs at a pace of around 0.35 m along the Katwa, 1.97 m on the left bank, and abundant channel oscillation and bar expansion downstream of the Katwa reach. Moreover, the right bank's Nabdwip urban area has slowed the channel's swift flow (Fig. 2.39). The annual rate of movement along Nabadwip is (Right: 0.96 m and Left: 1.36 m), while the rate of channel oscillation in its upstream area is around (Right: 8.63 m and Left: 15.94 m), and its downstream reach has a rate of (Right: 1.4 m and left: 4 m).

2.6.3.2 Brick Fields Development and Channel Morphology

Moreover, road-stream crossings and brickfield industries are examples of human interventions over the BRB. Especially noticeable is the abundance of brickfields along the riverbanks (Fig. 2.40). At now, the head reach has 41 brickfields (16 on the Left Bank and 25 on the Right Bank), the middle reach has 61 brickfields (32 on the Left Bank and 29 on the Right Bank), and the tail reach has 52 brickfields (18 on the

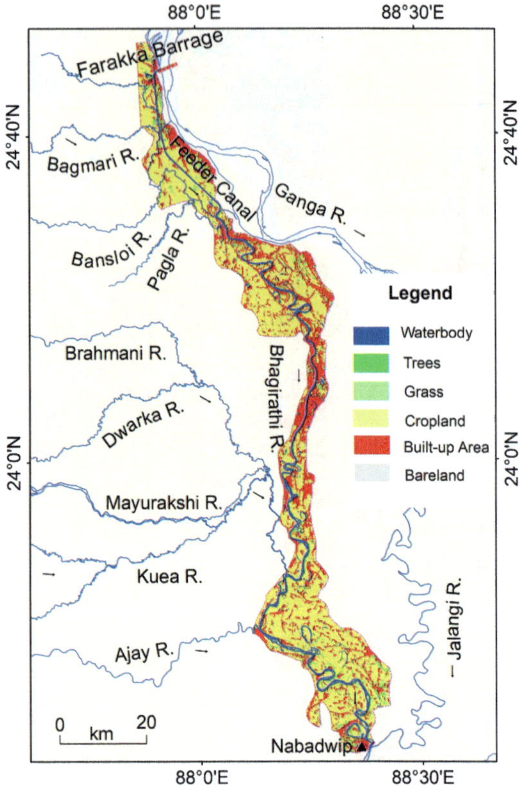

Fig. 2.35 Present-day land use and land cover of the BRB

Left Bank and 34 on the Right Bank). River bank erosion is usually exacerbated by brickfields (Das 2016). Because earth resources are extracted from the riverbank, a higher number of brickfields along a river correspond to a greater region of bank collapse (Fig. 2.41a). Clay or silt is the preferred material for brick kiln businesses. While the fluvial action at the Bhagirathi River accelerates bank collapse mechanisms by undermining the sandy layer at the bottom, brick kiln operators collect these silty materials from the higher layers of the soil. Furthermore, in order to gather silt during flood stages for later extraction for industrial uses, brick kiln operators frequently dig shallow depressions along the riverside (Fig. 2.41b). The channel width may eventually grow wider as a result of this procedure.

According to estimates by the DL and RO, Nadia (2009), a single kiln can move approximately 133,110 ft^3 of earth materials per year, extracted from within the river basin. Multiplying this figure by the number of brick kilns located adjacent to the Bhagirathi River, it amounts to

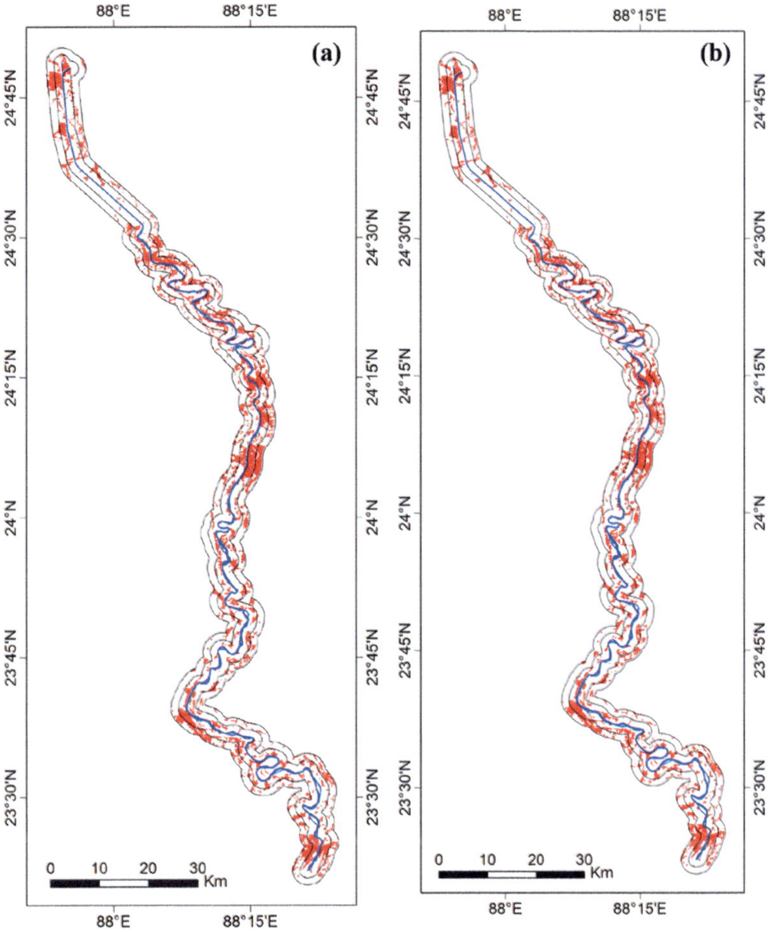

Fig. 2.36 Changing urbanization in the active belt of Bhagirathi River **a** 2005–2006, **b** 2015–2016

about 20,498,940 ft^3 of material movement in 2019, an increase from 6,522,390 ft^3 in 2000. This significant movement of earth materials by anthropogenic processes raises concerns when compared to the annual sediment budget (ranging from 0.5 to 1.0×10^6 m^3 per year) transported naturally by the Bhagirathi River. It underscores the influence of human activities on the sediment transfer process of the Bhagirathi river belt.

Moreover, the morphological alterations along a bank that is peppered with kilns are remarkable. Several cross-sections were drawn in order to evaluate the effect of the brick kiln industry on the morphological changes along the Bhagirathi River. Generally speaking, elevation rises landward from the riverbank or occasionally falls upon reaching the natural levee (Fig. 2.42a). Excavation of shallow depressions

(a)

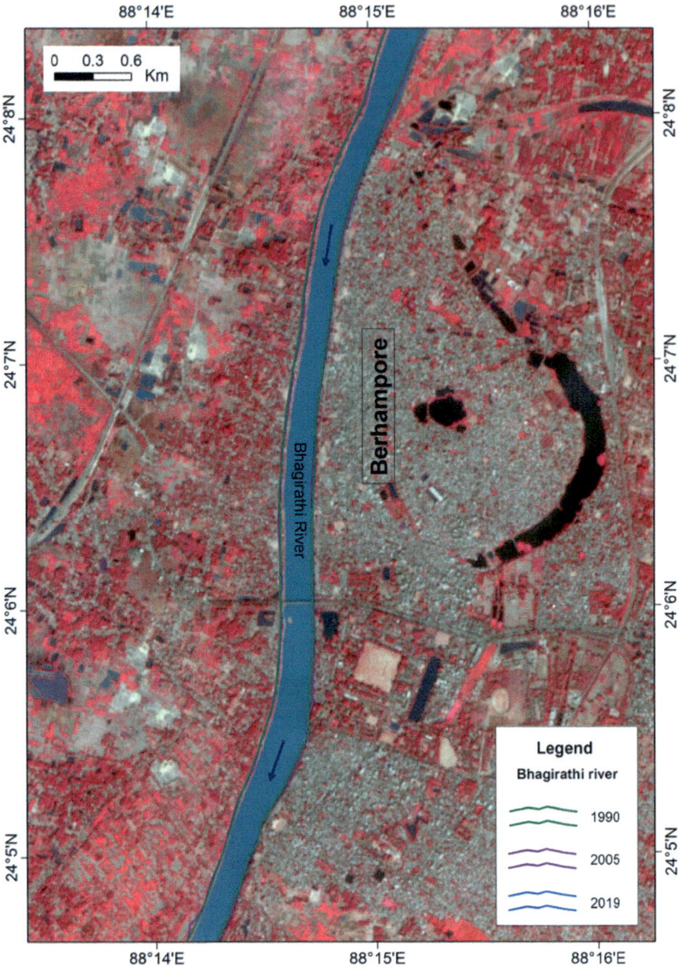

(b)

Fig. 2.37 Urbanization and channel stabilization adjacent to Berhampore city **a** Bank line shifting, **b** Concrete urban bank (Field photograph, 2019)

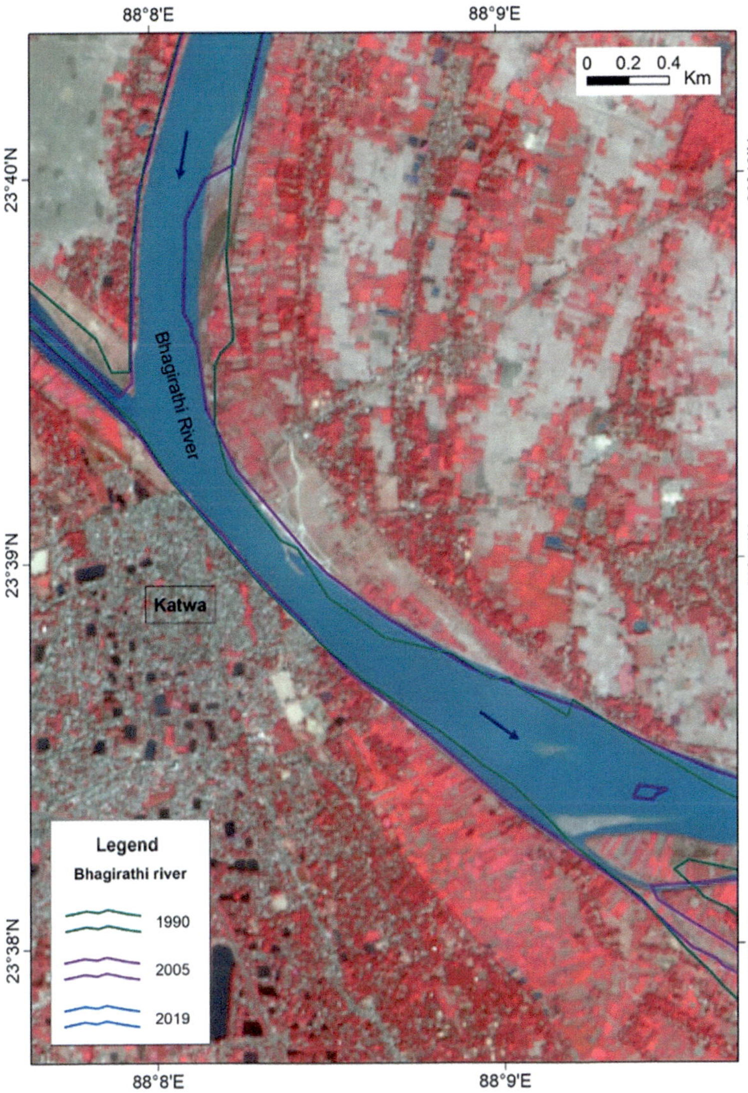

Fig. 2.38 Impact of urbanization on channel narrowing and stabilization near Katwa town (Islam et al. 2020)

to hold silt during floods disturbs the natural landscape. The cross-sections' characteristics show a sudden rise in areas without sediment cutting after a steep descent from the Bhagirathi River bank (Fig. 2.42b).

2.6.3.3 Road Stream Crossing and Channel Changes

There are four road crossings (two in the head reach and two in the middle reach) over the Bhagirathi. Hydrogeomorphological features

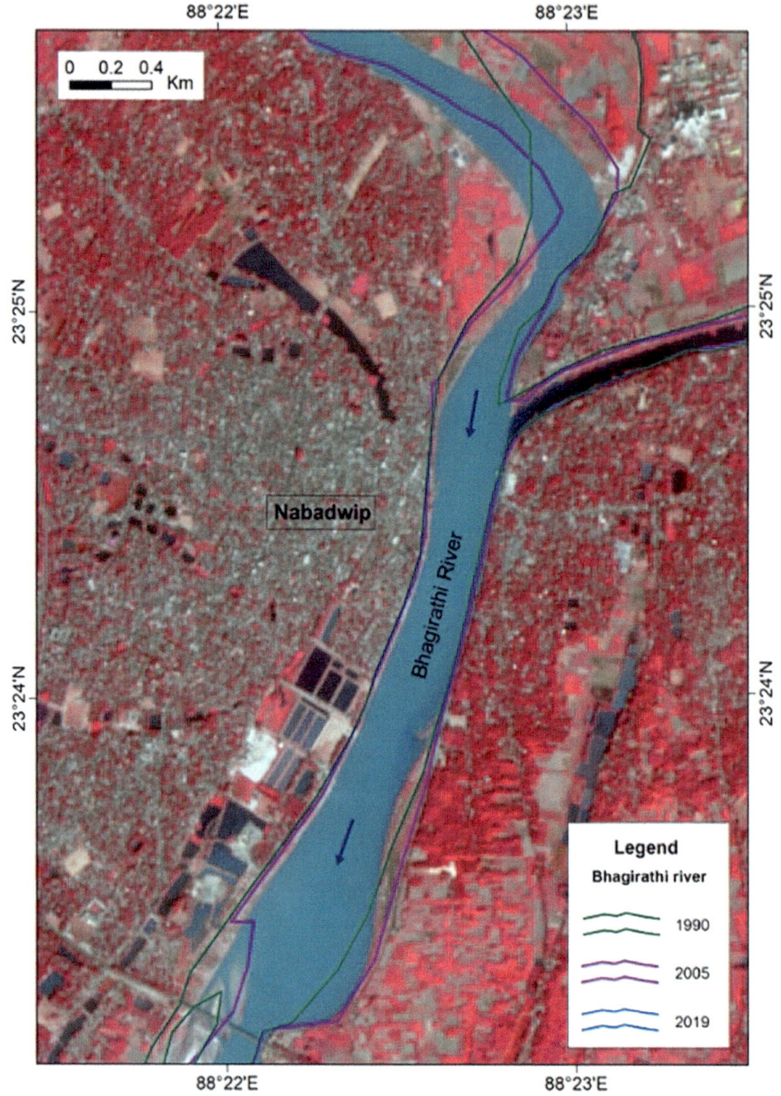

Fig. 2.39 Impact of urbanization on channel stabilization near Nabadwip town (Islam et al. 2020)

are also influenced by human activity, such as ship movements along rivers and the construction of bamboo fence across them. Road crossings across rivers are important because they control a channel's flow characteristics, which in turn affects morphological changes (Gregory and Brookes 1983). Cantilever bridges and suspension bridges are the two primary kinds of structures that are frequently utilized. Cantilever bridges immediately change the flow characteristics over the Bhagirathi River, whereas suspension bridges tend to narrow the channel. Most river bridges are cantilevered, with piers placed halfway between the two ends of the structure.

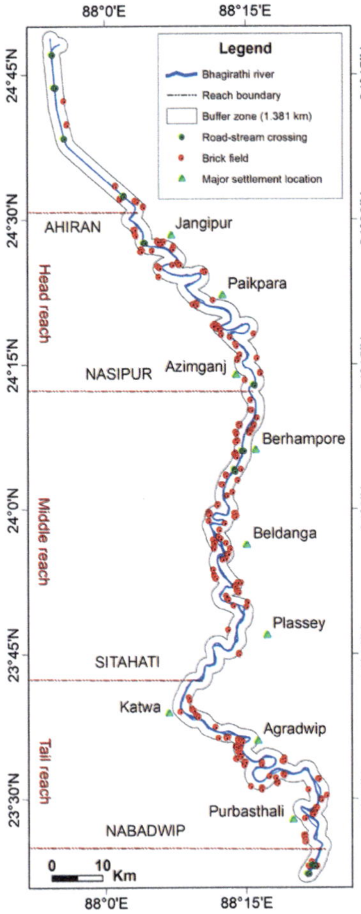

Fig. 2.40 Location of brick field and road stream crossing (*Source* Islam and Guchhait 2024a)

These piers play a key role in altering flow patterns because they impede the flow, which causes deposition between two piers and scouring on one side.

Morphological alterations were observed underneath the Berhampore Bridge (Fig. 2.43a). For example, thalweg direction was shown to be reversed between 1976 and 1994. As can be seen in Fig. 2.43b, the thalweg was orientated towards the left bank in 1976 and towards the right bank in 1994. The alternating pattern of deposition and scouring in the bed during 1976 is thought to have been impacted by the bridge piers. Road-stream crossings have an impact on channel morphology, but the hydraulic impact from the FBP has hastened bed scouring and increased the cross-sectional area at a faster rate. Furthermore, the bridge at Hazarduari, which was built in 2010 with the use of contemporary technology to support larger weights (Fig. 2.44a), has two piers situated immediately in the channel. With a broader bottom width tapering upwards, these piers have an impact on the local water level upstream.

Because of these piers, the effective water area has decreased, raising the upstream water level and causing afflux. Furthermore, during the height of the monsoon season, the flow is impeded by river training projects on both banks. In reality, the boulder-strewn surface is ready to limit channel oscillation and avert future bridge collapse. The Bhagirathi River‘s hydraulic behavior is altereds by the flow resistance provided by the man-made boulder bed. At Balarampur near Berhampore, another bridge (Fig. 2.44b) with three piers in the middle of the channel and another pier for extra services is presently being built.

2.6.3.4 Guide Bank and Sedimentation

A guide bank is constructed to assist a river's flow through a constricted area without endangering neighboring buildings. A bamboo guide bank was built in 2011 over the Bhagirathi River near Prachin Mayapur, Nabadwip. This structure's main objective is to reroute the Bhagirathi River‘s flow into a restricted channel on the right bank so that Inland Waterway Authority of India (IWAI) vessels can travel through it. The river is around 1 km wide in this section, and during the lean months, the flow is much less, at about 25,000 cusecs. Obviously, the distributed flow is not enough to facilitate vessel movement, which necessitates a minimum of two to

Fig. 2.41 **a** Bank failure due to cutting of soil from the upper horizons near Nabadwip (*Source* Field photograph, 2011); **b** Shallow depression in the brick kiln area for entrapping sediment during flood along the left bank of Bhagirathi river, near Beldanga

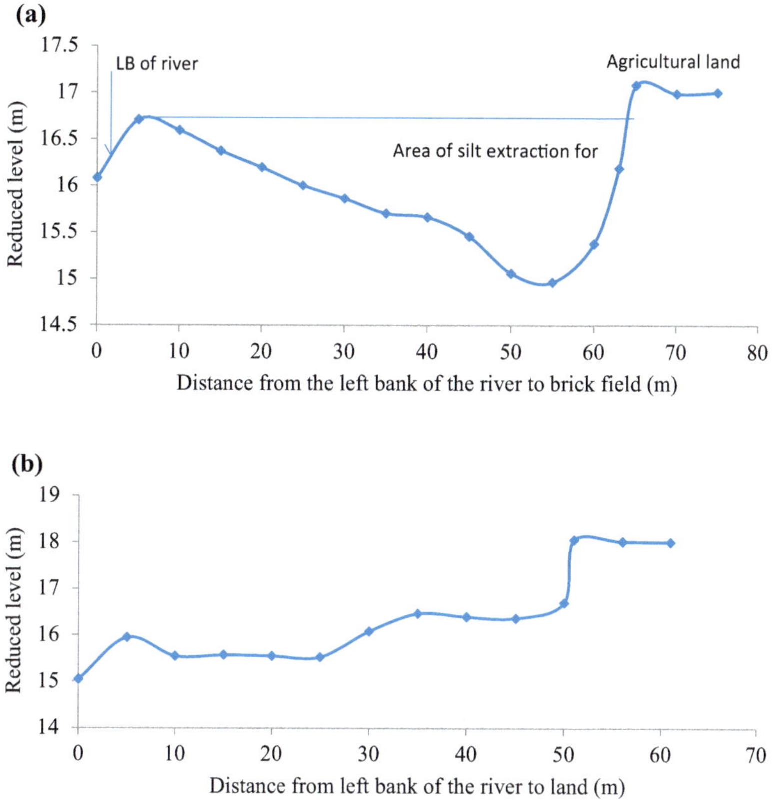

Fig. 2.42 Topographic expression is altered by brick kiln industries through: **a** Creation of artificial valleys resulting from silt excavation. **b** Formation of artificial spur-like features (*Source* Islam and Guchhait 2024a)

three meters of water. Therefore, rather than minimizing the canal to improve depth, man-induced sedimentation is initiated to restrict the flow across its whole length. The fencing's slanted placement catches the silt and deposits as a mid-channel bar (Fig. 2.45).

2.7 Concluding Notes

The BRB is a truly oscillatory river characterized by fluctuations in the river regime. Once historically active river decayed and disconnected from the main Ganga in the lean months due to the eastward titling of the Bengal basin. The river again rejuvenated through the FBP. The mythology, history, geography and environment are really spatially and temporally varied as depicted in this chapter. The work has eloquently depicted the various transitions to transformation in its trajectory. The naturally processing river is now under severe control of human intervention through the hydraulic inputs from the Farakka Barrage Project. The planform and cross-sectional morphology have exhibited profuse changes in the post-FBP period. Rapid progression in meander geometry in terms of the frequent channel cut-offs, and fluctuating meander geometry characterize the post-FBP platform. Similarly, higher *w*/*d* ratio, and

(a)

(b)

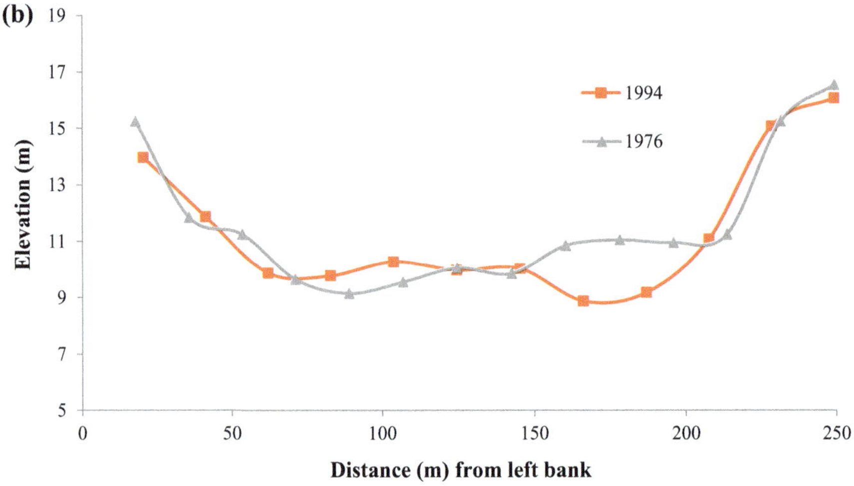

Fig. 2.43 Bed scouring caused by the presence of road piers at the Berhampore Bridge: **a** Location of piers. **b** Alterations in cross-sectional morphology (*Source* Ghosh 2000)

higher channel asymmetry are also prominent in the post-barrage situation. This regulated river regime has induced frequent riverbank erosion and channel oscillation in the Murshidabad and Nadia districts. Moreover, controlled hydrology from the FBP has created locally inundated areas along the banks of the Bhagirathi River due to the higher release of water from the FBP every 15 days as per the Indo-Bangkadesh water sharing treaty. These floods and bank erosion affect local agricultural ecosystems to a great extent causing an escalated level of community vulnerability. Apart from this mega hydraulic control, there are several other forms of anthropogenic interventions like brick fields, changes in the LULC with higher urbanization, bamboo fencing across the river, and frequent ship movement causing a significant change in the

Fig. 2.44 **a** Nasipur Railway Bridge near Hazarduari, Murshidabad (Field photograph, 2019); **b** New Bhagirathi bridge near Balaram under construction (Field photograph, 2019)

channel morphology. Moreover, human intervention has caused severe pollution of the river water that affects the water quality for domestic, drinking and irrigation purposes. Thus, a minimum river flow is maintained through the FBP, but water quality is getting depleted with time over the BRB. Thus, it needs attention from the various stakeholders for the sustainability of the riverine ecosystem. Like this river, other Nadia rivers like Jalnagi and Churni are facing these human-induced basin scale alterations. Thus, the next chapter is committed to addressing various issues of the Jalangi River.

Fig. 2.45 Bamboo-made guide bank triggering sedimentation near Nabadwip (Field Photograph, 2011)

References

Allison MA et al (2003) Stratigraphic evolution of the late Holocene Ganges-Brahmaputra lower delta plain. Sed Geol 155:317–342

Bagchi KG (1944) The Ganges delta. Calcutta University, Calcutta

Bandyopadhyay S (1996) Location of Adi Ganga Paleo Channel, S. 24 Parganas, West Bengal. Geogr Rev India 58(2):93–109

Banglalibrary (2025) রামায়ণ, কৃত্তিবাস ওঝা. Accessed 15 Jan 2025. https://www.ebanglalibrary.com/

Basu SR et al (2005) Meandering and cut-off of the river Bhagirathi. In: Kalwar SC (ed) Geomorphology and environmental sustainability. Concept Publishing Company, New Delhi, pp 20–37

Bhattacharya K (2000) Bangladesher Nadnadi o Parikalpana. Vidyoday Library Private Ltd., Kolkata

Brice JC (1964) Channel patterns and terraces of the Loup Rivers in Nebraska. Geological Survey Professional Paper 422-D

Chatterjee SN (1989) River system of West Bengal. River Research Institute, Govt. of West Bengal, Kolkata

Chin A (2006) Urban transformation of river landscapes in a global context. Elsevier, Amsterdam, pp 1–28

Colebooke RH (1801) On the courses of Ganges through Bengal. Asiatic Res 7:1–31

Das BC (2013) Changes and Deterioration of the Course of River Jalangi and Its Impact on the People Living on its Banks, Nadia, West Bengal, An Unpublished. PhD Thesis. Kolkata: Calcutta University

Das BC (2016) Impact of in-bed and on-bank soil cutting by brick fields on Moribund deltaic rivers: a study of Nadia river in West Bengal

District Census Handbook (2001) Murshidabad District, Directorate of Census operation, West Bengal, Part XII A & B Series 20

District Census Handbook (2011) Murshidabad District, Directorate of Census operation, West Bengal, Part XII B Series 20

Eck DL (2012) India: a sacred geography. New York: Harmony Books. pp 216–221

Hirst FC (1915) Reports on Nadia Rivers. Reprinted in River of Bengal (vol III, 2002). Gazetteers Department, Calcutta

Ghosh A (2000) A systematic study of the effect of farakka barrage on the morphological changes of the river Bhagirathi. PhD Thesis. University of Calcutta, Kolkata

Ghosh S, Hoque MM, Islam A, Barman SD, Mahammad S, Rahman A, Maji NK (2023) Characterizing floods and reviewing flood management strategies for better community resilience in a tropical river basin, India. Natural Hazards, 115(2):1799–1832

Goodbred SL Jr, Kuehl SA (1998) Flood plain processes in the Bengal Basin and the storage of Ganga-Brahmaputra River sediment: an accretion study using 137Cs and 210Pb geochronology. Sed Geol 121:239–258

Goodbred Jr SL, Kuehl SA (2000a) Enormous Ganges-Brahmaputra sediment load during strengthened early holocene monsoon. Geology 28(12):1083–1086

Goodbred Jr SL, Kuehl SA (2000b) The significance of large sediment supply, active tectonism, eustasy on the margin sequence development. Late quaternary stratigraphy and evolution of the Ganges-Brahmaputra delta. Sediment Geol 133(3–4):227–248

Gregory KJ, Brookes A (1983) Hydrogeomorphology downstream from bridges. Appl Geogr 3(2):145–159

Guchhait SK, Islam A, Ghosh S, Das BC, Maji NK (2016) Role of hydrological regime and floodplain sediments in channel instability of the Bhagirathi River, Ganga-Brahmaputra Delta, India. Phys Geogr 37(6):476–510

Islam A (2011) Variability of stream discharge and bank erosion—a case study on the river Bhagirathi. J River Res Inst River Behav Control 31(1):55–66

Islam A (2016) River bank erosion and its impact on economy and society a study along the left bank of river Bhagirathi in Nadia District West Bengal

Islam A, Guchhait SK (2015) Is severity of river bank erosion proportional to social vulnerability? A perspective from West Bengal, India. In: Life and living through newer spectrum of geography, p 35

Islam A, Guchhait SK (2017a) Analysing the influence of Farakka Barrage Project on channel dynamics and meander geometry of Bhagirathi river of West Bengal, India. Arab J Geosci 10(11):245

Islam A, Guchhait SK (2017b) Search for social justice for the victims of erosion hazard along the banks of river Bhagirathi by hydraulic control: a case study of West Bengal, India. Environ Dev Sustain 19(2):433–459

Islam A, Guchhait SK (2018) Analysis of social and psychological terrain of bank erosion victims: a study along the Bhagirathi River, West Bengal, India. Chin Geogr Sci 28(6):1009–1026

Islam A, Guchhait SK (2019) Social engineering as shock absorbing mechanism against bank erosion: a study along Bhagirathi river, West Bengal, India. Int J River Basin Manag 1–14

Islam A, Guchhait SK (2020) Characterizing cross-sectional morphology and channel inefficiency of lower Bhagirathi River, India, in post-Farakka barrage condition. Natural Hazards, 103(3), 3803–3836

Islam A, Guchhait SK (2024a) Riverbank Erosion: a Human-Induced Process. In Riverbank Erosion in the Bengal Delta: an Integrated Perspective (pp 99–143). Cham: Springer International Publishing

Islam A, Guchhait SK (2024b) Economic Vulnerabilities Induced by Riverbank Erosion. In Riverbank Erosion in the Bengal Delta: An Integrated Perspective (pp 201–247). Cham: Springer International Publishing

Islam A, Guchhait, SK, Hoque MM (2020) Anthropogeomorphology of the Bhagirathi River. In Anthropogeomorphology of Bhagirathi-Hooghly River System in India (pp 41–82). CRC Press

Islam A, Laskar N, Ghosh P (2012) An areal variation of fluvial hazard perceptions of various social groups—a perspective from Rural West Bengal, India. Indian Streams Res J 2(9):1–9

Majumder SC (1942) Rivers of Bengal delta. Calcutta University, Kolkata

Mirza MMQ (2004) The Ganges water diversion: environmental effects and implications—an introduction. In The Ganges water diversion: Environmental effects and implications (pp 1–12). Dordrecht: Springer Netherlands

Mukherjee RK (1938) The changing face of Bengal. Calcutta University, Calcutta

Ness R (2004) Belt width delineation procedures. Toronto and region conservation authority, Parish Geomorphic Ltd.

Oldham (1870) President's address. In: Proceedings, Asiatic Society of Bengal for February 1870, Calcutta

Parua P (1992) Stability of the banks of Bhagirathi-Hooghly River system. PhD Thesis. Jadavpur University, Kolkata

Parua PK (2009) Farakka barrage and its alleged impact on floods and bank erosion problems of Malda and Murshidabad Districts of West Bengal. In: Parua PK (ed) Some aspects about farakka barrage project (vol II, pp 30–44). Shilpanagari Publishers, Berhampore

Ray A (1999) Locational problems of the sixteenth century Bengal coast. Pratna Samiksha 6(8):121–134

Reaks HJ (1919) Report on the physical and hydraulic characteristics of the rivers of the Delta, Appendix-II of the report of the Hooghly River and its head waters. The Bengal Secretariat Book Depot, Calcutta

Rennell J (1788) Memoir of Hindoostan or Mughal Empire. London

Rudra K (2010) Banglar Nadikatha. Sahitya Samsad, Kolkata

Rudra K (2011) The encroaching Ganga and social conflict: the case of West Bengal, India. Unpublished paper, Kolkata

Sen H (2010) The drying up of River Ganga: an issue of common concern to both India and Bangladesh. Curr Sci 99(6):725–727

Sen S (2019) Ganges: the many pasts of an Indian river. Yale University Press. ISBN: 9780300119169. P 214–215

Sherwill WS (1858) Report on rivers of Bengal. Savielle Printing and Publishing Co Ltd., Calcutta

Umitsu M (1987) Late quaternary sedimentary environment and landform evolution in the Bengal lowland. Geogr Rev Japan 2(60):164–178

Willcocks W (1930) Lectures on the ancient system of irrigation in Bengal

ভট্টাচার্য, আশুতোষ (1940). রামায়ণ, কৃত্তিবাস ওঝা প্রণীত। অখিল ভারত জনশিক্ষা প্রচার সমিতি। কলকাতা | p-26

ebanglalibrary.com (2025). https://www.ebanglalibrary.com/topics/%E0%A7%A7%E0%A7%AE-%E0%A6%B8%E0%A6%97%E0%A6%B0-%E0%A6%AC%E0%A6%82%E0%A6%B6-%E0%A6%89%E0%A6%A6%E0%A7%8D%E0%A6%A7%E0%A6%BE%E0%A6%B0/ (retrived on 10.06.2025)

Jalangi: A Sarcastic Name of a River

3

Abstract

This chapter examines the Jalangi Rivers transformation and characteristics from both spatial and temporal perspectives. Formerly, the Jalangi River held significant importance among the 'Nadia Rivers,' stretching from Char Madhubona in the northeast of Nadia district to Swarupganj opposite Nabadwip, where it meets the Bhagirathi River. In the past, the river exhibited notable dynamism in its planform geomorphology, evident in various oxbow lakes and palaeo channels. Despite its current ironic name, historical records highlight its significant hydraulic role. Over time, the river has drastically declined, largely due to the eastward tilt of the Bengal Basin, post-Farakka reduced water levels of Padma River and human interventions. This chapter provides a comprehensive analysis of the river, covering hydro-geomorphic changes, declining water quality, environmental flow dynamics, and socio-economic shifts resulting from fluvial changes. The chapter also aims to foster sustainable management for the local community's development.

Supplementary Information The online version contains supplementary material available at https://doi.org/10.1007/978-3-031-85971-7_3.

Keywords

River decay · Water quality · Sialmari river · Bhairab river · Suti river · Anjana river

3.1 Etymology

The word 'Jalangi' is a united form of two Bengali words 'Jala' and 'angi'. "Jala" is derived from the Sanskrit word "जल" (jala), which translates to "water" in English. "Angi" is derived from the Sanskrit word "अङ्गी" (aṅgī), which signifies having a "body" or a "part of the body". In this compound term, "Jalangi" essentially conveys the idea of a body composed of water. The amalgamation of "Jala" and "Angi" encapsulates the essence of water, suggesting a form or entity of water. In the context of a river's name, it reflects the essence of a watercourse characterized by a significant volume and continuous flow of water, emphasizing the dynamic vigorous nature of the river.

The river bore a different name before being identified as 'Jalangi.' Originally it was known as 'Khoriya' or 'Khore'. On the right bank of the river Padma, there exists a village named 'Jalangi,' from where the 'Khoriya' River originated. Presently situated in Jalangi C.D. Block within the Murshidabad district, this village serves as a historical reference point. During

B. C. Das et al., *The Nadia Rivers*, Geography of the Physical Environment,
https://doi.org/10.1007/978-3-031-85971-7_3

the eighteenth century, Major James Rennell, engaged in surveying the Ganga River, began his survey in the autumn of 1764 (Lees 1896). It was during this survey that he possibly bestowed the name 'Jalangi' upon the river, drawing inspiration from the village of the same name. Despite this renaming, some residents, nestled within the river's catchment area, still refer to it by its original names of 'Khoriya' or 'Khore.' Different authors spelt the river Jalangi differently e.g. Jellinghy (Rennell 1781) or Jelenghee (Phillimore 1945), Jellinghi (Fergusson 1863), Jelingi (Willcocks 1930), Jellingi (Long 1846), Jellinghee (Mukharji 1932).

3.2 River in Epic, History, and Legend

The myth regarding the origin of the Bhagirathi River is: Bhagīratha (भगीरथ), a lineage of King Sagara from the Sūrya Dynasty, or Ikṣāku Dynasty played a pivotal role in the descent of the Ganga. Guiding Ganga (River Ganges) through mountains, foothills, the plains of India, and finally to the sea at Gangasagar, Bhagīratha facilitated the liberation of sixty thousand sons of King Sagara. Thanks to his instrumental role, the source stream of the Ganga earned the name Bhagirathi. However, in the Bengal delta, the right branch of the Ganga bears the same name Bhagirathi and the left branch is called Padma. The narratives of epics regarding the origin of Ganga, Bhagirathi, and Padma has been described in detail in Chap. 2 of this volume. The myth surrounding the Churni River falsely claims that it is not a natural river, rather an artificial canal created by Maharaja Krishnachandra (Das 2024; Sarkar et al. 2020). According to this belief, the Maharaja redirected a portion of the Ichamati River. This story originates from "Kshitish Vansabali Charita" (Ray 1876), which describes the construction of the royal palace at Shibnibas, including a moat and a canal that supposedly connected to the Churni. Chapter 4 of this volume provides a detailed account of this myth and possible truth regarding the Churni River.

Unlike the Bhagirathi or Churni rivers, the Jalangi River lacks any mythical or legendary story regarding its origin. However, a significant controversy surrounds the timeline of the Jalangi River's origin.

As Jalangi River is a distributary stemming from the Padma River, it appears reasonable to infer that Jalangi originated from the Padma when the Padma was at its full vigour, seeking an outlet for relief from its flood and coursing through Jalangi. Bose (1972) mentioned that the presence of numerous historical and contemporary hubs of art, culture, and industry along the Bhagirathi-Hooghly River suggests that, historically, the right arm of the Ganga Delta served as the primary channel of the Ganga River. However, the gradual shift of river discharges towards the left branch Padma-Meghna became more pronounced towards the close of the eighteenth century. Majumder (1941) proposed that the Jalangi and Mathabhanga rivers formed relatively recently after the Ganga underwent a diversion of its main flow from Bhagirathi to the Padma. However, the timeline of the significant shift in the flow of the Ganga through Padma remains contentious, with some sources suggesting it occurred in the fifteenth century (Majumder 1941).

In summary, different perspectives converge on the notion that the Ganga deviated from its primary course through Bhagirathi and shifted eastward via the Padma River, with potential occurrences in the 15th, 16th, or eighteenth centuries. Considering Jalangi as the distributary of the Padma, it is reasoned that its origin aligns with the redirection of the Ganga through the Padma, especially during periods of heightened flow. Opinions on the origin date of Jalangi vary, spanning the 15th (Fergusson 1863), 16th (Majumder 1978), 17th (Stevenson-Moor et al. 1919), and 18th (Mukeijee 1938) century. Fergusson (1863) asserted that Jalangi came into existence 300 or 400 years ago, with the earliest feasible calculation placing its emergence in 1463 AD, coinciding with the extinction of the Boyrub (Bhairab River). Majumder opined that during the end of the fifteenth century, the main flow of the Ganga shifted from Bhagirathi to the Padma (Majumder 1941).

Around 300 to 400 years ago, the Jalangi River is said to have originated, altering a segment of the Bhairab River's course and redirecting it southwestward. Eventually, it joined the Bhagirathi at Swarupganj, opposite Nabadwip (Majumdar 1978). Majumdar (1978) proposes that Jalangi likely had its beginnings in 1578, approximately 400 years before 1978. Despite differing historical opinions on the origin date of the Jalangi River, there is evidence of its existence in 1660, as documented by Van-Den-Broucke's map. Mukerjee (1938) notes that the eastward shift of the Padma River began in the sixteenth century, and by the middle or late eighteenth century, the Jalangi River, along with Mathabhanga-Churni, Tista, Jamuna, Kirtinasa, and Naya Bhangini, became prominent features in the Bengal landscape. He further mentions that the Gorai River, originating between 1820 and 1830, is the youngest among the delta rivers.

Among these observations, Van-Den-Broucke's account holds factual significance, as he witnessed and mapped the Jalangi River. Another historical reference supporting the existence of the Jalangi River can be traced to Chaitanya Bhagbata by Das (1535), which describes the life of Sri Chaitanya and the Nabadwip of that era. In terms of geographical importance, Brindaban Das's (1507–1589 CE) Chaitanya Bhagbat stands out from earlier hagiographies, as it not only designates Nabadwip as the birthplace of Sri Chaitanya but also provides additional geographical details about Nabadwip.

Following the Sena dynasty, Bengal fell into a state of chaos as Turk invaders, driven solely by the desire for plunder, disrupted the political, economic, social, and spiritual fabric of the region. Despotism prevailed over Bengal for approximately three centuries (Mukherjee 1984). Amid this tumultuous period, Sri Chaitanya emerged as a saviour in Nabadwip in 1486.

Alauddin Hussain Shah gained control of Bengal in 1494 and the dynasty reigned until 1538 (Eaton 1993; Markovits 2004). The constables of Chand Kazi (Mishra 2006), the then local administer of Hussain Shah, vandalized some of the Mridangas (earthen drums) during the Hare Krishna Sankirtana by followers of Sri Chaitanya. Upon learning of this, Sri Chaitanya initiated a civil disobedience movement in Navadwip, becoming a pioneer of such movements in India for the right causes. He organized a procession featuring a large crowd carrying thousands of Mridangas and Karatalas (hard cymbals), which traversed the streets of Navadwipa. Das (1535) describes the procession as follows:

গঙ্গা তীরে তীরে পথ আছে নদীয়ায়।
আগে সেই পথে নাচি যায় গৌর রায়।। (Couplet 298)
'আপনার ঘাটে' আগে বহু নৃত্য করি।
তবে 'মাধাইর ঘাঁট' গেলা গৌর হরি।। (Couplet 299)
বারকোনা ঘাটে নাগরিয়া ঘাটে গিয়া।
গঙ্গারনগর দিয়া গেলা সিমুলিয়া।। (Couplet 300)

....

নদীয়ার একান্তে নগর সিমুলিয়া।
নাচিতে নাচিতে প্রভু উত্তরিলা গিয়া।। (Couplet 348)

....

অনন্ত অর্বুদ লোক সঙ্গে বিশ্বম্ভর ।
প্রবেশ করিলা শঙ্খ বনিক নগর।। (Couplet 428)

....

এই মত সকল নগর শোভা করে।
আইলা ঠাকুর তন্তুবায়ের নগরে।। (Couplet 433)

....

সর্বমুখে হরিনাম শুনি প্রভু হাসে।
নাচিয়া চলিলা প্রভু শ্রীধরের বাসে।। (Couplet 436)

....

জল পানে শ্রীধরেরে অনুগ্রহ করি।
নগরে আইলা পুনঃ গৌরাঙ্গ শ্রীহরি।। (Couplet 494)

....

সর্ব নবদ্বীপে নাচে ত্রিভুবন রায়।
গাদিগাছা পাড়ডাঙ্গা মাজিদা দিয়া যায়।। (Couplet 498)

Gangā tire tire path āchhe Nadiyāy.
āge sei pathe nāchi jāy Gour rāy.
āponār ghāte āge bahu nritya kari
Tabe Mādahir ghāt gelā Gour hari.
Bārkonā ghāte Nāgoriyā ghāte giyā
Gangārnagar diyā gelā Simuliā. (Couplets 298–300)
P-799/Chap. 23/Madhya Khanda
Ananta arbud lok sange biswambhar
Prabesh karilā Shankhabanik nagar. Couplet-428/
P-805/Chap. 23/Madhya Khanda
Ei moto sakal nagar shobha kore
Aailā thakur tantrabāyer nagare. Couplet-433/
P-805/Chap. 23/Madhya Khanda
Sarba mukhe hari nām suni prabhu hase
Nāchiya chalilā prabhu sridharer bāse. Couplet-436/
P-805/Chap. 23/Madhya Khanda
Jal pāne Sridharere anugraha kori
Nagare āila puno Gouranga Srihari. Couplet-494/
P-809/Chap. 23/Madhya Khanda
Sarba Nabadwipe nāche tribhuban rāy
Gādigāchā Pārdāngā Mājidā diyā jāy. Couplet-498/
P-809/Chap. 23/Madhya Khanda

[Meaning: Along the banks of the Ganga River in Nadia, Goura Ray (Sri Chaitanya) leads a pioneering dance through the paths. Beginning with an exuberant dance at His own ghat, Goura Hari (Sri Chaitanya) proceeded to Madhai's ghat. Passing through Barkona Ghat, Nagariya Ghat, and Ganganagar, Sri Chaitanya continued to Simuliya. Dancing joyfully, Prabhu (Sri Chaitanya) reached Simuliya, located at the farthest end of Nadia (Nabadwip). Having triumphed over the Kazi, Sri Chaitanya retraced His steps back home, visiting various places along the way. Accompanied by a multitude of people, Biswambhar (Sri Chaitanya) entered the village of conch product sellers. Bringing joy to all the villages, Thakur (Sri Chaitanya) arrived at the weaver's village. Upon hearing hari nām from all, Prabhu (Sri Chaitanya) smiled and danced his way towards Sridhar's house. After drinking water from Sridhar and satisfying him, Gouranga Srihari returned to the town. The Lord of tribhuban danced throughout Nabadwip, traversing Gadigacha, Pardanga, and Majida].

During the era of Sri Chaitanya, Nabadwip was situated on the left bank of the river Bhagirathi, a fact supported by (Mukherjee 1984; Nakatani 2003) and confirmed by Van-den-Broucke's map from 1660. The confluence of Bhagirathi and Jalangi (referred to as De Galgatese Spruyt in the map) occurred at Amboa near Samudragarh, approximately 15 km downstream from the present day confluence. This geographical alignment remained almost unchanged in the late eighteenth century, with the only notable difference in Rennell's (1781) map, which depicted Nabadwip as encircled and trisected by the Jalangi and Bhagirathi rivers. Consequently, Sri Chaitanya's house in Nabadwip, Kazi's residence in Simuliya, and all other locations along the procession route were positioned to the east of Bhagirathi and north of Jalangi. This arrangement ensured that the procession did not need to cross either the Bhagirathi or the Jalangi during its course.

In addition to Van-den-Broucke's 1660 map, an earlier account by Das (1535) also provides evidence for the existence of the Jalangi River during the period of Sri Chaitanya. Historical records indicate that following his 12-day period of renunciation (Sann'yāsa), on 26 January 1510, Sri Chaitanya returned to Shantipur. Devotees from Nabadwip seeking to visit him reportedly arrived at a ferry point on a river. According to sources, the Bhagirathi River during Sri Chaitanya's time flowed west of Nabadwip (Chaudhury 2004; Van-den-Broucke 1660), placing both Nabadwip and Shantipur on the same bank. Consequently, there was no necessity to cross the Bhagirathi for travel between Nabadwip and Shantipur. This evidence suggests that the river crossed by the devotees was likely the Jalangi River. Such observations support the conclusion that the Jalangi River existed in the fifteenth century during Sri Chaitanya's time. Additionally, historical references indicate the deteriorating condition of the Jalangi at that time, as illustrated in subsequent textual descriptions.

অনন্ত অর্বুদ লোক হৈল খেয়া ঘাটে।
খেয়ারী করিতে পার পড়িল সঙ্কটে।।
কেহ বান্ধে ভেলা কেহ ঘট বুকে করে।
কেহ বা কলার গাছ ধরিয়া সাঁতরে।।
Ananta arbuda lōka ha'ilō khēẏā ghāṭē
khēẏārī kōritē par parialō saṅkaṭē
kēhō bāndhē bhēlā kēhō ghāṭa bukē kōrē
kēhō bā kālara gacha dhariẏā satrē
সহস্র সহস্র লোক এক নায়ে চড়ে।
কথ দুর গিয়া মাত্র নৌকা ডুবি পরে।।
sahasra sahasra lōka ēka naẏē cārē
katō dūr giẏā mātra naukā ḍubi pārē
যে না জানে সাঁতরিতে সেহ ভাসে সুখে।
ঈশ্বর প্রভাবে কুল পায় বিনি দুঃখে।।
jē nā jānē santritē sēhō bhāsē sukhē
īśwara prabhabē kul pa'y bini duḥkhē. (Das 1535).

[Meaning: A vast multitude of Sri Chaitanya's visitors gathered at the ferry ghat, presenting a significant challenge for the boatman in facilitating their river crossing. Consequently, some individuals resorted to unconventional means, using a makeshift raft crafted from tied-together banana trees (Musa Paradisica), and empty earthen pots for support. Despite a boat sinking, through the benevolence of God, the visitors were able to safely reach the opposite bank without any swimming experience].

This narrative implies that the state of the river was likely in decline, with diminished water levels resulting in shallow depths. Consequently, this allowed the followers of Sri Chaitanya to cross the river with ease, without requiring swimming. Thus, it may be inferred that the river Jalangi was extant by at least the 15th century, as noted by Fergusson (1863).

The presence of the river Padma at its full strength is a prerequisite for the emergence of Jalangi in the fifteenth century or prior. To substantiate the redirection of the primary flow of the river Ganga through the Padma before the fifteenth century, it is necessary to provide evidence. Examining the Mahabharata epic as a historical resource for reconstructing the history of the Bengal delta reveals indications of the river Padma's existence much earlier, possibly contemporaneous with the Bhagirathi if not predating it. As in the second chapter of this book, the narrative of Bhagiratha and his sixty thousand ancestors in the Mahabharata suggests, according to Willcocks (1930), that the Bhagirathi originally served as an irrigation canal from the Ganga. However, the same narrative could be interpreted to suggest that both the Bhagirathi and Padma Rivers existed simultaneously. If we trace the epic back to the Vedic period, around c. 1500–c. 500 BC (Winternitz 1981) or around c. 900 or c. 800 BC (Brockington 2022), it implies that both the Bhagirathi and Padma must have existed at least during the period of c. 1500–c. 500 BC. However, the gradual shifting of the main flow from Bhagirathi to the Padma has been evident from studies of several scholars (Allison et al. 2003; Allison 1998; Goodbred et al. 2003).

Evidence from a 1550 map by De Barros, indicates that the Padma river was recognized as the wider channel of the Ganges (Bhattacharyya 1974). Sixteenth-century accounts from Abul Fazl and other foreign travellers, as recorded in *Akbarnama* (Vol. III, p. 46 ff.), refer to the Padma as a significant part of the Ganges river system. Moreover, in the fourteenth century, the traveller Ibn Battuta described entering the city of Sadkawan, a substantial settlement near the sea, where the Ganges—sacred to Hindu pilgrims—merged with the river Jun before flowing into the ocean (Gibb 1971). Yule later identified Sadkawan as Chittagong and equated the Jun with the Jamuna River. Assuming these identifications are correct, this suggests that even in the fourteenth century, the Padma served as a primary channel of the Ganges.

In van-den-Broucke's map, the River Jalangi, known as *De Galgatese Spruyt*, exhibited a highly dynamic existence (Fig. 3.1). As depicted in historical maps, the Jalangi originated from the Ganges near *Bilgassi*, flowing in a south-southwest direction through an intricate network of channels and settlements. The river's course was particularly significant as it traversed regions like *Neddia* (modern-day Nabadwip), a prominent centre of cultural and religious activity, before merging with another river near *Amboa*. This hydrological dynamism was further emphasized by the Jalangi's proximity to geographically notable markers, such as its position north of the *Tropicus Cancri* (Tropic of Cancer).

However, historian Dr. R. C. Majumdar, drawing on the Stevenson-Moore Committee's report, posits alternative identifications: he suggests that Sadkawan could correspond to Satgaon, with the Ganges and Jun represented by the Bhagirathi and Jamuna, respectively. Nevertheless, evidence from *Āin-i-Akbari*, the *Ramacharita*, and other sources supports the conclusion that the main branches of the Ganges—now known as the Bhagirathi and Padma—existed as distinct channels from at least the beginning of the Christian Era, if not earlier (Bhattacharyya 1974) (Table 3.1).

Between 1764 and 1915, spanning 150 years, the course of the river Jalangi underwent remarkable changes. However, in the lower section from Patkabari to Krishnagar, the formation of meanders has been well-preserved (Reaks 1919). The alterations observed appear minimal when accounting for slight inaccuracies in Rennell's chart mapping. This section presents the appearance of a river with a well-formed and established channel, likely in stiffer bank materials that are less prone to erosion. In 1796–1797, the Kulkuli or Kulkooli River (spelt as Culcullia River on the map) served as the primary tributary of the Jalangi River. Additionally, two other inlets were feeding the Jalangi (also known as Jellinghy) located north of Hooglebarya (Fig. 3.4).

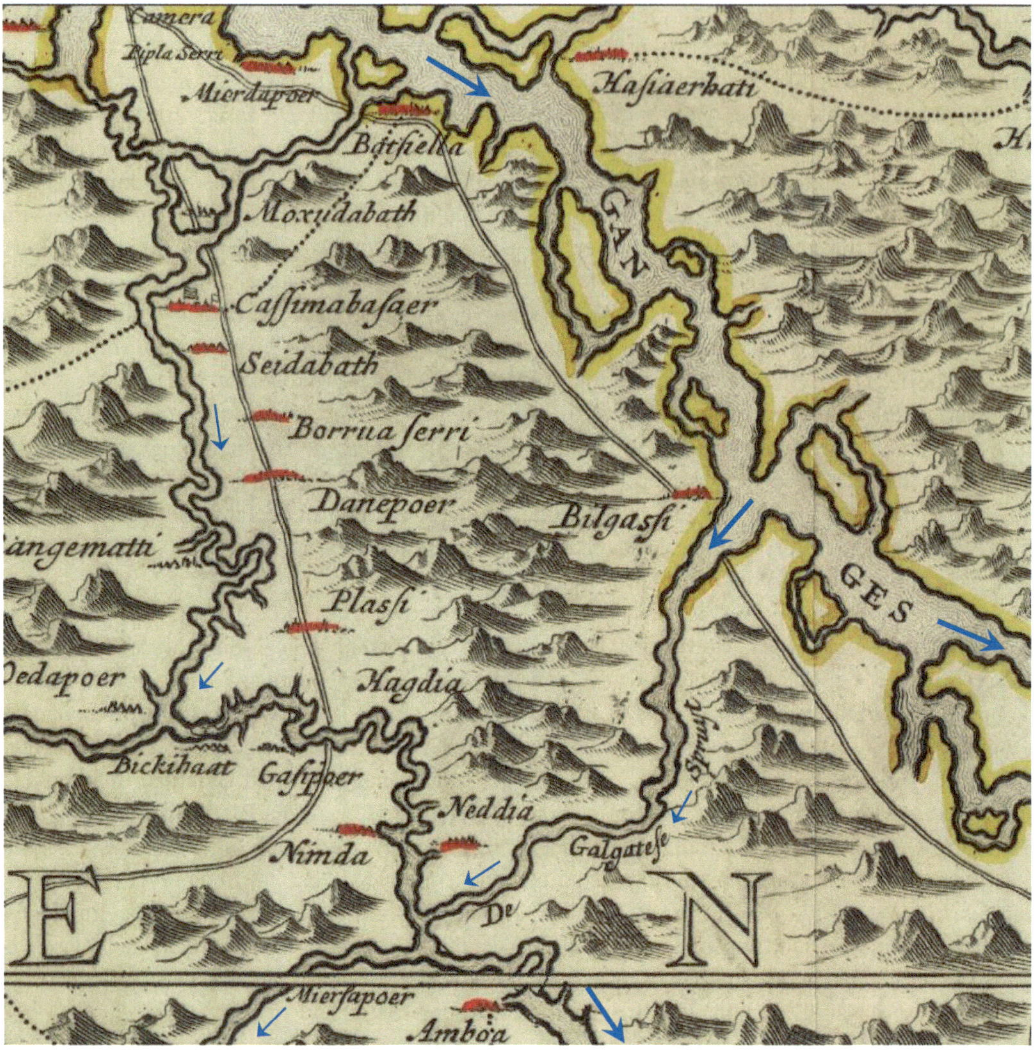

Fig. 3.1 In 1660, De Galgatese Spruyt' (Jalangi) takes off from Ganges at 'Bilgassi' and runs south-southwest and joins another river (Bhagirathi) down 'Neddia' (Nabadwip) near Amboa near Mierjapoer, located north of 'Tropicus Cancri' (Tropic of Cancer). To see the full map, please click on the link retrieved on 18.10.2024. https://www.raremaps.com/gallery/detail/75478/bangladesh-nieuwe-kaarte-vant-koninckryk-bengale-door-la-valentijn

New Bhairab opened in 1874 and appropriated the entrance of the river Jalangi. Old Jalangi entrance completely silted up. During Moor's committee, the old Jalangi channel flowed only during the flood season. It was opined that the old Jalangi channel may resume its activity with any deterioration of Bhairab (Stevension-Moor et al. 1919, p. 55).

3.3 Course of the River

Jalangi is a distributary of the river Padma and a tributary of the river Bhagirathi. The Jalangi originates from the Padma in the far north of the Nadia district at coordinates 24°05′26″ N and 88°41′53″ E. After taking off from Padma, the river takes a winding path along

Table 3.1 Timeline of remarkable events

Time	Events	Source of information
15th Century or before	Origin of the river Jalangi	Das (2021)
1463	Fergusson (1863) reported that the R. Bhairab was extinguished around 300 to 400 years ago, (i.e. between 1563 to 1463) by the R. Jellinghy, which altered course as the delta's eastern slope lessened, redirecting its flow southward to join the Bhagaruttee at Nuddea, once known as Nyadwipa—a newly formed island when the area was a tidal swamp or sea. During the end of the fifteenth century, the main flow of the Ganga shifted from Bhagirathi to the Padma (Majumder 1941, p. 54).	Fergusson (1863)
1510	The winter river was shallow enough to walk across because twelve days after His renunciation (Sanyas) on January 26th, 1510, Sri Chaitanya returned to Shantipur and His visitors from Nabadwip reached at a ferry ghat on the river Jalangi and many among them forded the river on foot.	Das (1535, 2013, p. 116)
1660	Van-den-Broucke's. '*De Galgatese*' (present daypre-sent-day Jalangi) takes off from a larger river at '*Bilgassi*' and runs south-southwest and joins another river (Bhagirathi) down '*Neddia*' (Nabadwip) at Amboa, north of '*Tropicus Cancri*' (Tropic of Cancer). The river was more vigorous than that of the river Bhagirathi, because it was drawn wider than Bhagirathi in Broucke's map	Van-den-Broucke (1660); Das (2013, p. 119)
1666	On 6th January 1666, traveller 'Bernier was going overland from Rajmahal to Cossimbazar as the river route was impracticable' through the Bhagirathi River. However on van-den-Brouck's map, Jalangi was drawn wider than Bhagirathi River	Mukharji (1932)
1757	The Battle of Plassey in 1757, between Siraj-ud-Daulah and Lord Clive, marked the beginning of British dominance in Bengal. Since the British acquisition of the region, maintaining the navigability of the Nadia rivers has posed significant challenges	Captain Lang (1848)
1764	Shifting of Bhagirathi-Jalangi Confluence from Samudragarh (near Amboa) to Swarupganj andf shifting of Bhagirathi from west of Nabadwip to East	Van-den-Broucke (1660); Rennell (1781); Das (2013)
1764	The offtake of the river Jalangi was at Jalangi. Rennell named Khariya or Khore River as Jalangi	Reaks (1919, p. 47); Majaumder (1985)
1765	Jalangi and Mathabhanga Rivers had separate offtakes from a bypass secondary channel of the river Padma	Biswas (2001). Map No. 2, Part-I
1803	Captain Colebrooke proposed measures for maintaining the navigability of the Cossimbazar (Bhagirathi) or Jellinghy (Jalangi) River during the dry season	Colebrooke (1803)
1832	In the flood of 1832, the entrance of the river shifted 5 miles to the north of the then offtake	Reaks (1919, p. 50)
1838	In January 1838, entrance depth was improved from 3.5 feet to 4.5 feet	Reaks (1919, p. 51)
1840	During 1840–50, Jalangi and Mathabhanga had a common entrance from the river Padma (lower reach of the Ganges or Ganga downstream of Farakka barrage)	Reaks (1919, p. 55)
1840	Both Jalangi and Mathabhanga had separate offtakes from a southward bend of the river Padma (Fig. 3.2)	Tassin (1840), from Rudra (2010)
1841	1841–1850 Mathabhanga shared a common entrance with the river Jalangi	Reaks (1919, p. 52)

(continued)

Table 3.1 (continued)

Time	Events	Source of information
1850	Mathabhanga River was taking off from the river Jalangi at Bansmari (Bausmari). In November 1850, the primary flow of the Ganges (Ganga) diverted into a bend of the Mathabhanga above Jutpara, creating a separate offtake of Mathabhanga	Reaks (1919, p. 46)
1850	In 1850–51, Jalangi had two entrances	Reaks (1919, p. 51)
1853	Two large sandbars in the Ganges (Ganga), positioned above and below the river Jalangi's entrance, merged, effectively sealing off the river Jalangi's offtake completely	Reaks (1919, p. 51)
1862	The inception of the railway connecting Kolkata to Kustia under Eastern Bengal Rail, marked a significant shift in the Nadia's transportation dynamics. This probably marked a decline in the significance of river transport, including the Nadia Rivers. In 1877, the government took control of the railway company, renaming it the 'Eastern Bengal State Railway.' By that time, a network of 100 miles of railway tracks had been constructed in the undivided Nadia with 20 stations	সান্যাল (১৯৭৩)
1874	During the 1870s, the original offtake of the Jalangi River at Jalangi had become entirely silted up, only flowing during the flood season. In 1874, a new Bhairab emerged from the Ganges (Ganga) at Akhriganj, tracing its path towards the south to converge with Jalangi at Char Moktarpur. Rennell's records provide no hint of the presence of any river in this course. This new Bhairab became the main feeder of the river Jalangi.	Reaks (1919, p. 54–55)
1905	The first railway bridge over the river Jalangi at Krishnagar was completed and became operational in 1905. As of the date of publication of this volume, the bridge has been in use for 120 years. However, till date (June 2025), there are two highway bridges and two railway bridges are in operation. A new railway bridge (3rd railway bridge) downstream to the first (oldest) one is under construction.	Pringle and Kemm (1928); Mukharji (1932)
1909	Reaks (1919) reports that "About 6 years ago a natural cut-off took place at the Nawpara loop just above the discharge observation at Panditpur" and the Nawpara Oxbow lake originated in 1909 due to the natural neck cut-off of the meander	Reaks (1919, p. 56); Stevension-Moor et al. (1919, p. 56)
1914	In 1914, Bhairab appropriated the upper course of the Kulkuli River and took off from the Ganges 10 miles downstream from Akhriganj	Reaks (1919, p. 55)
1915	A cut-off in 1915, as reported by Reaks (1919) was imminent just below the Patkabari, which is probably the present-day Swardanga *Beel* at 23°49′30″ N 88°30′00″ E	Reaks (1919, P. 56)
1918	Jalangi and Mathabhanga rivers had a common offtake above Bausmari, near Jalangi	Biswas (2001). Map No. 2, Part-I
1960s	The Karimpur-Jalangi State Highway (SH-11) was constructed across the remnants of the river channel, representing a final human intervention to close off the channel in that area	Das (2013)
2010	Formation of two shoals at 731 m to 1175 m downstream from Palashipara Bridge	Google Earth Pro. Image Dated-22.11.2010
2024	The two shoals emerged as islands 731 m to 1175 m downstream from Palashipara Bridge (Fig. 3.3)	Google Earth Pro. Image Dated-27.04.2024

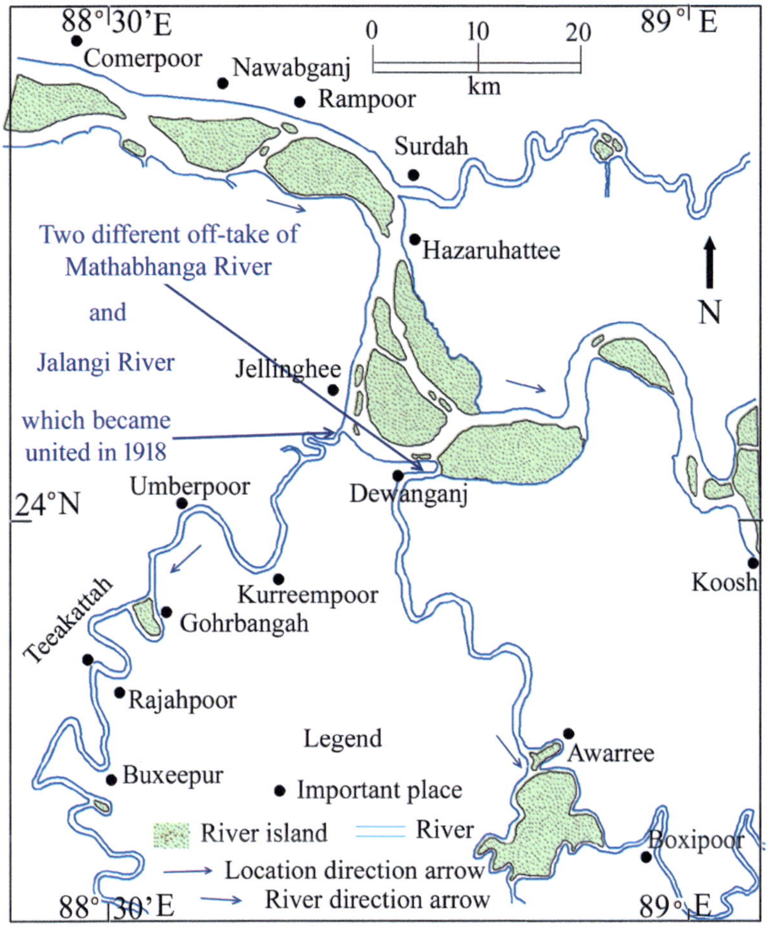

Fig. 3.2 Tassin's map (1840) shows two separate entrances of the rivers Jalangi and Mathabhanga during 1840. These two offtakes became united in 1918

the northwest boundary of the Nadia district for 51 km until it reaches Char Moktarpur village. This stretch is inactive due to the absence of a connection with the Padma River, its former water source (Fig. 3.5). From Moktarpur, the river flows for 45 km to Gopinathpur and Sahebnagar villages in C. D. Block Tehatta-I and C. D. Block Tehatta-II, respectively, marking the border between Murshidabad and Nadia districts (Table 3.S2). Subsequently, the river enters Nadia district and remains within its confines until it merges with the Bhagirathi River at Swarupganj opposite Nabadwip (Fig. 3.6 and Table 3.2).

Passing through Gopinathpur and Sahebnagar villages, the Jalangi continues its meandering course southward through Palashipara, Tehatta, Bara Andulia, Chapra, and Doier Bazar until it reaches Krishnanagar, the district headquarters of Nadia. From Krishnanagar, the river changes its course to the west, covering a total distance of 233 km before joining the Bhagirathi River at Swarupganj (at coordinates 23°29′23″ N and 88°28′57″ E), opposite the town of Nabadwip, where Sri Chaitanya was born. The combined flow of the Bhagirathi and Jalangi rivers downstream of Nabadwip, known as the Jalangi–Bhagirathi confluence, forms the Hooghly River.

Out of the total course of 233 km, 96 km (51 + 45 km) runs along the borders of Nadia and Murshidabad districts, while the remaining 137 km traverses Nadia district, separating various villages

Fig. 3.3 Mid-channel bar emerged during the early decades of the twenty-first century. Soils are being cut and transported (yellow arrows) to the brick kiln altering river channel morphology. Floating and rooted algae (red arrows) along with grassed mid-channel bar indicating shallow, degraded, flowless state of the river.

across seven C. D. Blocks (excluding C. D. Blocks Karimpur-I and II, where the river serves as a natural boundary between Murshidabad and Nadia) on both the left and right banks. The Jalangi can be considered a "river of Nadia" in the truest sense, as it flows entirely either along the district's border or through its territory, unlike the Bhagirathi and Mahabhanga–Churni rivers.

It's crucial to note that the Annual Flood Report-2016, published by the APPE & M Cell, I & WD, Government of West Bengal, raises concerns about the unclear delineation of the flow path of the Jalangi River. The report specifically notes that—

> On the left bank of the Bhagirathi river system, the Bhairab-Jalangi-Sealmari group of rivers originate from Ganga-Padma at Akherigunj in Murshidabad district and meet the Bhagirathi at Swarupgunj in Nadia District.

However, Bhairab, Jalangi, and Sealmari (Sialmari) do not constitute a single river. Instead, Bhairab and Sialmari serve as tributaries to the Jalangi River. Consequently, their respective points of origin differ, and they do not share a common starting point like "Akherigunj," from where the Bhairab River used to branch off from the Padma River. Once again, it was stated—

> The river Jalangi originates from the right bank of the river Padma in Murshidabad district, 165 km. downstream of Farakka. Jalangi is dead for all purposes except during the periods of heavy rain when it receives water from Padma. The river ends its journey by finally outfalling into the river Bhagirathi near Nabadwip town of Nadia district. The major tributary of Jalangi is river Bhairab which starts its journey from the river Ganga near Lalbag of Murshidabad district. It is now almost a dead channel but during rainy season it receives water from Padma.

First, the original off-take point of the Jalangi River at Char Madhubona near Bausmari in C.D. Block Karimpur-I has ceased to exist, erasing it permanently from the landscape. It no longer receives any supply from the Padma River. This offshoot of the Jalangi River is approximately 132 km downstream from Farakka. Secondly, it is important to clarify that Akherigunj or Ankhriganj serves as the point of off-take for the Bhairab River, not the Jalangi River.

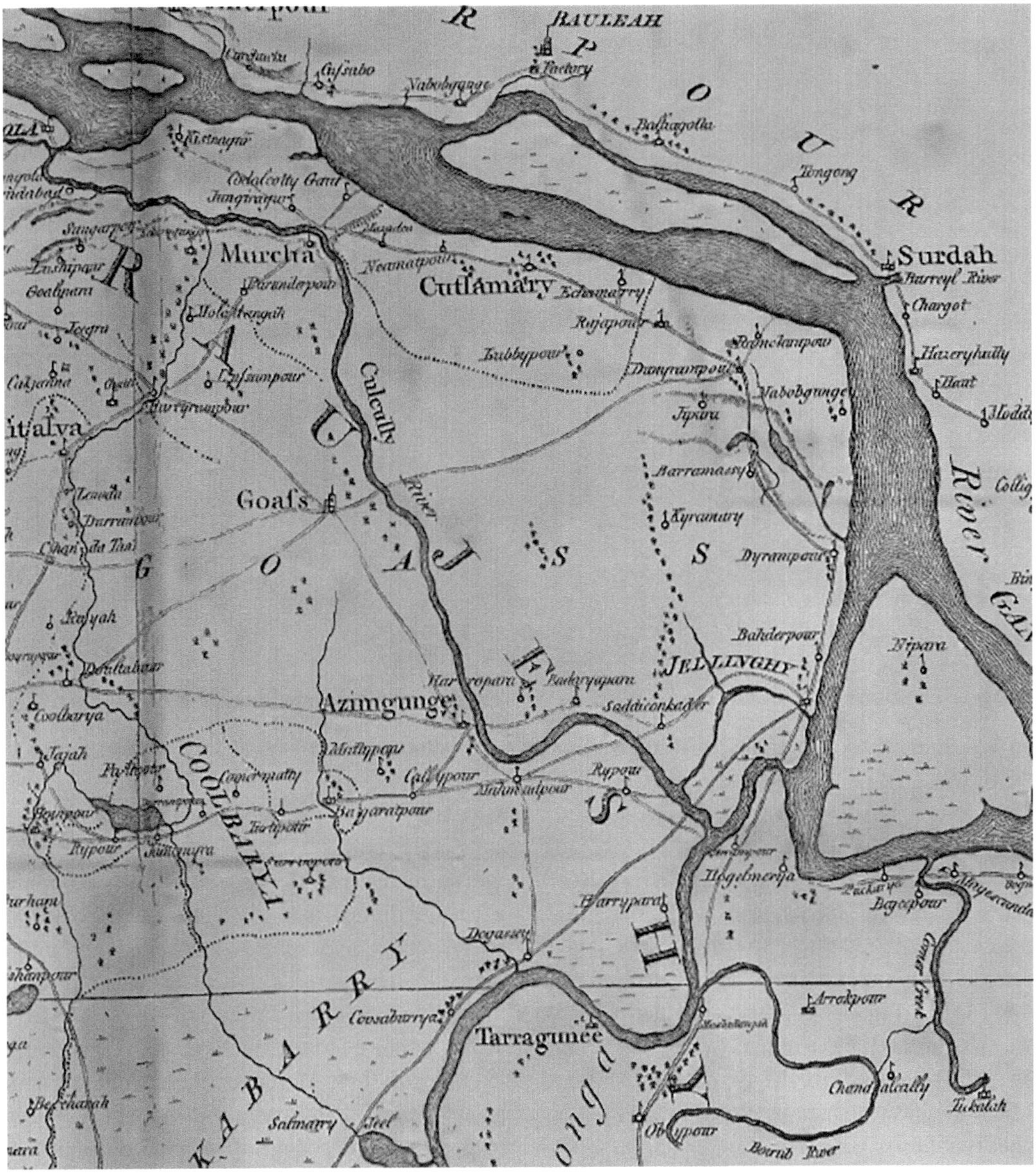

Fig. 3.4 The Culcullia River was the main feeder of the river Jalangi. There were two more inlets of the river Jalangi (Jellinghy) north of Hooglebarya. Map of the Cossimbuzar Island, prepared by Rennell (1780) (Fig. 3.S2). The figure is also represented in the 'Chart of the river Ganges from Colgong to Hurrisonker exhibiting the state of its islands and sand bars during the dry season of 1796–1797, by R. H. Colebrooke, Surveyor General [Fig. 3.S1 in supplementary document]. (Colebrooke 1803)

3.4 Jalangi Basin

Precisely delineating the drainage basin of a deltaic river like Jalangi is more challenging than difficult. However, an approximation was attempted using various sources, including the Google Earth image of the flood-affected area in 2000, a contour map (Bagchi 1978), a flood slope map (Biswas 2001, plate-19), and transport networks. The estimated basin area

Fig. 3.5 The bamboo bridge is made on the river Jalangi immediately before the confluence of Bhairab-Jalangi. The wide course of the river flowing right to left before the confluence (right) is Bhairab and after the confluence (left) is Jalangi. The channel has been narrowed by putting earthbags in bamboo cages from both banks

is 2815.33 km^2 (Das and Bhattacharya 2020). The geographical extension in terms of latitude ranges from 23º24′00″N to 24º27′00″N and in terms of longitude, it spans from 88º12′00″E to 88º44′15″E (Fig. 3.7). The basin has a length of 110 km and an average width of 45 km. According to the circularity index (Das et al. 2022), the basin's shape is elongated, with a circularity index of 0.47.

3.5 Spills from the Right Bank

Tributary spills-play a pivotal role in regulating the flow of the Jalangi River, especially considering that Jalangi's direct connection with the Padma River has been virtually obliterated. These tributaries contribute significantly to the Jalangi by not only draining marshes but also through the infusion of seepage water from the basin. During floods, these spills distribute excess sediment-laden water from the river to backswamps and relieve the sediment burden there. In lean periods, when the water level in the channel is much lower, silt-free water from swamps returns to the river and maintains partial lean season flow. Moreover, these spills recharge groundwater of the basin part of which returns to the river as seepage flow and contributes to lean season flow.

3.5.1 Sialmari River

Among right-bank spills, the Sialmari River is the first contributing channel to the Jalangi, in terms of proximity to the original offtake of the river Jalangi. Originating from the Padma River, approximately 5 km east of Hursi in C. D. Block Raninagar-I or 7 km north of the

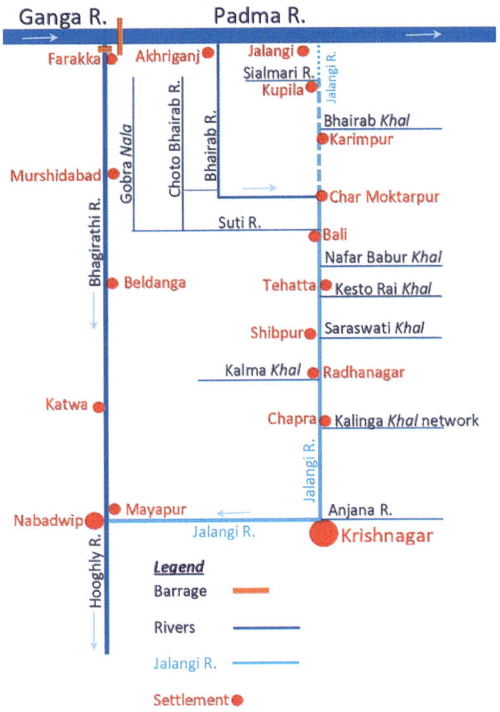

Fig. 3.6 The wide course of the river flowing right to left before the confluence (right) is Bhairab and after the confluence (left) is Jalangi. The channel of the Jalangi has been narrowed by putting earthbags in bamboo cages from both banks

village Malibari in C. D. Block Raninagar-II, of Murshidabad district, the Sialmari River derives its name from the nearby village "Sialmari." The river follows a winding course towards the south, passing through Raninagar, Kasba Goas, Domkal, and Raipur, ultimately merging into the Jalangi River at Kupila on the right bank (Fig. 3.7). Notably, the offtake of the Sialmari River from the Padma has also dried up, resulting in a cessation of the direct supply from the Padma. Consequently, the Sialmari River and the upper reaches of the Jalangi have suffered significant degradation. At present, the Sialmari River contributes to the Jalangi by draining the Bhairab–Sialmari inter-fluvial tract of the Murshidabad district. During Rennell's time, the Sialmari River had a well-developed channel (Reaks 1919, p. 58).

3.5.2 Bhairab River

Before delving into the Bhairab River, it's important to note the presence of three distinct Bhairab Rivers within the Jalangi-Mathabhange-Churni system. In this present volume, we distinguished their name as (i) Bhairab River, (ii) Choto Bhairab, and (iii) Bhairab Khal (in Bengali, the literal meaning of Khal is an artificial canal. However, colloquially the term is also used for small natural rivers and Bhairab Khal bears this second meaning). Of these three, the Bhairab River serves as the primary tributary to the Jalangi River. Choto Bhairab, accompanied by Gobra Nala and Suti River, contributes to the Jalangi at Bali-Tungi, upstream of Palashipara. The third one, Bhairab Khal, takes off from the Jalangi river at Karimpur, winding its way over the Jalangi-Mathabhanga interfluve to join Mathabhanga opposite Raghunathpur, near Subalpur bridge, downstream of Kapasdanga in Bangladesh, at 23°34′26″ N and 88°44′38″ E.

Bhairab River, positioned second in terms of proximity to the closed offtake of the river Jalangi, emerges as the primary contributor to the Jalangi River. The river supplies the largest volume of water to the Jalangi and became the main tributary among all the tributaries. Functioning both as a distributary of the Padma River and a contributor to the Jalangi, the Bhairab River plays a crucial role in maintaining the flow of the Jalangi, particularly during one or two monsoon months. Originating from the Padma River at Ankhriganj (Fig. 3.7), approximately 5 km upstream from Hursi, along the border of C.D. Block Bhagwangola-II and C.D. Block Raninagar-I in the Murshidabad district, the Bhairab River follows a southward course until it reaches the tri-junction point of C.D. Block Raninagar-I, C.D. Block Jalangi, and C.D. Block Berhampore. At this point, the river bifurcates, with the eastern branch, known as the main Bhairab, continuing its southerly journey through Islampur, Bhagirathpur, Swaruppur, Garibpur, and Dakshin Jitpur, eventually merging into the Jalangi River at Char Moktarpur in

Table 3.2 Important opinions/observations about the river

Authors	Period	Comments
Holwell	1756	While travelling by boat, he was detained at Santipur due to shallow water below the confluence of the Bhagirathi and Jalangi Rivers
Rennell	1781	The Bhagirathi was almost dry from October to May. In contrast, the Jalangi maintained its flow throughout the year but sometimes in the driest months, Jalangi was found to be a non-navigable channel
Colebrooke	1797	The Mathabhanga-Churni River was surveyed in 1795 and found that the river was navigable throughout the year for the boats of moderate burden. The Bhagirathi and the Jalangi Rivers could not maintain their navigability during the dry season
Capt. Sherwill	1857	The Bhagirathi, the Mathabhanga, and the Jalangi Rivers were not navigable throughout the year while the Gorai was a very active river
Prestige Franklin	1861	The Bhagirathi and the Jalangi Rivers became disconnected from the river Ganga and the then government took a vital initiative to keep the mouth of the Mathabhanga River open in order to have good water communication between the Ganga and the Bhagirathi River throughout the year
Major Hirst	1914–1915	The decay process of the Bhagirathi-Hooghly River was so powerful that no artificial measure would be capable enough to rescue the condition of the river and it would create a menace to the then port of Calcutta (presently Syama Prasad Mookerjee Port) (Hirst 1916).
Major Long	1948 and 1954	The off-take of the Bhagirathi and the Jalangi Rivers were non-navigable
Man Singh Committee	1952	The condition of the Hooghly River between Nabadwip and Kolkata was found to deteriorate

C.D. Block Karimpur-II of the Nadia district. The Bhairab River serves as the natural border for C.D. Blocks of Berhampore–Raninagar-I, Berhampore-Domkal, Hariharpara–Domkal, and Nawda–Domkal in the Murshidabad district. After the initial bifurcation from the tri-junction point of Raninagar I-Jalangi-Berhampore, the right branch of the Bhairab River follows a westerly course along the border of C.D. Blocks Jiaganj and Berhampore for approximately 7 km. It then turns southwards, traversing Berhampore and Hariharpara C.D. Blocks. About 1–2 km north of Protappur in the southern part of Hariharpara C.D. Block, this branch bifurcates again, with the left branch (hard to trace now) rejoining the main Bhairab near its confluence with the Jalangi. Simultaneously, the right branch continues its southward trajectory, meeting Gobra Nala near Goghatia in C.D. Block Nawda.

3.5.3 Choto Bhairab

This is the third tributary. It drains low-lying areas like Mirpur, Dhanaipur, and Gangaprasad. The river flows through Hariharpara and eventually joins the Gobra–Suti Nala at Trimohini near Jhowbona on the Beldanga–Amtala road at coordinates (23°55′26″ N and 88°22′06″ E).

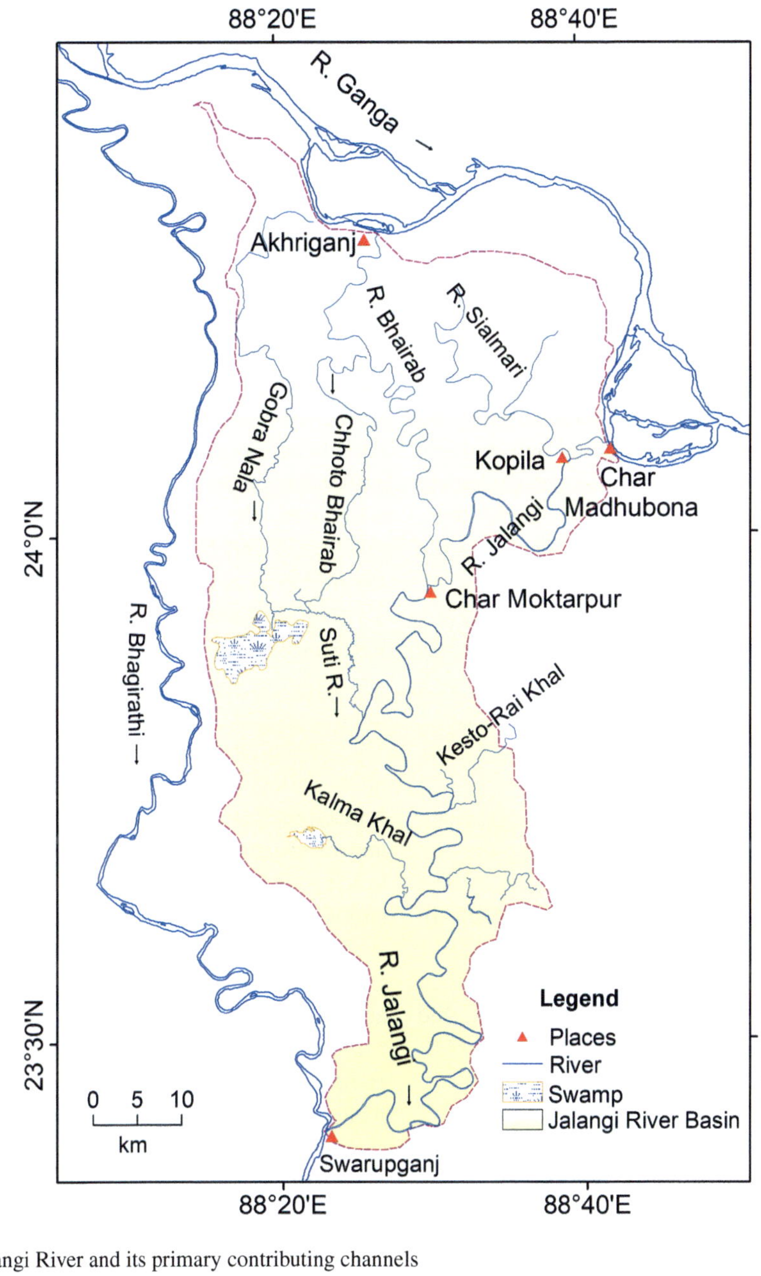

Fig. 3.7 Jalangi River and its primary contributing channels

3.5.4 Gobra-Suti Nala

Once this tributary of the Jalangi River used to get its water from the Padma River, 2 km upstream from the offtake of the Bhairab River. It flows westward up to Bhander–Kismat, south of Bhagwangola. Crossing the Bhagwangola–Hursi–Raninagar road, it continues south through C.D. Blocks of Bhagwangola-II, Jiaganj, and Berhampore, reaching 2 km north of Tarakpur, passing through Mrigi, Jiaganj, Lalbagh, and Begamnagar. From there, it follows the border of C.D. Blocks Berhampore–Hariharpara and Beldanga-I–Hariharpara, branching into multiple channels in the Kālāntar depression. Collecting water through various channels in the Kālāntar back swamp, it reunites as Suti Nala and joins the Jalangi River at Bali-Tungi.

During 1st decade of the twentieth century, the river, a sluggish one with occasional depths of 20–50 feet at Bhandardaha Nala, became beheaded by the Ranaghat-Lalgola railway line.

3.5.5 Kalma Khal

This smaller contributing channel drains *Beel* Kumari and Pon *Beel* in C.D. Block Tehatta-II, adding water to the river Jalangi during rainy months. The confluence with the Jalangi is at geographical coordinates 23°47′13.26″ N and 88°29′31.75″ E in the village Radhanagar, C.D. Block Nakashipara, Nadia.

3.6 Spills from the Left Bank

There are six noteworthy spill channels or channel systems on the left bank of the river, playing crucial roles in maintaining flow of the river Jalangi. During floods, these spills carry silt-laden water into the back swamps or low-lying floodplains, raising their levels. When the river's water level decreases, the reverse flow of clear, silt-free water through these spills sustains the lean season flow, helping scour the loose slurry bed deposits. The identifiable spill channels are discussed below.

3.6.1 Bhairab Khal

Historical accounts (Mukharji 1932; Reaks 1919; Stevension-Moor et al. 1919) suggest that the Bhairab River, previously discussed under Sect. 3.5.2 as the prime tributary and the spill Bhairab Khal was originally a single river until the eighteenth century before the Jalangi and Mathabhanga rivers emerged. However, the unconventional position of the Bhairab Khal's offtake from Jalangi, far upstream from the Bhairab-Jalangi confluence raises questions regarding the above narrative. Moreover, the presence of the river Jalangi during the fifteenth century or even earlier (Majumder 1941; Das 1533, 2013) also contradicts the abovementioned narrative that the Jalangi and Mathabhanga rivers bisected the Bhairab River in the eighteenth century.

Bhairab Khal, takes off from the Jalangi River at 23°59′10.16″ N and 88°37′31.85″ E in Uttampur, near Karimpur Śmaśāna Ghāṭa (cremation ground), C.D. Block Karimpur-I, Nadia. The winding 19 km stretch, immediately after Pattabuka Bridge, from Natna through Jamserpur, Haripur, Kechuadanga, Santoshpur, and Mojlishpur back to Natna has now become a relict channel on the left bank (Fig. 3.8). A straight cut of 800 m has bypassed the loop, straightening the course. It then runs through C.D. Block Karimpur-I and enters Bangladesh at Kutubpur of Meherpur district, Bangladesh. Moving due south from Meherpur, it reaches Ballavpur at 23°39′45″ N and 88°36′45″ E, where it makes an eastward turn towards Jagannathpur. The river adopts a zigzag course, disregarding the general regional slope. After Gopalpur, the river flows southeast and joins the Mathabhanga River opposite Raghunathpur, near Subalpur bridge, Bangladesh, at 23°34′26″ N and 88°44′38″ E. The total length of the Bhairab Khal is 138 km.

Reaks (1919) mentioned that the region between Jalangi and Mathabhanga is intersected by innumerable number palaeo-channels, trending in every direction. The overall slope is south-easterly, and the Bhairab Khal joins the Mathabhanga at a significant angle. Most of the Mathabhanga spill flows through

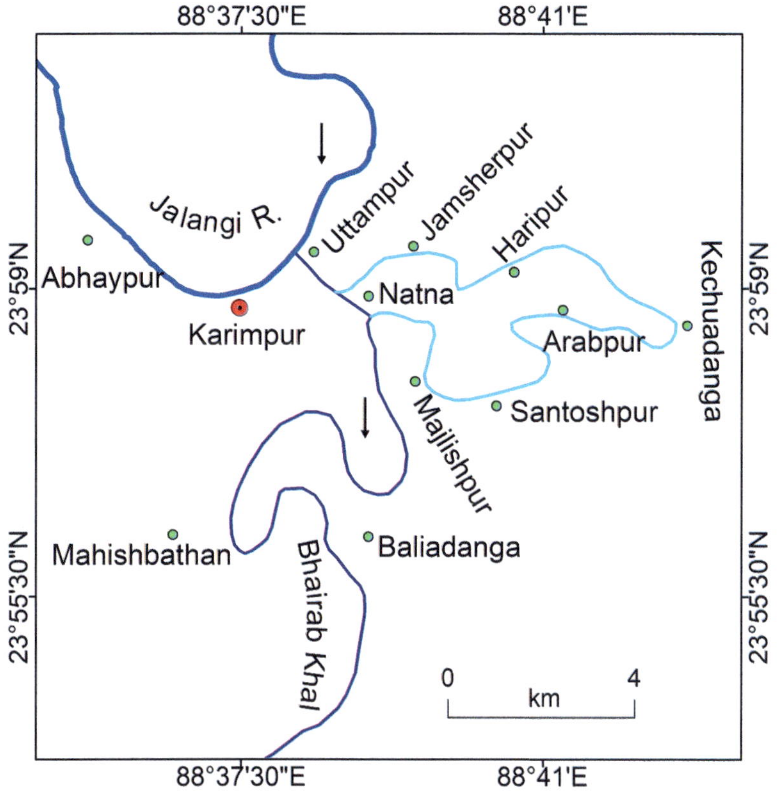

Fig. 3.8 The upper loop of Bhairab Khal, shown in a light blue line by the villages of Jamserpur, Haripur, Kechuadanga, Santoshpur and Majlishpur, was rejected by a cut of ~ 800 m at Natna

Kumar. In 1820, an attempt was made to close Kumar off-take, but it led to another outlet in the lower reach. Since the early nineteenth century, no successful attempts have been made to draw a significant volume of water through Mathabhanga, with 70% of its volume directed to Jessore spills before entering only 30% to Hooghly River via Churni River.

The Bhairab Khal is considered the oldest river of Nadia District (Islam 1993) and is witness of the old history of the region. Khan Jahan Pir of Gour is documented to have reached Bagerhat by navigating the Bhairab River. In 1598, Mansingha sought shelter in the palace of Bhabananda Majumder in Bagoan and crossed the Bhairab River to successfully defeat Pratapaditta (Islam 1993). The river is associated with several important historical, cultural, social, and economic traits of the regions through which it flows. Till the nineteenth century, the river served as an important navigation route between different trade centres like Meherpur, Damurhuda, Chuadanga, Barbazar, Jessore, Jibannagar, Khalishpur, Kaliganj, etc. Several indigo factories were there (e.g. Kapasdanga) along the river (Islam 1993).

3.6.2 Nafar Babur Khal

The Khal serves as a waterway linking the Jalangi River to the Chapagara *Beel*. It meets the Jalangi River at coordinates 23°44′38″ N and 88°31′23″ E, flowing nearly parallel to the Tehatta-Shyamnagar road through the village of Jitpur. The channel then turns eastward, crossing the road near Pulpara Primary School, Jitpur

(23°45′11″ N and 88°31′09″ E). The Chapagara *Beel* to the east of Tehatta-Shyamnagar Road is drained through the Khal.

3.6.3 Kesto-Rai Khal

This spill leaves the Jalangi River at 23°44′19.20″ N and 88°31′31.55″ E near Tehatta Ferry Ghat in C.D. Block Tehatta-I. It flows northwest towards Gobri *Beel* and Chand *Beel* in C.D. Block Tehatta-I and Margangni in C.D. Block Karimpur II, eventually reaching Meherpur in Bangladesh. The khal is also known as Gobri and Chand *Beel* in-draw channel.

3.6.4 Saraswati Khal

The khal relieves flood water of the river Jalangi towards the east across the Indo-Bangladesh border up to the west of Bagoan. However, draining the lowlands of Meherpur district in Bangladesh, during post-monsoon months, this spill meets the Jalangi River at 23°42′29.20″ N and 88°31′11.55″ E near Putimari in C.D. Block Tehatta-I. Inglis (1909) noted that Saraswati Khal "opens from the Jellenghy near Putimari and flows across the whole Nuddea district, emptying itself into the Matabhangah near Khalbolia". Probably during that period reported by Inglis (1909), the Saraswati Khal being interconnected with other channels, used to connect Jalangi with Mathabhanga. However, at present, there is no such channelized link through Saraswati Khal between Jalangi and Mathabhanga.

3.6.5 Kalinga-Kumri-Thagboge-Jhor-Palda-Damodar

Kalinga River (now Kalinga *Beel*) via Kumri Khal—Thagboge Khal—Jhor Nadi—Palda River—Damodar *Beel* and other depressions makes a connection between rivers Jalangi and Mathahanga during rainy months (Fig. 3.9). During the past days, when river routes were the major mode of transport of goods and passengers in the region, these water courses played significant roles in the economy, society, and culture of the region. Once the Thagboge Khal had significant importance. Daily necessities such as rice, wheat flour, coal, oil, tins, cement, sugar, etc. were dropped in at Thagboge Ghat (northeast of present day Asannagar bridge on Krishnagar-Majdia road) and then sent to different villages by boat. The crops produced in those villages used to be loaded and brought down to Thagboge Khal by boats (Biswas 2024) for sending to different markets of the district and beyond.

3.6.6 Anjana River

Originating as a distributary from the feeder river Jalangi at Krishnanagar, River Anjana flows south and south-easterly course (Fig. 3.10). The Anjana River once played a vital role in connecting the major rivers Jalangi and Churni, weaving a network of waterways in the densely populated agrarian region of the Moribund Deltaic West Bengal (Das and Das 2020). Collaborating with rivers Bhagirathi, Jalangi, and Mathabhanga-Churni, it contributed to the formation of the lower delta, supplying water to the land for crop and plant growth. The river served multiple purposes: irrigating fields during dry months, preventing inundation by efficiently channelling floodwaters, providing fish for the diet of those living along its banks, creating water routes for transportation, and offering a water source for both drinking and daily life. Unfortunately, the river's current state is far from its past glory. It has been transformed into a stagnant, segmented, shallow, and narrow ditch in rural areas and a mere sewage channel in urban areas, rendering it incapable of fulfilling its previous basin-serving functions.

The Jalangi River feeds the Anjana River at Krishnagar and empties into the Churni River at Byaspur. It's worth noting that the river bifurcates into two branches, with the minor branch, Haler *Khal*, merging with the Churni River at Hanskhali (Fig. 3.10). The river's basin is located between

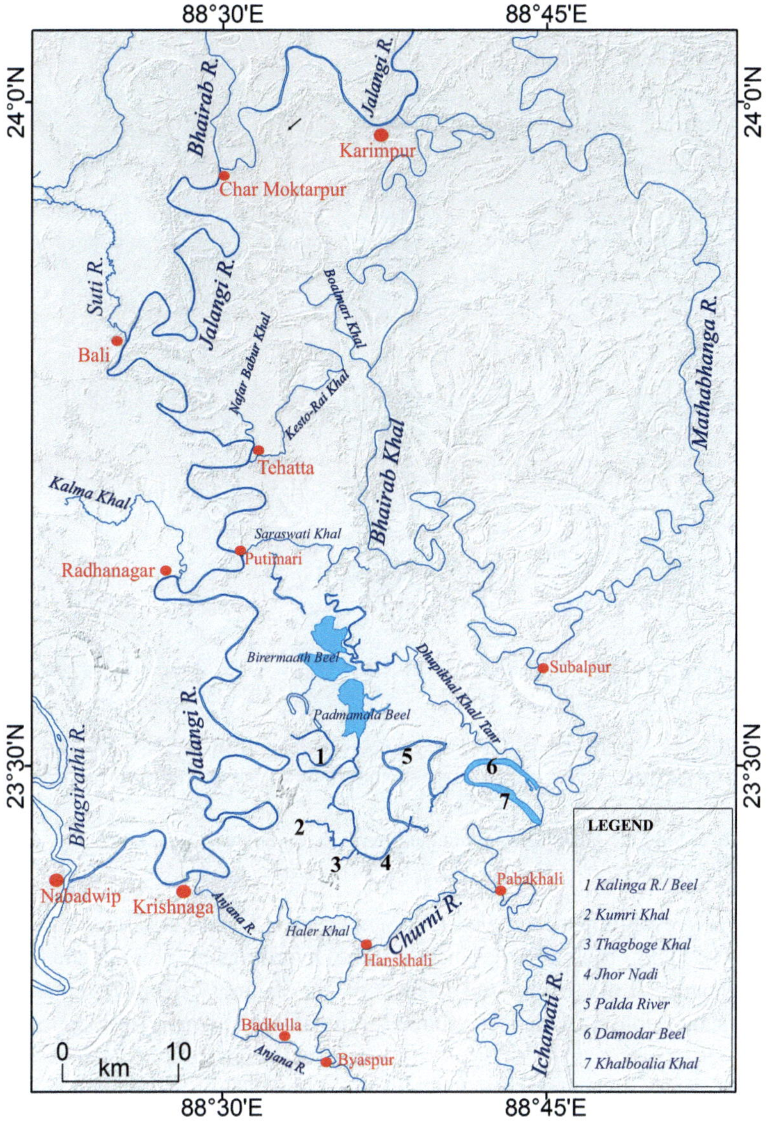

Fig. 3.9 Spills on the left bank and relict channels between Jalangi and Mathabhanga-Churni interfluvial tract. Namable of them are Kalinga *Beel*, Kumri Beel, Thagboge *Khal*, Jhor *Khal*, Palda *Beel,* and Damodar *Beel.*

latitudes 23°16′10″ to 25′02″ N and longitudes 88°27′39″–88°36′46″ E. To learn more about the hydrogeomorphology of the Anjana River, refer to the book chapter titled "The Anjana: A Journey from River to Canal" (Das & Das 2020).

3.7 Wetlands

Numerous strokes of oxbow lakes, back swamps, and relict channels are prominent features in this basin, each representing a unique

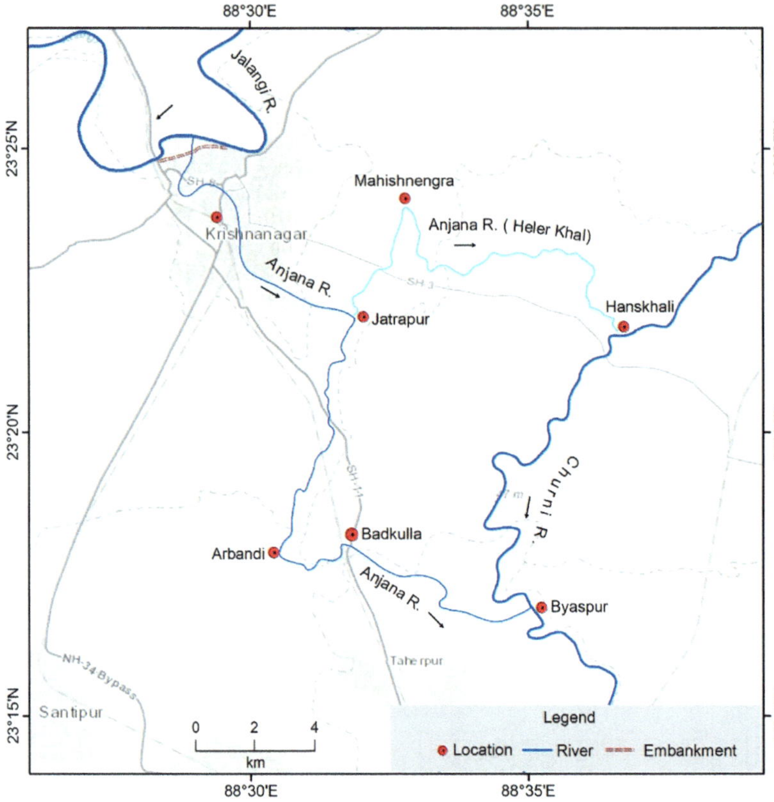

Fig. 3.10 Anjana River links the Jalangi River with the Churni River (Source: Das 2024).

phase in the river's evolutionary history (Das et al. 2020). These landforms provide insight into the hydrodynamic shifts within the basin and serve significant ecological and economic roles.

The formation of oxbow lakes in the Jalangi River basin is primarily attributed to lateral migration and meander cut-offs of channels necks. As the river meanders across the sediments of the delta, hydraulic forces and sediment deposition create crescent-shaped water bodies disconnected from the main channel. Over time, these oxbow lakes evolve, filling with organic-rich sediments and transitioning into wetlands. Back swamps, typically found in the floodplain's low-lying areas, form due to overbank flooding during monsoon seasons. These water-logged depressions are frequently replenished by riverine floodwaters, accumulating fine-grained sediments, organic matter, and nutrients. Their persistence is closely linked to the hydrological regime and sediment supply of the Jalangi River system. Relict channels, remnants of former river courses, illustrate the temporal mosaic of channel abandonment of the basin. These features often result from avulsion events, where the rivers shift courses due to reaching the thresholds of morphodynamic events. Relict channels in the Jalangi basin exhibit a range of ages, offering a stratigraphic record of palaeo-hydrological conditions and climatic variations. Inglis (1909) mentioned that it is highly unlikely that all the channels were ever simultaneously active. As one channel deteriorated, another was likely formed. Thus numerous relict channels of varied ages featured the Bhagirathi-Jalangi-Churni basin.

These geomorphological features hold immense economic value. Oxbow lakes, rich in biodiversity, support local fisheries and provide irrigation water, contributing significantly to agriculture in the delta (Das et al. 2015). Back swamps, with their nutrient-rich sediments, enhance soil fertility, promoting the cultivation of crops like rice, a staple in the region. Furthermore, these wetlands act as natural water storage systems, aiding flood mitigation and groundwater recharge (Chatterjee et al. 2020). Relict channels, often influence land use and settlement patterns (Das 2024). Their fertile banks are prime locations for agriculture. Their stratigraphic records are valuable for understanding sedimentary processes and evolutionary history (Greb et al. 2006), which can inform water management practices in the flood-prone delta.

A summary of wetland types is provided in the table (Table 3.3). Wetlands characterized by an arc-shaped fundamental component, as described by Weihaupt (1977), are categorized as oxbow lakes. Wetlands exhibiting a linear configuration with a length at least five times their width are identified as relict channels, while those with irregular gently depressed forms are designated as back-swamps (Fig. 3.11).

3.8 Changes Over Time and Space

Reconstructing the historical trajectory of the River Jalangi's spatio-temporal changes and deterioration requires reliance on archival maps, records, and other historical documents. Therefore, to piece together the history of these changes, we have divided the timeline into distinct phases based on the availability of such resources (Fig. 3.12).

3.8.1 Changes Occurring Before 1780

Changes and deterioration before 1780 are divided into two periods based on the availability of Van-Den-Brouck's map (1660) and Rennell's survey from 1764–1777.

3.8.1.1 Changes from Its Inception Until 1660

Literature from epic, historical accounts, and legend, as presented in the section 'River in Epic, History, and Legend' and illustrated by Das (2013), has confirmed the River Jalangi's existence as far back as the fifteenth century or earlier. However, no surviving documents from the fifteenth century to 1660 provide a sufficient basis for commentary on the geographical history of the River Jalangi. Like the precise date of the river's origin, its condition immediately following its formation, and whether its course deteriorated or improved by 1660 remain entirely unknown. The only inference that can be drawn from Van-den-Brouck's map (1660) is that the river had reached a state more robust than the Bhagirathi River, as depicted wider than the Bhagirathi in his map.

3.8.1.2 Changes Took Place from 1660 to 1780

In Van-den-Broucke's map, only one off-take of the river Jalangi (De Galgatese Spruyt) is depicted, flowing into the river Bhagirathi opposite Amboa. Majaumder (1985) disputes this, arguing that De Galgatese Spruyt is not Jalangi as its off-take and confluence do not align with historical references. However, the present researcher suggests that Van-Den-Broucke's outdated and imprecise map may still hold relevance. It indicates Jalangi's debouchment into the Bhagirathi at Amboa, possibly solving the route dilemma for visitors of Sri Chaitanya from Nabadwip to Shantipur. Following his renunciation on 26 January 1510, Sri Chaitanya returned to Shantipur, where devotees from Nabadwip reportedly crossed a river to visit him (Dev 2004). Given that the Bhagirathi River then flowed west of Nabadwip, situating both Nabadwip and Shantipur on the same bank (Chaudhury 2004; Van-den-Broucke 1660), the river crossed was likely the Jalangi (Das 1535).

Between 1660 and 1780, Jalangi underwent significant course changes, with Nabadwip initially situated east of the Bhagirathi. By 1764, Nabadwip was encircled and dissected by

Table 3.3 Relative positions, formation period and dimension-related attributes of the oxbow lakes along Bhairab-Jalangi

Sl. No	Name of the oxbow lake/ nearest locality	Relative position	Location	Plan form	Probable formation period
B1	*Damus Beel,* Char Shib-nagar	Left Bank of Bhairab	24°03′05″ N & 88°30′16″ E	Oxbow Lake	1921–1967
B2	Nayankhali *Beel*	Left Bank of Bhairab	24°01′37″ N & 88°30′10″ E	Oxbow Lake	Before 1849
B3	Char Sahajadpur *Beel*	Right Bank of Bhairab	24°00′30″ N & 88°29′09″ E	Oxbow Lake	1949–1967
1	Sahebrampur *Beel-1*	Right bank	24°06′23″ N & 88°39′35″ E	Oxbow Lake	Before 1849
2	Sahebrampur *Beel-2*	Right Bank	24°05′37″ N & 88°39′43″ E	Irregular	1766–1854
3	Karua *Beel,* Bajitpur	Left Bank	23°58′00″ N & 88°38′00″ E	Relict Channel	1766–1854
4	Tengramari *Beel*	Left Bank	23°57′30″ N & 88°37′00″ E	Back swamp	1766–1854
5	Kishorpur *Beel*	Left Bank	23°53′26″ N & 88°32′15″ E	Oxbow Lake	Before 1849
6	Topla *Beel*	Left Bank	23°52′56″ N & 88°31′00″ E	Oxbow Lake	Before 1849
7	Damuk *Beel, Fatepur*-An-andanagar	Left Bank	23°50′03″ N & 88°30′24″ E	Oxbow Lake	1766–1854
8	Sardanga *Beel*	Left Bank	23°49′59″ N & 88°29′27″ E	Oxbow Lake	1916–1967
9	Argarini *Beel*	Left Bank	23°48′03″ N & 88°29′00″ E	Relict channel	1766–1854
10	Banur *Beel,* Nischintopur	Left Bank	23°46′45″ N & 88°29′16″ E	Oxbow Lake	Before 1849
11	Chanderghat *Beel*	Right Bank	23°41′22″ N & 88°29′21″ E	Oxbow Lake	*Beel*
12	Sholua *Beel*	Left Bank	23°41′48″ N & 88°31′00″ E	Oxbow Lake	1766–1854
13	Mathabhanga *Beel,* Mota-Baragachi	Right Bank	23°35′30″ N & 88°30′06″ E	Oxbow Lake	1766–1854
14	Badelangi *Beel*	Left Bank	23°31′35"N & 88°32′46"E	Oxbow Lake	Before 1849
15	Chulkani *Beel,* Chapra	Left Bank	23°31′36″ N & 88°32′46″ E	Oxbow Lake	Before 1849
16	Gokhurapota *Beel*	Left bank	23°32′06"N & 88°30′37"E	Back swamp	1766–1854
17	Nawpara *Beel*	Right Bank	23°31′00″ N & 88°29′22″ E	Oxbow Lake	1849–1921
18	Hansadanga *Beel*	Right Bank	23°27′14″ N & 88°27′30″ E	Oxbow Lake	Before 1849
19	Harkhali *Beel*	Left Bank	23°24′46"N & 88°26′37"E	Back swamp	Before 1766
20	Chamta *Beel*	Left bank	23°23′53"N & 88°26′9"E	Back swamp	Before 1766

(continued)

Table 3.3 (continued)

Sl. No	Name of the oxbow lake/ nearest locality	Relative position	Location	Plan form	Probable formation period
21	Alakananda *Beel*, Gangabas	Left Bank	23°24′20″ N & 88°25′06″ E	Oxbow Lake	1660–1766
22	Bachhamari *Beel*, Miyapur	Left Bank	23°23′04″ N & 88°23′54″ E	Oxbow Lake	1660–1766
23	Khayrar *Beel*, Simulgachi	Left Bank	23°22′38″ N & 88°23′35″ E	Oxbow Lake	1660–1766

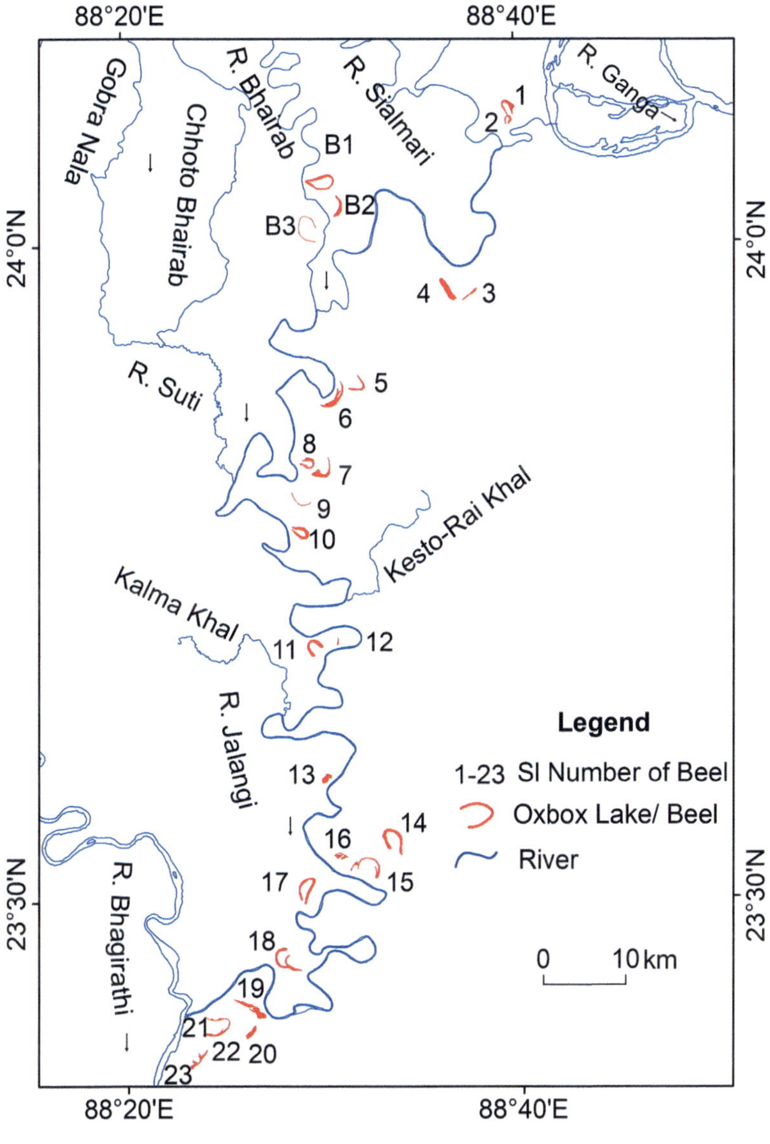

Fig. 3.11 Oxbow lakes and other notable wetland along the banks of the river Bhairab and Jalangi

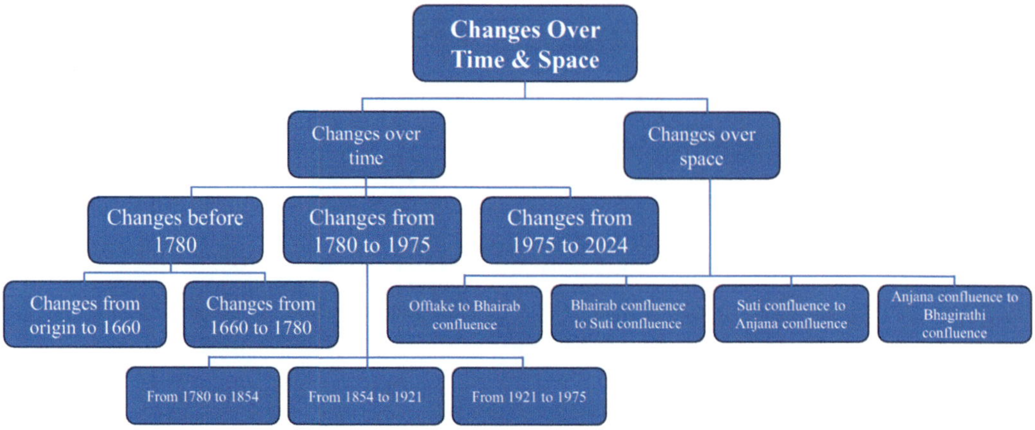

Fig. 3.12 Scheme of temporal scale for portraying changes in the river Jalangi

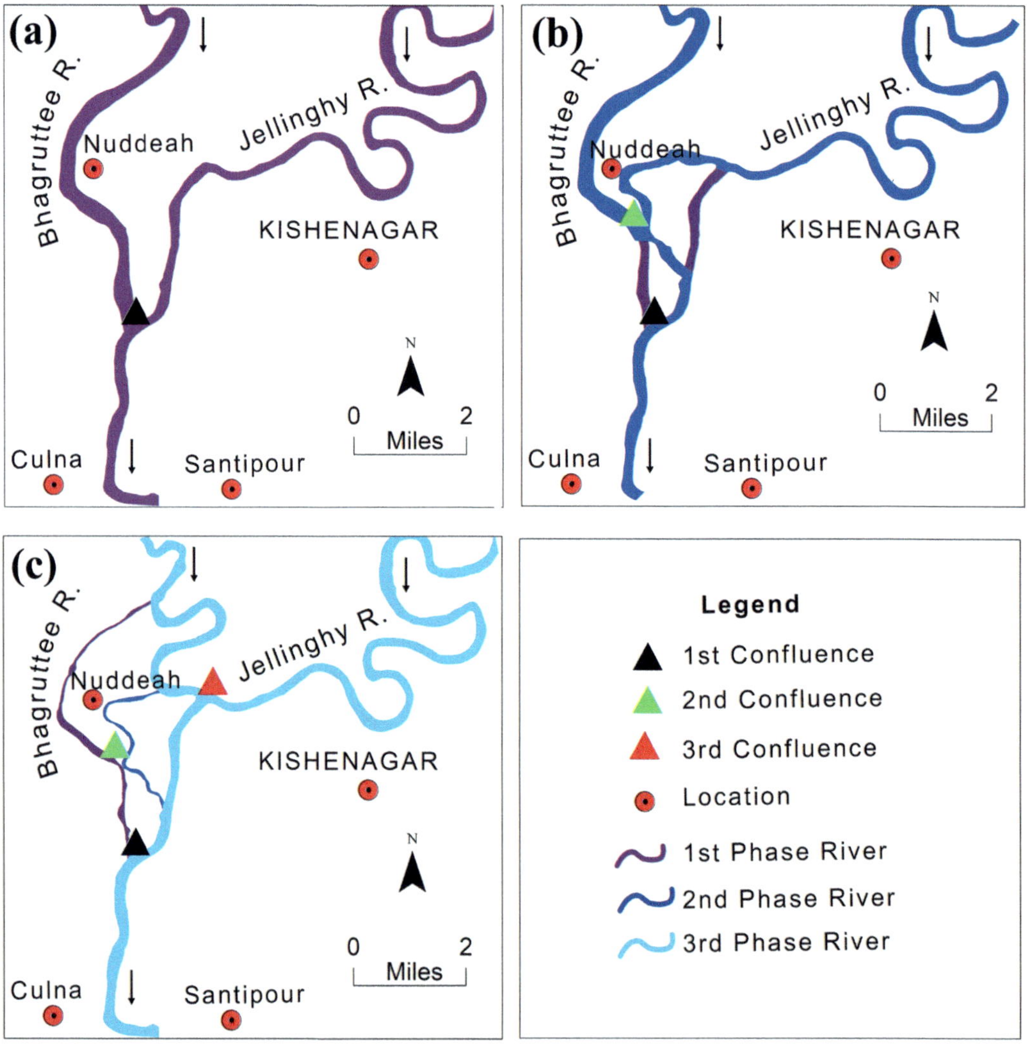

Fig. 3.13 Three-panel geographical map showing a probable sequence of events illustrating the 15 km upstream shift in the confluence of the Bhagirathi and Jalangi Rivers, as the Bhagirathi captured the course of the Jalangi due to the disproportionate wandering of their meanders. The depiction is based on Rennell's map

Table 3.4 Changes in the position of off-take of the river Jalangi

Period	Position of off-take	Latitude and longitude	Source of information
1660	East of Bilgassi from R. Ganga	–	Map of Van-den-Broucke
1764	At the village Jalangi of Murshidabad district, from R Padma	24°06′00″N 88°22′00″E	Map by James Rennel
1852	Bausmari, from R Mathabhanga	24°02′30″N 88°45′45″E	Revenue Survey Map, by Thullier (1852)
1910	Near Bausmari, from R. Padma or Ganga	24°11′00” N 88°43′00″E	J. H. E. Garrett, Bengal Dist. Gazetteer (1910, p. 256)
1930	From R. Padma, 3 km south of Jalangi	24°05′40″N 88°21′50″E	Mukharji (1932). Map showing important rivers in district of Murshidabad From 1766–1930, Scale: linch to 5 miles Appendix-XIX
1994	The off-take is not tracable from R. Padma, east of Bausmari village	24°04′30″N 88°41′54″E	D. P. M. S, Nadia Survey of India (1994)
2024	The off-take is not tracable from R. Padma, east of Bausmari village	24°04′30″N 88°41′54″E	Googl earth Pro image, as on 10.04.2024

rivers, with the Jalangi's confluence shifting about 15 km upstream. This shift, likely occurring in the early eighteenth century, altered the Nabadwip's geography and left former Bhagirathi river's channels as stagnant swamps (Paltar Khal) west of Nabadwip. Geological formations, nature of substrate, channel slope, and hydraulics, influence the deltaic river's course evolution. The Jalangi, characterized by low volume and velocity, meanders extensively due to its negligible slope of 0.000043 (Stvenson-Moor 1919) in deltaic plains, leading to increased course length and meander amplitude, especially at the confluence. During Rennell's time (1764–1780), Jalangi's sinuosity increased significantly (2.23) which was much lesser (1.14) during Van-den Brouck's time (1660). The disproportionate meandering of the Bhagirathi and Jalangi rivers has resulted in the Bhagirathi capturing the course of the Jalangi. Consequently, the confluence of these rivers has migrated approximately 15 km upstream. A potential sequence of events leading to this phenomenon is illustrated in Fig. 3.13. Following the catastrophic flood of 1770, the Bhagirathi progressively abandoned its western channel, shifting its primary flow to the eastern course (Chaudhury 2004) (Table 3.4).

During the late seventeenth century, the ‘Khoriya’ or ‘Khore’ River used to take off from Padma at Jalangi village, leading to its renaming by Rennel as Jellinghy.

Historical records, such as Tavernier's account (1666, pp. 125–126), describe Bernier's journey on January 6, 1666, from Murshidabad (Kāsimbāzār) to Hooghly (Hugli) via land due to the low water levels and obstructions at the Bhagirathi River's offtake. A similar situation might have affected the Jalangi River as well. Hedges' journey through the river in 1682 (Barlow 1887), highlights the river's navigability during the post-monsoon season. His journey provides a sequential account of riverine transit, reaching Krishnagar on the 16th, Andulia on the 17th, and Bali at the Suti-Jalangi confluence on the 18th. This account underscores the river's strategic importance as a navigable route during the period. The ability to traverse these points sequentially implies consistent water levels and the river's role as a vital corridor for travel and commerce, particularly in post-monsoon conditions when the flow was likely favourable for such endeavours.

Rennell depicted the Bhagirathi-Jalangi confluence as dividing greater Nabadwip and

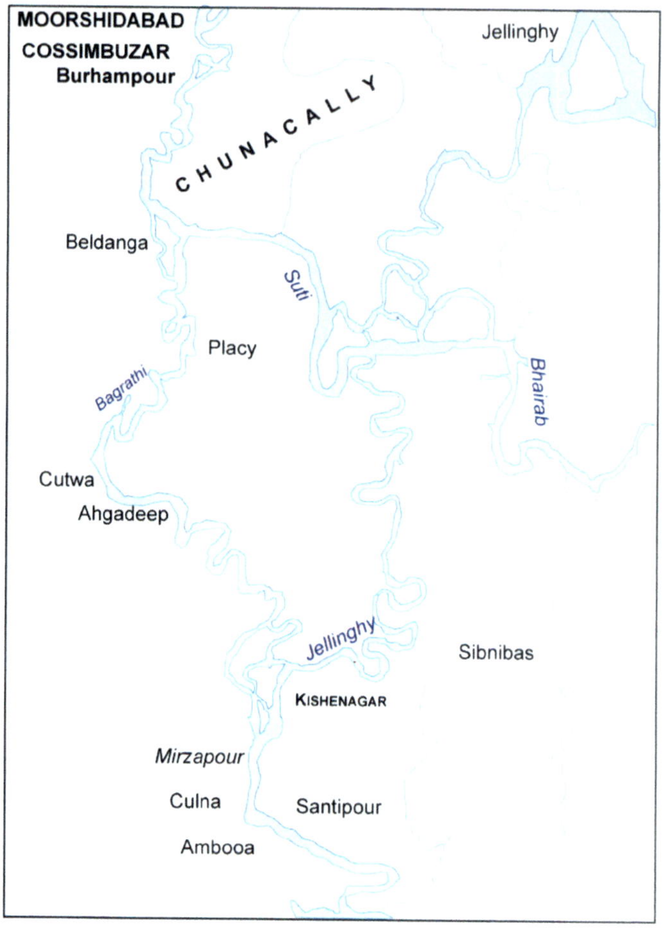

Fig. 3.14 The course of the river Jalangi during Rennell's time

its surroundings, forming a network of channels. This configuration encircled at least three islands within the Bhagirathi-Jalangi system (Fig. 3.14). Although Rennel, India's first Surveyor General, crafted a map in 1764 depicting Jalangi as narrower than Bhagirathi, by 1787, he noted that the 'Cossimbazar river was almost dry from October to May and that Jalangi was sometimes unnavigable for two or three months during the dry season' (Reaks 1919). Both Bhagirathi and Jalangi were impassable during this period. However, Jalangi was navigable during this period for most of the year, unlike Bhagirathi. These indicate that by the late eighteenth century, Jalangi had become more important for navigation than Bhagirathi.

3.8.2 Changes Occurring from 1780 to 1975

Under this section, the time span of 195 years has been subdivided in to three periods spanning from 1780–1854, 1854–1921, and 1921–1975 to discuss on spatio-temporal changes and deterioration in the course of the river Jalangi. The decay of the Jalangi, along with the Bhagirathi and the Mathabhanga–Churni rivers, is generally attributed to the eastward shift of the Ganga's main discharge (Morgan and McIntire 1959). Despite the decline of the Bhagirathi since the early seventeenth century, both the Jalangi and Mathabhanga remained navigable (Rudra 2018).

3.8.2.1 Changes Took Place from 1780 to 1854

Rennell found the Jalangi navigable for around nine months of the year, while in 1797, the Mathabhanga was more easily navigable than both the Bhagirathi and the Jalangi (Colebrooke 1801). In 1796, Bhagirathi was navigable during the dry season, but by 1797, it was no longer feasible for boats to travel to Kolkata. During the first half of the nineteenth century, the Garai River experienced a resurgence at the cost of the Jalangi and the Mathabhanga (Rudra 2018). The changes in the course of the Jalangi River between 1780 and 1854, spanning 75 years, can be divided into four spatial segments: (i) From the source of Jalangi to the Bhairab confluence at Moktarpur, (ii) From the Bhairab confluence at Moktarpur to the Suti confluence at Tungi, (iii) From the Suti confluence at Tungi to the Anjana confluence at Krishnagar, and (iv) From the Anjana confluence at Krishnagar to Swarupganj.

(i) From offtake to Bhairab confluence

During the 1832 flood, the Jalangi River experienced a significant alteration, shifting its off-take approximately 8.05 km (5 miles) northward. This change is evident from various historical maps, including the tricolour map (Mukharji 1932), as well as the (i) Revenue Survey of 1849–55 of India, District Nadia, and (ii) Revenue Survey of 1852–55 of India, District Murshidabad, surveyed under the direction of Major H.L. Thullier. These sources collectively illustrate the notable adjustments in the Jalangi River's course during this period.

Rennel's survey depicted only one primary offtake of the Jalangi River. However, the Revenue Survey map shows two offtakes at Jalangi village, approximately 6.44 km (4 miles) apart. Additionally, two other channels were supplying the Jalangi River: one originating from Dhuinchapara (between the two main offtakes, 3.22 km west of Jalangi village), and the other about 8.05 km north of Dayarampur from Padma. The southernmost offtake was originally the old course of the Mathabhanga River. Two offtakes are also shown in 1854 (Fig. 3.15).

The confluence of Sialmari remained almost unchanged. Furthermore, the Bhairab *Khal* at Karimpur branched off from Jalangi through two channels, one at Takipur and the other at Ukunpur, resulting in a reduced width compared to Rennel's 1764 map but remaining in good condition. All feeder channels such as Bharat *Beel*, Vikramkhali *Beel*, Kalitala *Beel* channel, river Sialmari, and river Bhairab were in good condition.

During this period, the "old Coolcooly" diverted a significant volume of water from the Padma River, flooding the area and directing a large flow downstream through the Bhairab River. The Bhairab subsequently split into two branches, the Bhairab River and Choto Bhairab, which converged into the Jalangi River at Moktarpur. The Coolcooly discharged the remaining water into the Jalangi through the Sialmari channel. At its eastern end, the Jalangi received a considerable influx from the Padma, causing flooding along both of its banks (Mukharji 1932). In 1854, a sandbar measuring approximately 3.22 km (2 miles) in length formed along the right bank of the Jalangi River near the village of Jagabandhupur in C.D. Block Karimpur-2. Furthermore, three islands (chars) or mid-channel bars surrounded by wet channels were documented on maps: one southwest of Jagabandhanpur, measuring 0.97 km (0.6 miles) in length; another north of Diar Kartikpur, measuring 0.64 km × 0.48 km (0.4 miles × 0.3 miles); and a smaller one just upstream of the Bhairab confluence, measuring 0.32 km × 0.16 km (0.2 miles × 0.1 miles) (Thullier Revenue Survey map, 1849–1855).

(ii) From the Bhairab confluence to the Suti confluence

A notable spatial transformation observed between 1780 and 1854, spanning approximately 75 years, occurred along the stretch from the Bhairab confluence at Moktarpur to the Suti confluence at Tungi. This period witnessed the formation of Topla Beel and the Margangnee-Paikpara Khal. In 1766, the Jalangi River underwent a bifurcation and rejoining near Topla and

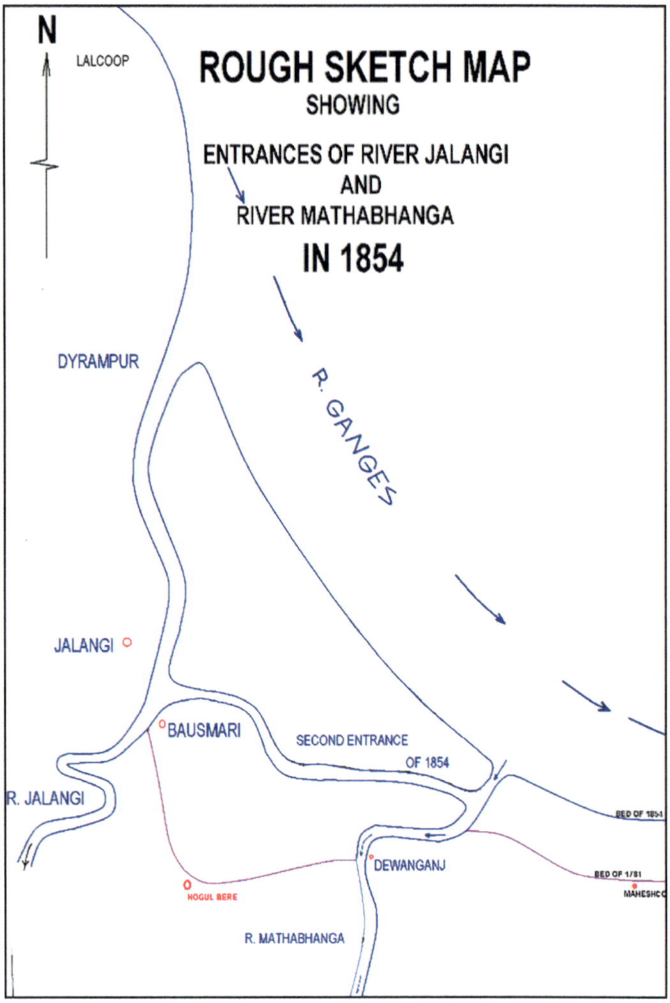

Fig. 3.15 Offtake of the river Jalangi in 1854 (*Source* Annual Report on Nadia Rivers, 1925–26. Govt. of Bengal. pp. 12–13)

Mrigi, resulting in the encirclement of Mrigi as an island. This enclosed area measured 8.05 km in length, and 4.03 km in width, and covered a total area of 32.44 km^2. By the time of the Revenue Survey Map of 1852–54, the river had shifted its course to the right channel, leaving the left channel to evolve into oxbow lakes—Topla Beel and Margangnee-Paikpara Khal (Fig. 3.16). Furthermore, during this period, the Hansadanga Beel was formed as a cut-off meander of the Jalangi River.

(iii) From the Suti confluence to the Anjana confluence

Under this sub-section, changes in the stretch from the Suti-Jalangi confluence at Tungi to the Anjana-Jalangi confluence at Krishnagar have been noted. In 1766, Fatepur *Beel* on the left bank and Chanderghat *Beel* (an ox-bow lake) on the right bank were absent but observed in 1854. Anjana Khal, a left bank tributary and spill, had already separated from the Jalangi River, with its off-take closed by Raja Rudra Ray of Krishnagar in 1676 (Mallik 1910). Furthermore, the Bad-Elangi ox-bow lake, located north of Chapra, which was absent in Rennell's map was documented in the Revenue Survey map of 1854. A uniquely shaped ox-bow lake, the Hansadanga *Beel*, situated north of Krishnagar, absent in Rennell's map was delineated in the Revenue Survey Map of 1854. Hence, it can be inferred that within this ~ 100 years span

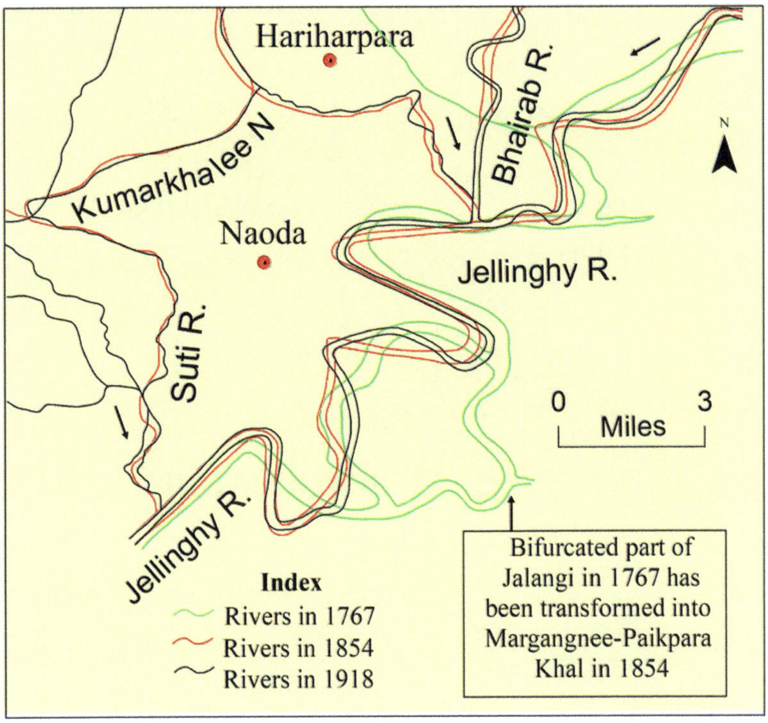

Fig. 3.16 The village of Mrigi was surrounded by a divided channel of the Jalangi (Jellinghy) River, as mapped by Rennell in 1767. The left channel was later abandoned, leading to the creation of the Margangni-Paikpara Khal, which was documented on maps in 1854 (Das 2013)

(1766–1854), two ox-bow lakes, the Bad Elangi *Beel* and the Hansadanga *Beel*, were formed as cut-off meander channels within this reach of the Jalangi River due to the cut-off of its meander necks.

(iv) From Anjana confluence to Bhagirathi Confluence

The changes, observed along the stretch from the Anjana confluence at Krishnagar to Swarupganj, were pivotal in reshaping the regional fluvial dynamics. The river underwent significant transformations, particularly around its confluence with the Bhagirathi River (Fig. 3.17).

The Bhagirathi began flowing predominantly through the captured course of the Jalangi River, leaving its own path to the west of Nabadwip town significantly weakened and likely stagnant during the dry season. During James Rennell's tenure in Bengal (1764–1777), he documented that near the Bhagirathi-Jalangi confluence, the Jalangi eroded a stretch approximately 2.42 km (1.5 miles) wide.

Between 1777 and 1782, the river's encroachment caused the birthplace of Sri Chaitanya to be engulfed by the Bhagirathi (Dev 2004). This event led Maharaja Krishna Chandra of Krishnagar to donate lands in Baidik Palli to the displaced Brahmins from the submerged areas of Nabadwip (Chaudhuri 2004). By 1784, although the river briefly flowed through the western channel, a sharp eastward avulsion near northern Nabadwip resulted in severe bank erosion. This erosion affected areas such as old Mayapur and Nidayar Ghat, with large portions of the old town being submerged into the Bhagirathi.

A pivotal moment occurred after the inundation of 1820 when heavy siltation caused the western channel of the Bhagirathi to close permanently, establishing its eastern course as the main flow (Chaudhury 2004). These dynamic hydrological changes significantly influenced the region's geography, history, and socio-cultural fabric (Das 2023a).

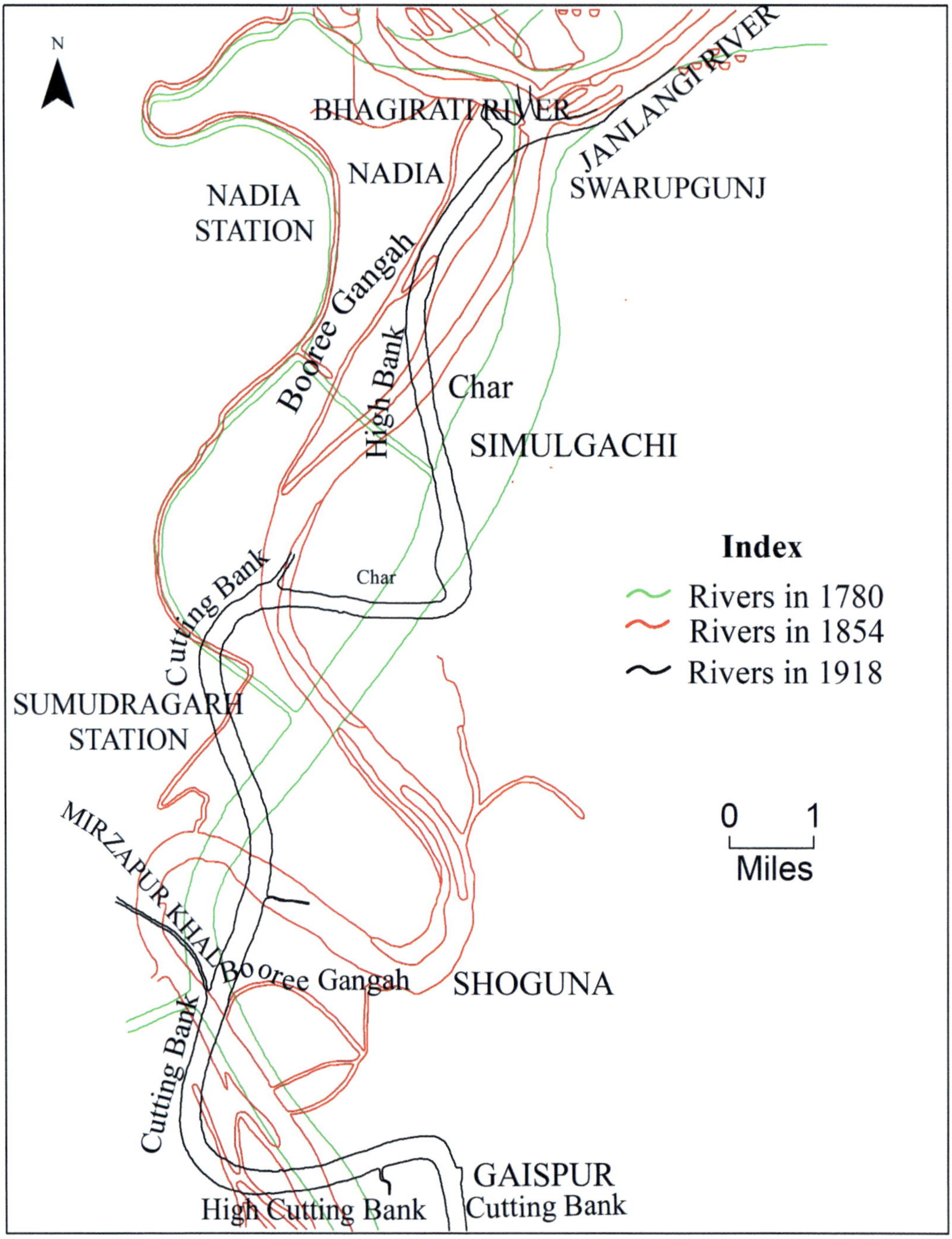

Fig. 3.17 Dynamics of Bhagirathi-Jalangi confluence

3.8.2.2 Changes Took Place from 1854 to 1921

The nineteenth century marked a period of decline for Nadia Rivers. Following the flood of 1874, the Bhairab entrance from the Padma River expanded into a significant distributary, diverting its water into the Jalangi River approximately 40 miles or 64.37 km further south. Consequently, the channel of the Jalangi River above this junction began to narrow (Hunter, 1877). Captain Brame noted, "*The Jalangi was used for main line of steamers proceeding from Calcutta to Goalundo or Assam during the rains from 1858 to 1866. It deteriorated about the later date, and the Mathabhanga was then used and was in use when I joined the inland service in 1882. It (Mathabhanga) was then quite a good river and in 1884 I went up with the 'Shillong' one of our steamers towing three fiats. It was never used for downward traffic, being too narrow and winding. It (Mathabhanga) afterwards gradually deteriorated while the Jalangi improved and by 1890 it became so bad that we abandoned it (Mathabhanga) in favour of the Jalangi and it has never been used. .... In 1891 we sent our Assam main line steamers through the Jalangi. There would appear to be some kind of oscillation between these two rivers, one improving while the other deteriorated. We used Jalangi for main line steamers towing fiats to Goalundo and Assam upto 1896. Subsequently, We continued to use the Jalangi for small craft proceeding to the Ganges districts up to 1907 (Stevension-Moor et al.* 1919*)*.

The above accounts indicate that, throughout most of the nineteenth century, except for occasional setbacks, Jalangi remained navigable even for steamers. However, remarkably, within approximately 50 years, the river became entirely disconnected from its feeder river, the Padma, and became essentially inactive, except for the downstream portion to the Bhairab confluence, which continued to be sustained primarily by the Bhairab's flow.

(i) From offtake to Bhairab confluence

To the east, the Coolcooly diverted water from Padma, flooding the area and directing water towards Bhairab, splitting into the Bhairab River and Chhoto Bhairab (Mukharji 1932). The Coolcooly discharged the remaining water via Sialmari to Jalangi, which further accumulated water from Padma, flooding its banks. However, by the early twentieth century, due to the eastward shift of Padma, Jalangi's thalweg level at its off-take rose above Padma's water level, abandoning Jalangi's own offtake.

Despite being in better condition than Bhagirathi in 1889–90, Jalangi gradually decreased over time. By 1932, it shared a common offtake with Mathabhanga near Bausmari for water supply. The river was in a livelier past compared to its current state (Annual report of Nadia River Division 1932, p. 25). The Jalangi was livelier than other Nadia Rivers but certainly poorer than what it was in the past (Mukharji 1932).

(ii) From the Bhairab confluence to the Suti confluence

Within this stretch, the river underwent minimal changes, except for a shift of approximately half a mile northward at Srinagar (J.L. No. 44) and a slightly larger eastward shift of over half a mile at Gopinathpur (J.L. No. 43) in C.D. Block Tehatta-I. The Topla Ox-bow Lake became significantly separated from the river, with a pathway (kutcha road) created through the cut neck to connect Narayanpur and Paranpur villages. Although water flowed from the river during floods, the Margangnee—Paikpara khal remained detached from Jalangi. It encircled the village of Mrigi from all sides except the east (Toposheet No. 79A/9, 1917).

(iii) From the Suti confluence to the Anjana confluence

During its active phase, Gobra Nala facilitated ample floodwater supply to Jalangi, benefiting the entire Bhagirathi-Jalangi interfluve. Historical maps, such as those by Rennell, illustrate the effective drainage system of the region, with Bhagirathi floodwaters originating near Lalitakuri and flowing through Gobra Nala's depression towards Jalangi. This floodwater journey continued through Bhanderdaha

Beel to Suti and Kalantar, with a significant portion flowing to Jalangi, while the remainder collected in the Kalantar depression. This livelier condition of Jalangi, even than Bhagirathi, may be due to the latter finding its path through the former via Bhandardaha Beel near Beldanga (Mukharji 1932; Fig. 3.14). During this period, no significant alterations of the channel occurred within this stretch, aside from slight reductions in the sinuosity index. Menders the Radhanagar loop shifted approximately 0.75 km northward, while the Nawpara loop expanded further, nearing a neck-cut stage to form the Nawpara *Beel*. Hansadanga *Beel* became noticeably separated from the river, with no spill channel connecting it to the Jalangi, even during flood events.

(iv) From Anjana confluence to Bhagirathi Confluence

From Krishnagar, the river flows due west with occasional bends. Downstream to the Railway Bridge at Krishnagar, the course shifted 300 m northward from 1854 to 1925. The Anjana offtake became entirely disconnected from Jalangi. Notably, Krishnagar Municipality was established in 1864 and the urban developmental processes might have impacted the detachment of Anjana from Jalangi. Micro-topographic features of Krishnagar town suggest that the Anjana once flowed through a different route which is illustrated in detail by Das and Das (2020). During 1917–18, the Anjana River underwent excavation to enhance drainage. North of the village Ruipukur (J.L. No. 52, C.D. Block Krishnagar-I), Jalangi shifted approximately 250 m northward. In C.D. Block Nabadwip, Bhagirathi flowed through the eastern channel, which was formerly Jalangi's channel captured by Bhagirathi in the eighteenth century, while the western Bhagirathi channel also existed, albeit in a declining state.

3.8.2.3 Changes Took Place from 1921 to 1975

During this phase of its life cycle, the river Jalangi has reached a stage where it lacks the energy required to erode its banks and alter its course. Moreover, the river's degradation was hastened by both natural processes and human interventions.

(i) From offtake to Bhairab-Jalangi confluence

The detachment of Jalangi from its feeder river Padma occurred only after the first quarter of the twentieth century (Mukharji 1932). By 1932, the river was receiving its supply through an off-take from Mathabhanga at Bausmari (Annual Report on Major Rivers, Nadia River Division, Bengal Irrigation Department, Govt. of Bengal, 1932, p. 25), a connection that became untraceable by 1975. The upper reach of the Mathabhanga-Jalangi, extending up to Bausmari, had been eroded by the Padma River, with Mathabhanga originating from Padma at the village Char-Mahishkundi.

(ii) From the Bhairab confluence to the Suti confluence

In 1971, a longitudinal bar stretching 1.5 km long and with a maximum width of 100 m was observed immediately downstream of the Bhairab confluence at Char Moktarpur (Topographical map No. 78D/4 of 1971). However, by 1975, this bar had been completely washed away, with no trace remaining. (ii) The longest bar in this reach, extending 4.5 km from Rajapur (J.L. No. 63) to Dakshin Char Chandpur (J.L. No. 114), Tehatta-I, experienced a reduction in length by 0.65 km. (iii) Topla *Beel* near Char Brindabanpur, C.D. Block Karimpur-II, diminished in size from over 0.5 km^2 in 1917 (measuring 4 km long and 130 m wide) to only 0.36 km^2 (measuring 3 km long and 120 m wide). (iv) The Margangnee Khal at Mrigi also underwent a significant reduction in length from 14.5 km to ~2 km. The course of the Margangnee Khal from north of the Mrigi village has been entirely erased.

(iii) From Suti confluence to Anjana confluence

In 1921, there was no sand bar observed along the right bank immediately downstream of the Suti confluence. However, in the late twentieth century, there exists a 650 m long sand bar with an average width of 50 m. (ii) The connector of Banur *Beel* near Nischintapur in C.D. Block Tehatta-I has been effectively blocked by a village road. (iii) The length of the Rudranagar longitudinal bar, situated opposite Banur *Beel*, has increased from 5.25 km in 1971 to 5.80 km. (iv) Approximately 40 acres of cultivable land on the right bank were engulfed by the river at Panditpur-Jhitkipota. Additionally, a road stretching about 5.80 km has been submerged into the Jalangi River.

(iv) From Anjana confluence to Bhagirathi Confluence

During this period, no significant alterations of the channel has been recorded within this stretch.

3.8.3 Changes Took Place from 1975 to 2024

In 1975, a significant turning point occurred in the hydrology of the Bengal delta with the commissioning of the Farakka barrage on the Ganga River. This barrage was constructed to maintain a minimum flow in the Bhagirathi River, ensuring navigability and sustaining the port of Kolkata. Chapter 2 of this volume delves into the details of the Farakka barrage and the water-sharing agreement between India and Bangladesh.

However, the construction of the barrage has led to a decrease in downstream water levels, particularly below the thalweg level at the entrance of the Bhairab and Jalangi rivers. Consequently, the Bhairab River no longer, except few days, receives its water supply from the Padma River, resulting in a decline in hydraulic parameters such as gauge height, velocity, discharge, etc., and the transformation of the ecosystem from lotic to lentic.

At Jayrampur, the Jalangi River's bed level measures 15.66 m, while at Gopalpur Ghat, the river has nearly disappeared, with a bed level of 17.76 m. However, the average highest water surface level of the Padma River during the monsoon months of July to October between 2007 and 2009 was only 16.75 m, which is 1.01 m below the Jalangi's bed level necessary for river entry. This difference prevents water from the Padma from feeding into the Jalangi (Das 2013). The Jalangi's entrance via Bhairab is effectively severed from the main source of the Padma, given that the Padma's water level is lower than that of the Jalangi's bed. Consequently, water from the Padma cannot consistently supply the Jalangi's main channel.

During this period, at lower reach, channel shifting has been minimal, with localized bank erosion pockets being the primary concern. Left bank erosion has destroyed a village road, agricultural lands spanning 3.5 acres, and five houses at Ruipukur-Natun Shambhunagar. Additionally, at Bahadurpur on the right bank, the average annual rate of bank erosion was 400 m^2, leading to an estimated annual loss of Rs. 15.68 lakhs in 2004, as reported by the Department of Irrigation & Water Ways (Table 3.5). At Ruipukur-Natun Shambhunagar, left-bank erosion has caused considerable loss to local infrastructure and livelihoods. This illustrates not only the magnitude of physical changes in the channel but also the economic vulnerability of communities dependent on the eroding bank. Conversely, deposition processes have been observed on the left bank at Char-Shambhunagar, directly opposite Bahadurpur. Here, the formation of a crescent-shaped sandbar signifies a dynamic depositional environment. The sandbar's surface area expanded notably over several decades, growing from 0.55 km^2 (200 m wide and 2750 m long) in 1971 to 0.725 km^2 (250 m wide and 2900 m long) in 2009. This growth indicates sediment accretion likely influenced by hydraulic changes in the channel, sediment supply, and flow patterns. The construction of the second highway and railway bridges at Krishnagar has significantly altered the local-scale channel morphology. As of 2025, the third railway bridge is under construction, further impacting the channel morphology and aquatic ecology.

Table 3.5 Changes in the position of the confluence of the river Jalangi-Bhagirathi

Period	Position of confluence	Latitude and longitude	Source of information
1660	Amboa north of Mirsapore, near present-day Samudragar, Hooghly	–	Map of Van-den-Broucke
1764	Mirzapur, Hooghly (Fig. 3.17)	23°20′00″N 88°00′00″E	Map by James Rennel
1852	Swarupganj, opposite Nabadwip	23°23′30″N 88°23′00″E	Map, R. S. by Thullier (1852)
1910	Opposite Nabadwip	23°24′00″N 88°23′00″E	J. H. E. Garrett, Bengal Dist. Gazetteer (1910, p. 256)
1971	Swarupganj, Opposite Nabadwip	23°24′00″N 88°24′00″E	Topographical map N0. 79A/7
2024	Swarupganj, Opposite Nabadwip	23°24′41″N 88°22′49″E	Googl earth Pro image, as on 10.04.2024

3.9 Meander Belt Dynamics

Mapping the meander belt poses minimal challenges with regular and straightforward meanders; however, irregular and compound meanders present a more intricate task (VTANR 2004). Ness (2004) discusses several approaches for determining the width of complex meander belts. Notably, the Jalangi river showcases striking irregular meanders (Fig. 3.18). Consequently, four methodologies, adapted from Ness (2004), have been employed to ascertain belt widths of Bhagirathi and Jalangi, revealing the intricacies of the meander pattern (Islam et al. 2020). These techniques are outlined as Eqs. (3.1)–(3.4):

$$MB_{w1} = 2a_m + w \quad (3.1)$$

$$MB_{w2} = 2d_a + w \quad (3.2)$$

$$MB_{w3} = 2d_p + w \quad (3.3)$$

$$MB_{w4} = (2a_m + w) \times FS + M_{ax}100, \quad (3.4)$$

where MB_{w1} to MB_{w4} denotes different levels of meander belt width based on method; w is channel width; a_m is meander amplitude from the meander axis M_{ax}; d_a is the distance of the farthest point of the current channels from the meander axis; d_p is the distance of the farthest point of the paleochannels from the meander axis; FS stands for the Factor of Safety $= 1 + \frac{M_{100}}{MB_{w\_max}}$; $M_{ax}100$ is the distance of migration or absolute shift of the Channel centre line in 100 years and MB_{w_max} is the maximum meander belt width of the present channel (Table 3.6).

By incorporating these variables, the formula provides a comprehensive method to estimate the maximum width of the meander belt, considering factors such as channel migration and the presence of paleochannels.

Meander-Belt width on the consideration of present channel geometry: The meander belt width (MB_{w1}) of the Jalangi River in its lower reach was computed for 1990, 2004, and 2014. In 1990, the average belt width of the Jalangi was 2.91 km. By 2004, it measured 2 km. The meander belt width (MB_{w2}) of the Jalangi River has remained almost constant over time, indicating that it is virtually an inactive river. However, historically, the Jalangi was a significant river, comparable to or even surpassing the Bhagirathi in the eighteenth century (Das 2013). Due to the eastward tilting of the Bengal Basin, the Jalangi now functions primarily during the freshet period (Sen 2010).

Meander-Belt width on the consideration of past channel geometry: For the Jalangi River, the meander belt width (MB_{w3}) derived from the palaeo-channel study is 10.2 km for the year 1990 (Fig. 3.18). The channel retains its original position, as it is now virtually inactive. The

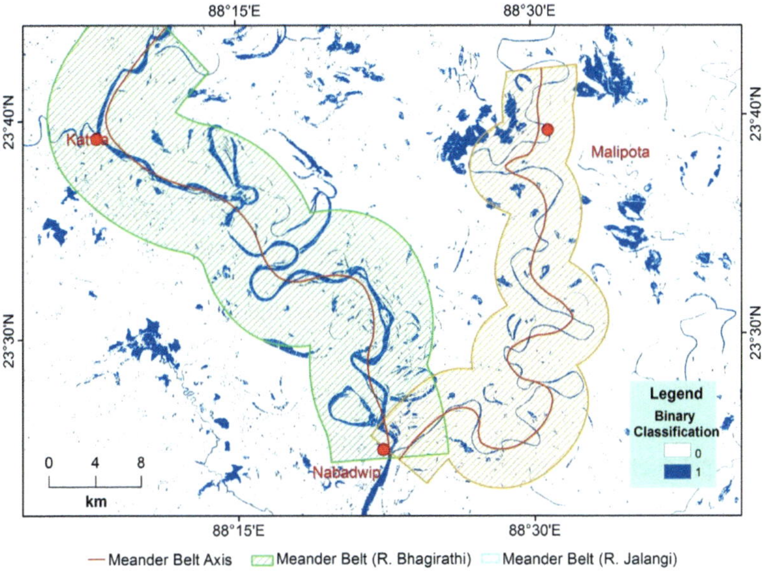

Fig. 3.18 Meander belts of river Bhagirathi and Jalangi (Source: Islam et al. 2020)

Table 3.6 Share of the overlapped area between Bhagirathi and Jalangi

Year	MB_{w2}			MB_{w3}			MB_{w4}
	1990	2004	2014	1990	2004	2014	2014
Total	13.75	9.18	8.89	65.82	84.48	76.93	17.7
Share of Bhagirathi (km^2)	10.6	7.02	6.85	50.77	65.16	59.77	13.68
Share of Jalangi (km^2)	3.14	2.16	2.04	15.05	19.32	17.16	4.02
Share of Bhagirathi (%)	77.09	76.47	77.05	77.13	77.13	77.69	77.29
Share of Jalangi (%)	22.84	23.53	22.95	22.87	22.87	22.31	22.71

Source: Islam et al. (2020)

delineation process used NDWI, pixel-based classification, field verification, and average channel width to resolve overlapping issues.

Meander-Belt width on the consideration of future channel characteristics (MB_{w4}): This analysis involves forecasting the factor of safety over 100 years by considering the migration rates of the channel and the meander belt axis. These migration rates were calculated using shifting data from the past 24 years, which were then extrapolated linearly to span a century. Observations indicated that the Jalangi channel will remain stationary over the 100-year projection. The 24-year data extrapolation revealed an average annual shift of 41.02 m for the Bhagirathi channel, however, the Jalangi channel exhibited consistent stability or near immobility over time, as confirmed by NDWI analysis (Table 3.7).

The contrast between Bhagirathi-Hooghly and Jalangi is striking. Bhagirathi-Hooghly, controlled by the Farakka Barrage, remains active, while Jalangi follows a more inactive moribund course. Their meander belt widths highlight this difference. In 2004 and 2014, MB_{w1} is similar for both rivers. However, due to Bhagirathi's straighter channel, MB_{w2} shows a wider belt for

Table 3.7 Computation of factors of safety

Parameters	Bhagirathi	Jalangi
Migration of channel axis, 100-year average (m)	1381.14	0
Migration of MB axis, 100-year average (m)	4102	0
Belt width of the present channel (m)	5920	6560
Factor of Safety	1.233301	1
Mean Channel width (m)	290	80
Final belt width (m)	11,760.8	6640

Source Islam et al. (2020)

Jalangi. In MB_{w3}, paleo-channel studies indicate Bhagirathi is wider. After accounting for a 100-year safety factor, Bhagirathi's final belt width (11.76 km) nearly doubles Jalangi's (6.64 km). This difference reflects fluvial dynamics and hydraulics, with Bhagirathi experiencing higher discharge, stream power, shear stress, and suspended sediment, which affects its channel movement more than Jalangi's. On the other hand, Jalangi is almost a stagnant water body except few days during rainy season.

3.10 Geomorphology and Landforms

3.10.1 Sedimentology

The pedology of the Jalangi basin, influenced by vegetation, climate, and human activity, gives rise to distinct soil regions, primarily entisols and inceptisols (Majumder 1978; Survey of India 1994). These soils, including Aerie Hapiaquepts, Typic Ustochrepts, and Typic Fluvaquents, vary based on depositional environments (National Bureau of Soil Survey & Land Use Planning, 1992). They exhibit different characteristics, such as varying textures like sandy loam and loamy sand (USDA Soil Taxonomy), affecting their susceptibility to erosion. On the banks, the predominant soil texture is coarse, characterized by sandy loam (Das 2013). However, patches of silty clay soil are observed in the northwestern part of Karimpur-I, Raghunathpur of C.D. Block Tehatta-I, and in the Kalinagar loop of C.D. Block Krishnagar-II. Loam soil dominates in C.D. Block Tehatta-I and II, as well as Krishnagar-I, while loamy sand is found on the banks of Nakasipara. In general, due to river velocity and hydraulic patterns, the texture on concave banks tends to be coarser compared to convex banks (Majumder 1978). One of the main drivers behind the environmental consequences of riverbed mining is the extraction of sediment at a rate faster than natural replenishment. Consequently, there's a pressing requirement for studies to assess the replenishment of riverbed sand deposit, aimed at mitigating the adverse effects caused by excessive extraction. Conducting annual replenishment assessments across the river is crucial for accurately gauging the potential sand reserves (DSR 2022). However, in the survey of potential sand reserves in riverbeds and banks, DSR (2022) possibly overlooked reporting on the Jalangi River due to its inability to generate, transport, and deposit sediment along its banks. Since 2013, the river has been largely disconnected from the Padma River, except for brief periods totalling only 2/3 days per year. Consequently, what was once a highly oscillating river, carrying a significant sediment load (Das 2013), has now stabilized within its established channel, formed several decades ago. The sediment introduced to the river now mainly comes from agricultural runoff and organic matter decomposition, particularly from hyacinth and other aquatic plants. However, the river has lost its transport capacity, leading to rapid deterioration as in-situ depositions occur. Aquatic plants (both rooted and floating) and algae are proliferating rapidly, exacerbating the degradation processes (Das et al. 2024a, b). Soil properties, particularly the textural pattern (Fig. 3.19), influence bank erosion and subsequent changes in the river course. Coarser textures correlate with higher vulnerability to bank erosion (Das 2013). However, other factors such as soil cutting on the banks and from the riverbed (Das 2014) can also trigger bank erosion

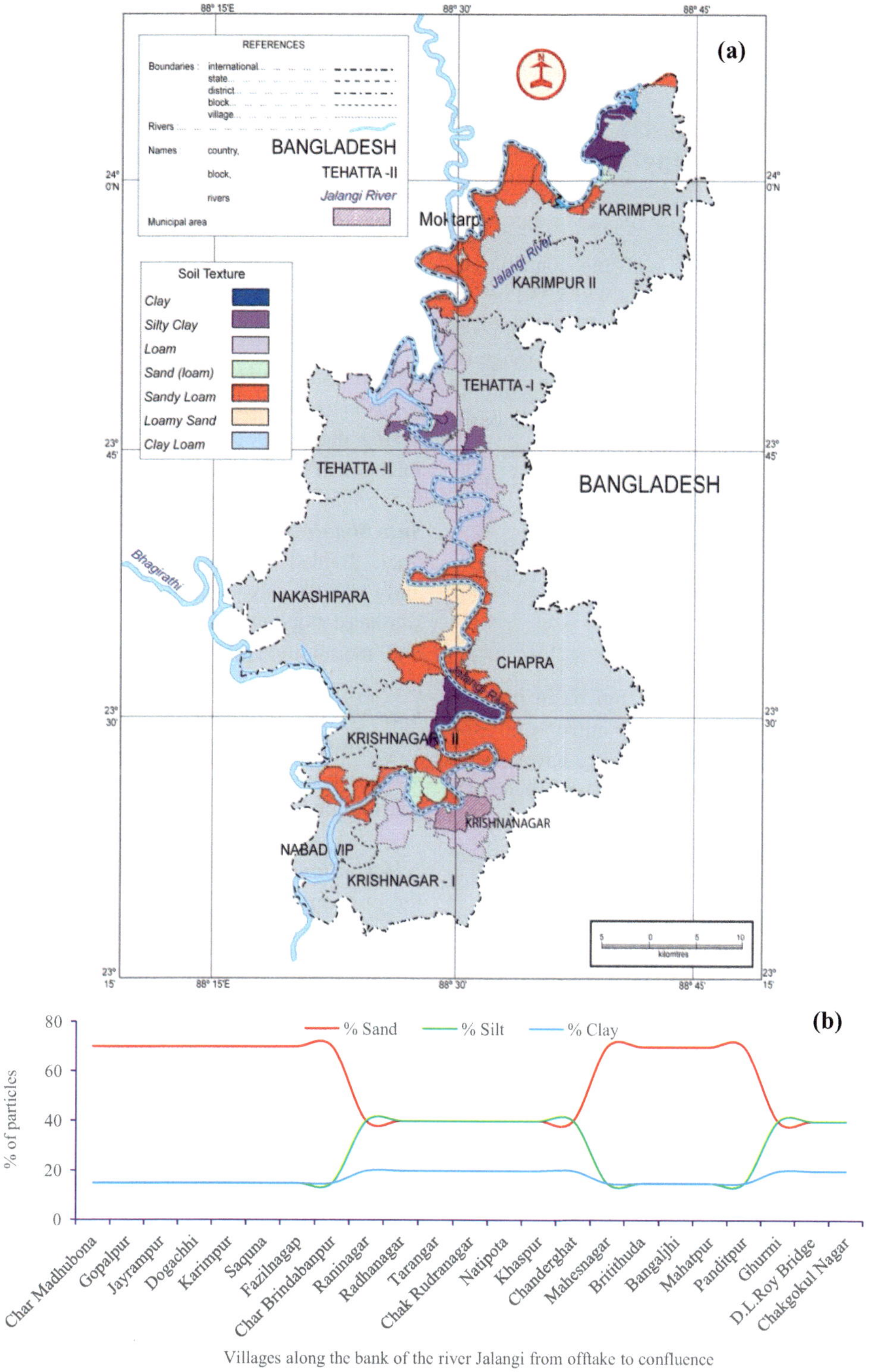

Fig. 3.19 Textural distribution of soil along Jalangi River banks from upstream to downstream. **a** Soil textural along the banks of the river Jalangi and **b** Particle distribution from upstream to downstream. (Source: Action plan 1999 & 2000–2001, Principal agricultural office, Nadia District, Krishnagar, West Bengal., p. 13, 26).

and consequent changes in channel profiles (Das 2019), even in areas with finer soil textures.

3.10.2 Extra-Channel Geomorphology

3.10.2.1 Relief

The Jalangi basin, along with the contributing spill basins, spans an area of 2815.33 km^2 in the districts of Nadia and Murshidabad. Over 63% of this area lies within an altitude range of 15.1 m to 27.00 m. Merely 1.91 km^2 (0.07%) of the lower region of the Suti and Jalangi basin is situated below the 5-m altitude mark. The medium-altitude zone, ranging from 5.1 m to 15.0 m, encompasses 1035.46 km^2 (36.78%) of the northern areas of the Gobra, Bhairab, and Sialmari basins. Altitude gradually diminishes from north to south.

3.10.2.2 Slope

The general slope of the Jalangi basin inclines towards the south, a pattern mirrored by its tributary channels and the Jalangi River itself. Analysis based on satellite imagery under ArcGIS software indicates that 93.83% of the basin, equivalent to 2641.79 km^2, boasts a slope lower than 4.385°. Conversely, only 6.17% of the area exhibits a slope greater than this value. The basin's landscape is characterized by predominantly gentle slopes, with nearly 57% of the total area featuring slopes of less than 1.67°.

The concept of flood slope, determined by the contour pattern directing floodwater movement during the inundation, reveals that the Jalangi basin's flood slope follows a southward trajectory. Flood slopes were identified using maps from Biswas (2001) and topographical maps (Fig. 3.20). In particular, the Jalangi-Bhagirathi interfluvial tract experiences a southward slope, with floodwaters from regions like Krishnagar converging towards the lower reaches of the Jalangi River, leading to annual inundation in specific areas. The catchment area, encompassing 2815.33 km^2, slopes towards the south and is divided into six sub-catchments:

Padma-Jalangi-Sialmari Interfluvial Tract: This region's slope descends from north to south, with surface runoff feeding into the Jalangi River either directly or via the Sialmari River.

Sialmari-Bhairab Interfluvial Tract: This area slopes southward, with water primarily flowing overland between and through the Sialmari and Bhairab rivers before reaching the Jalangi (Fig. 3.20).

Bhairab-Suti Interfluvial Tract: The Suti Nala contributes significant water flow, particularly during the monsoon months, draining into the Jalangi at Tungi.

Suti-Bhagirathi Interfluvial Tract: This tract, divided by National Highway-34 (presently NH-12), experiences north-to-south and southeast slopes, affecting the flow of water into both the Jalangi and Bhagirathi rivers.

Suti-NH-12 Intermediate Tract: The flood slope in this region is from north to south and southeast, with overland flow contributing to the yearly inundation in specific areas.

NH-12-Bhagirathi Intermediate Tract: Like the previous tract, this area experiences a flood slope from north to south and southeast, leading to occasional inundation, particularly in locations such as Mayapur (headquarter of ISKCON) and Swarupganj.

Furthermore, the left bank of the Jalangi River slopes southward and southeastward, with overbank flow predominantly moving in those directions. Only a portion of surface runoff on the left bank reenters the river through designated spill channels during post-monsoon months.

3.10.2.3 Landforms

The extra-channel geomorphic features include older deltaic plains (CosOri), older alluvial and flood plains (FluOri), back swamps, and intra-meander belt shallow depressions. These

(a)

(b)

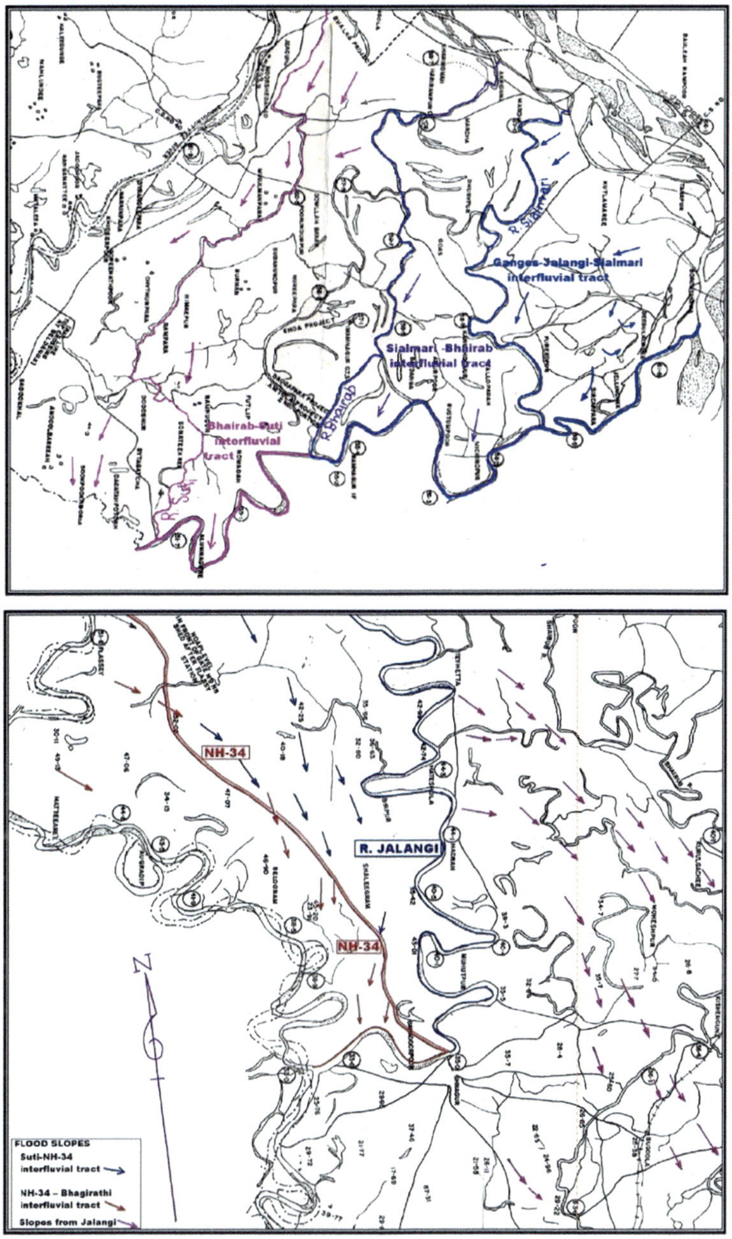

Fig. 3.20 Flood slope of the Jalangi basin. Arrows show flow directions. **a** Padma-Jalangi-Sialmari Interfluvial Tract; Sialmari-Bhairab Interfluvial Tract; and Bhairab-Suti Interfluvial Tract. **b** Suti-NH-12 Intermediate Tract; and NH-12 - Bhagirathi Intermediate Tract. (*Source* Biswas 2001, Plate 18 & 19)

features reflect the historical and geological evolution of the region (Fig. 3.21).

Older Deltaic Plains (Coastal Origin): These areas, marked in yellow, represent relict deltaic deposits formed by the river's past activities. The stratigraphic records in these regions indicate a decline in active sediment deposition due to upstream control structures like the Farakka Barrage. Their extensive spread suggests that the Jalangi was historically a more dynamic and active river system.

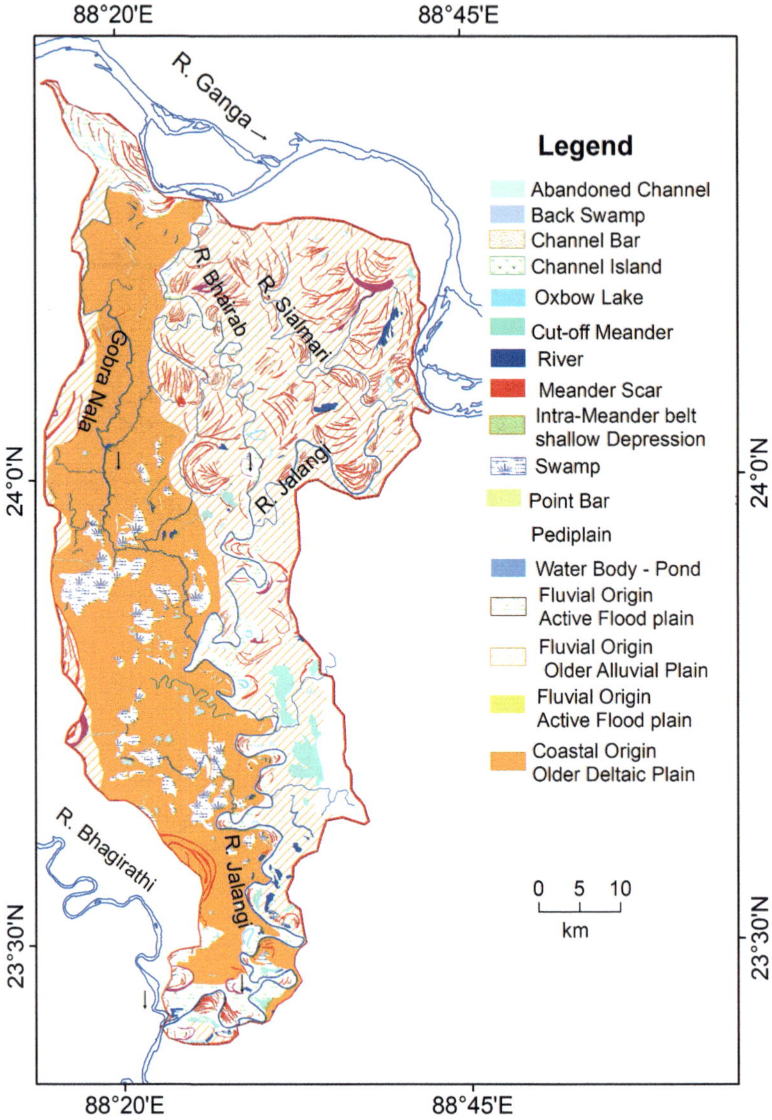

Fig. 3.21 Geomorphology of Jalangi River Basin

Older Alluvial Plains (Fluvial Origin): The green-marked areas indicate depositional features formed during periods of reduced channel energy. These plains serve as evidence of fluctuating discharge and sediment transport, showcasing the episodic nature of fluvial activity.

Back Swamps: Found adjacent to the channel, back swamps are indicative of poor drainage and a lower gradient. These areas accumulate fine sediments, such as clays and silts, during flood events, offering insights into past hydrological conditions.

Cut-Off Meanders and Oxbow Lakes: These features highlight the historical sinuosity and lateral migration of the Jalangi. Their occurrence demonstrates active meander dynamics in the past, which have now diminished due to reduced flow energy and sediment supply.

Intra-Meander Depressions and Swamps: These features reflect the interactions between fluvial processes and local topography. They often act as sediment traps and are key indicators of historical avulsion or channel shifting.

Meander Scars: Extensive meander scars dominate the floodplain, particularly in the central and southern reaches. These scars suggest historical episodes of lateral migration, but their lack of activity points to reduced channel dynamism caused by diminished discharge and sediment load.

3.10.3 Channel Geomorphology

The channel geomorphic features reveal the river's hydraulic behaviour, historical evolution, and anthropogenic impacts. Spatial observations from the Fig. 3.21 underline the following.

Channel Bars and Islands: Bars and islands appear sparsely within the active channel zone, indicating reduced sediment deposition due to upstream flow limitation from the feeder river. Their limited development is observed along straighter stretches of the river, reflecting subdued hydrodynamic variability.

Point Bars: Point bars are observed at the inner bends of meanders, with prominent examples in the southeastern stretch of the river. Their composition, likely fine sandy, reflects high-energy/moderate-energy sediment deposition during historical flood events, though their current morphology appears muted.

3.10.3.1 Cross-Channel Geomorphology

The study focused on examining the cross-sectional characteristics of the Jalangi River by conducting multiple cross-sections from its source to its confluence. The morphometric data derived from these sections shed light on various aspects.

Width (w): The width of the river channel varies based on its flow conditions. The average width is 162.58 m, with extremes observed at different locations along the river. For instance, at Moktarpur–Velanagar, where the river has a flat slope and meets the Bhairab River, the width reaches a maximum of 224.21 m due to lateral erosion. Conversely, at Moktarpur—Chandpur, just upstream of this wide section, the width decreases to 75.89 m. This reduction is attributed to both natural deterioration since the early twentieth century and anthropogenic activities such as river course alterations for infrastructure projects, leading to a severely degraded section.

Mean Depth ($\bar{d}$): Mean depth, calculated as cross-sectional area divided by channel width to width, varies across different sections. Average depths range from 0.82 m at Jitpur-Saguna to 7.74 m at Sibpur-Hatishala showing a downstream increasing trend i.e. Bed level = $-0.063 \times D + 13.875$ with $R^2 = 0.52$ (Fig. 3.22). Locations with deeper pools, like Sibpur-Hatishala, contrast with shallower reaches, such as the dead reach at Jitpur-Saguna. However, lowest bed level shows a decreasing trend downstream i.e. Average depth = $0.026 \times D + 0.860$ with $R^2 = 0.6373$ (Fig. 3.22).

Cross-sectional Area (A): The product of channel width and mean depth determines the cross-sectional area. Measurements show significant variations, with Krishnagar-Mayakol recording the largest area (265,042.70 m^2) and Jitpur-Saguna the smallest (2106.36 m^2).

Width to Mean Depth Ratio ($w/\bar{d}$): This ratio, expressing channel width relative to mean depth, indicates the shape of the channel. For instance, Jayrampur–Chanderpara exhibits a wide shape ($w/\bar{d} = 136.16$) due to shallow depths and silt accumulation from nearby agricultural activities.

3.10.3.2 Planform Geomorphology

The section examines meander patterns of the Jalangi River, revealing insights into its curvature and sinuosity. Major F. C. Hirst mentioned the abnormal winding of the Nadia rivers (Reaks 1919) especially if one considers the river Bhairab-Jalangi.

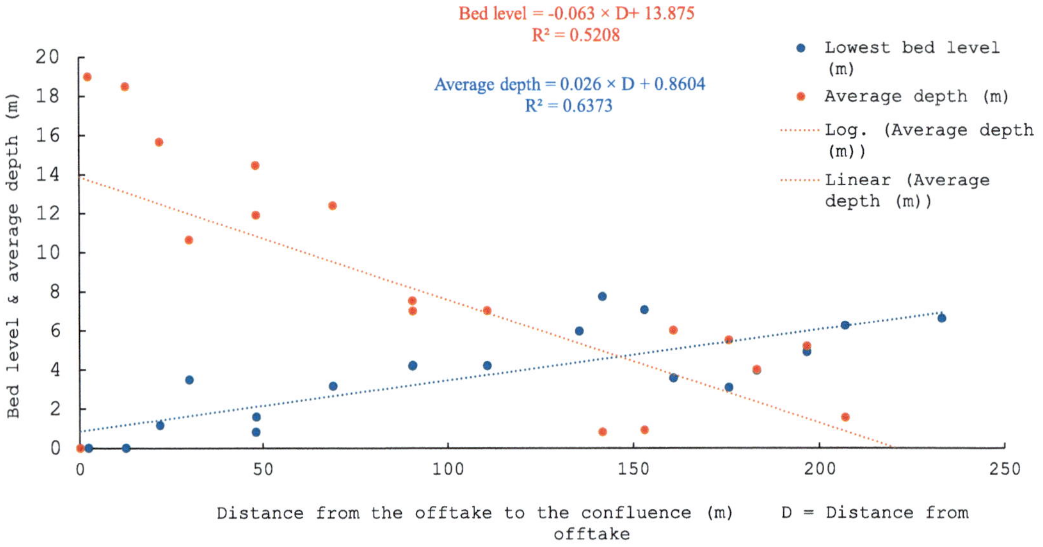

Fig. 3.22 Downstream variation of lowest bed level (m) and average depth (m) of the river Jalangi from offtake to confluence

Wavelength (λ): Wavelength refers to the distance between successive bends of a meander. It varies from 1500 m at Jayrampur to 10,000 m at Karimpur, with an average of 3646.74 m. The reasons for acute bends like at Jayrampur and broad ones like at Karimpur remain unclear, especially considering the absence of dynamic river processes in the dead reaches.

Bend amplitude (a_m): This measures the perpendicular height of meander bends from their axis. The Jalangi River exhibits varying amplitudes, with the Radhanagar loop (Fig. 3.23) having the highest (8500 m) and Jayrampur the lowest (1250 m). The high meander amplitude of the Jalangi River, now almost a dead river, signifies its past vitality and the dynamic geomorphological processes that once shaped its course.

Radius of Curvature (r_c): This indicates the curvature of meander bends, with values ranging from 250 m at Jayrampur to 4250 m at Karimpur. Despite downstream width increases, no significant trend in curvature is observed. Studies of different river types show that radius of meander bend (r_c) about 2–3 times the channel width at which the migration rate in bends peaks (Nanson and Hickin 1986). Hey (1976) found no consistent ratio of bend radius to channel width in two British rivers. The average ratio of r_c to width of the river Jalangi is around 6.88. Williams (1986) reported that 42% of the bend confirms the value 2-3 with an average of 2.43. About one-third of the values were below 2, a range where major changes in flow and sediment patterns can occur (Hickin and Nanson, 1975, 1984).

Sinuosity Index (*SI*): The river's sinuosity index averages 2.67, reflecting its highly meandering nature. Some meanders, like Radhanagar and Panditpur loops, exhibit necks ready for cut-off, though the river lacks the energy to complete the process (Das 2013).

Meander Shape Index (S_mI) is a ratio of the amplitude (a_m) to the radius of curvature (r_c) and reflects the openness or tightness of a meander (Das 2013, 2014). Values greater than 1 indicate open meanders, those between 0.5 and 1 denote regular meanders, and values below 0.5 represent acute meanders, where smaller values

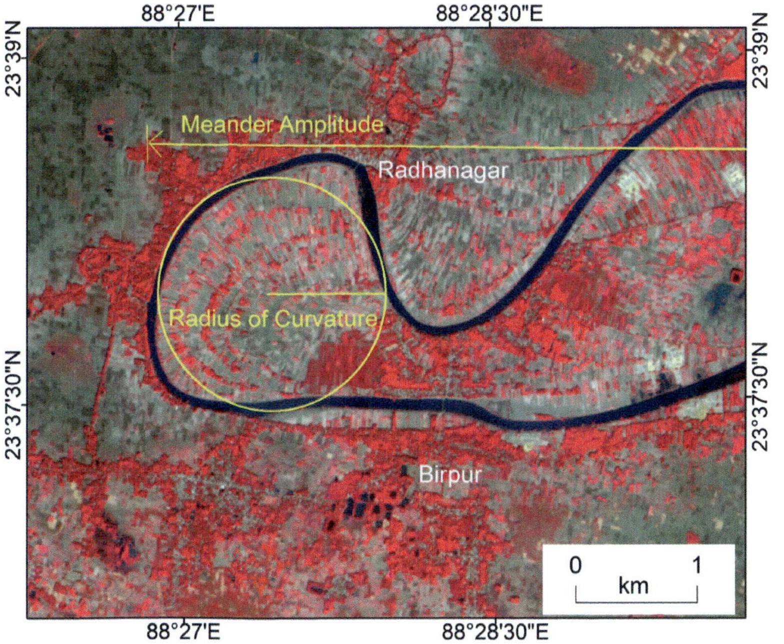

Fig. 3.23 The high meander amplitude of the Jalangi River, now almost a dead river, signifies its past vitality and the dynamic geomorphological processes that once shaped its course

suggest more intense meanders. The study found S_mI values ranging from 0.03 to 0.89, with a median of 0.14, indicating that most meanders are more sinuous than a regular sine curve, with over half showing acute characteristics. No significant downstream trend in S_mI was observed.

Meander Form Index (F_mI) measures meander curvature intensity by comparing the amplitude (a_m) to the meander wavelength (λ) (Das et al. 2020). A higher F_mI value denotes more intense meanders, with specific thresholds indicating open (<0.5), regular (0.5), intense (>0.5), or acute meanders. The study recorded F_mI values between 0.48 and 4.57, with an average of 1.44 and a median of 1.17, suggesting that more than half of the meanders exhibit acute characteristics.

Relationships between Cross-sectional parameters and Meander Properties: Meander characteristics like wavelength (λ), amplitude (a_m), and radius of curvature (r_c) all increase with channel width (w). These relationships are quantified, with the wavelength (λ) being approximately 24 times the channel width ($\lambda = 23.73$ w), and the amplitude being about 29 times the channel width ($a_m = 29.08$ w). Studies on meandering rivers from different environment suggests that the meander wavelength is about 10 to 14 times the channel width (Knighton 1998). radius of curvature channel width(r_c).

The spatial variability of older deltaic plains, back swamps, and cut-off meanders emphasizes the geomorphic dormancy of extra-channel features. This stagnation reflects reduced sediment flux and hydraulic activity in the Jalangi basin, resulting from upstream flow restriction. The channel morphology demonstrates a subdued fluvial system with limited active meandering, sparse bar formation, and static reaches. The historical dynamism observed through meander scars and oxbow lakes has been curtailed, indicating a shift from an active to a passive river

Fig. 3.24 Aerial view of a landscape where two rivers converge at Swarupganj, the Bhagirathi with muddy water and the Jalangi with clear blue water converge at Swarupganj. The blue-domed Chandradoy temple of ISKCON of Mayapur is at a rare view. Nabadwip is to the left side of the picture and Swarupganj is to the right side of the picture. Photo source: District Magistrate's Office, Nadia

system due to anthropogenic and climatic influences. This dual-layered analysis, combining spatial and geomorphic insights, underscores the need for detailed sedimentological studies and hydrodynamic modelling to better understand the Jalangi River's ongoing evolution.

The rivers' confluence of the Bhagirathi and Jalangi (Fig. 3.24) highlights contrasting hydrogeomorphological characteristics. The Bhagirathi, with its muddy water, carries a higher sediment load, due to intense upstream erosion driven by a larger catchment area and dynamic hydrological conditions. In contrast, the Jalangi, with its clearer water, indicates a slower flow and negligible sediment load, suggesting minimal upstream erosion or sediment deposition along its course. This contrast influences the confluence dynamics, where Bhagirathi's strong sediment-laden flow (Fig. 3.24) deposit material, alters the channel morphology and potentially causes bank erosion and the formation of bars. The Jalangi, with its gentler flow, acts as a stabilizing element, creating a balance in the sediment and flow interaction. These processes significantly shape the surrounding landscape, affecting the region's cross-channel geomorphology and ecological systems.

3.11 Hydrology and Water Resources

3.11.1 Gauge Level

The gauge height of the Bhairab River and the present offtake of the river Jalangi at Akhriganj from 1890 to 1920 show a declining trend (Fig. 3.25a). The linear trends of Low Water Level (LWL) and Highest Flood Level (HFL) show a similar downward slope, suggesting a consistent decline. While there is a general downward trend, significant variability exists due to natural fluctuations in rainfall and other factors. The difference between LWL and HFL

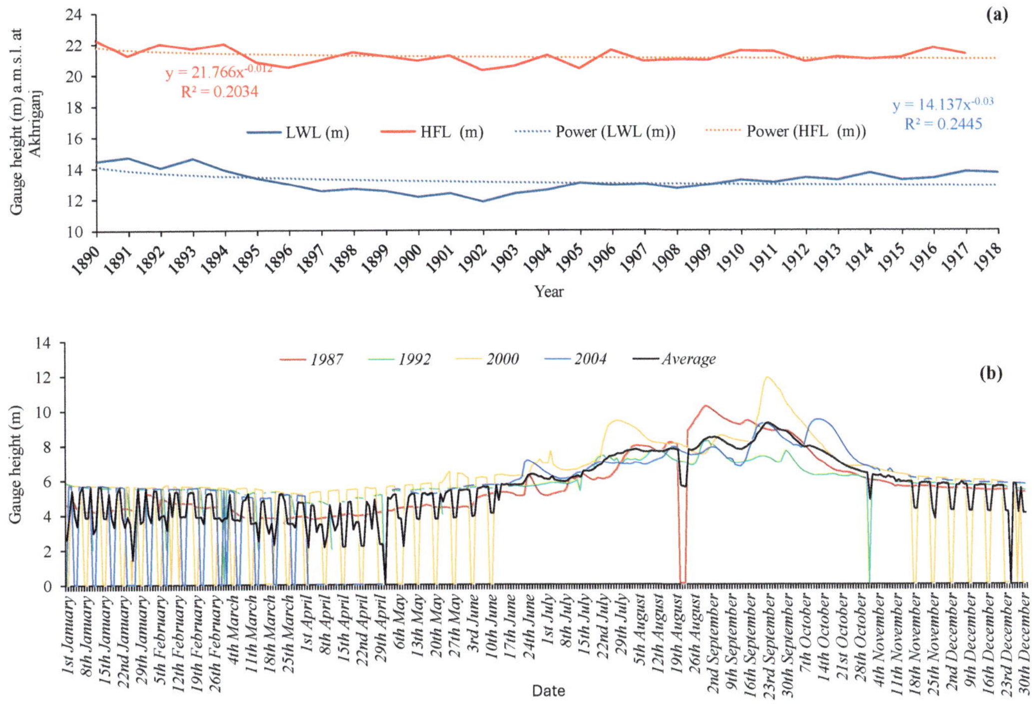

Fig. 3.25 **a** Gradual decline in the gauge height of the river Bhairab at its entrance at Akhriganj. Source: Reaks 1919, P. 104. **b** Daily gauge height at Swarupganj for the years 1987, 1992, 2000, 2004 and their average. Graphs touching the baseline indicate the absence of data

has remained relatively constant, suggesting no significant change in the overall range of water levels. The gauge height pattern of the River Jalangi for 1987, 1992, 2000, and 2004 shows clear seasonal and temporal variation (Fig. 3.25b). During the dry months (January to May), the water level remains low and stable. A gradual rise starts from June with the onset of the monsoon, peaking between August and mid-September, reflecting maximum rainfall and runoff. Among the years, 1987 and 2004 show higher peaks, indicating stronger monsoon impacts. From late September, the water level starts declining rapidly, reaching the lowest by December, marking the post-monsoon recession. The average hydrograph captures this typical seasonal cycle, while sharp drops, notably in 2000, suggest data gaps (For data table refer to Table 3.S1 in supplementary file).

Historical changes in the depth of the river Jalangi during the dry season are shown (Fig. 3.26) at two specific locations: the entrance and downstream from the entrance. The data reflects significant variations over time, with distinct trends observable for both locations. The entrance (blue line) exhibits marked fluctuations in depth during the early years (1840s–1850s), followed by the absence of data from 1854 to 1902. However, a sharp decline in depth is evident in the early twentieth century. Conversely, the depth below the entrance (orange line) remains relatively shallow and stable throughout the period, with a gradual increase but consistently lower values than at the entrance.

The large fluctuations in entrance depth during the 1840s–1850s may reflect natural processes such as sedimentation, erosion, or channel shifts. The sharp decline in the early twentieth century could indicate silting and reduced upstream water flow. The consistent disparity between the entrance and below-entrance depths points to sediment deposition downstream, likely reducing the channel's carrying capacity. While the entrance shows more

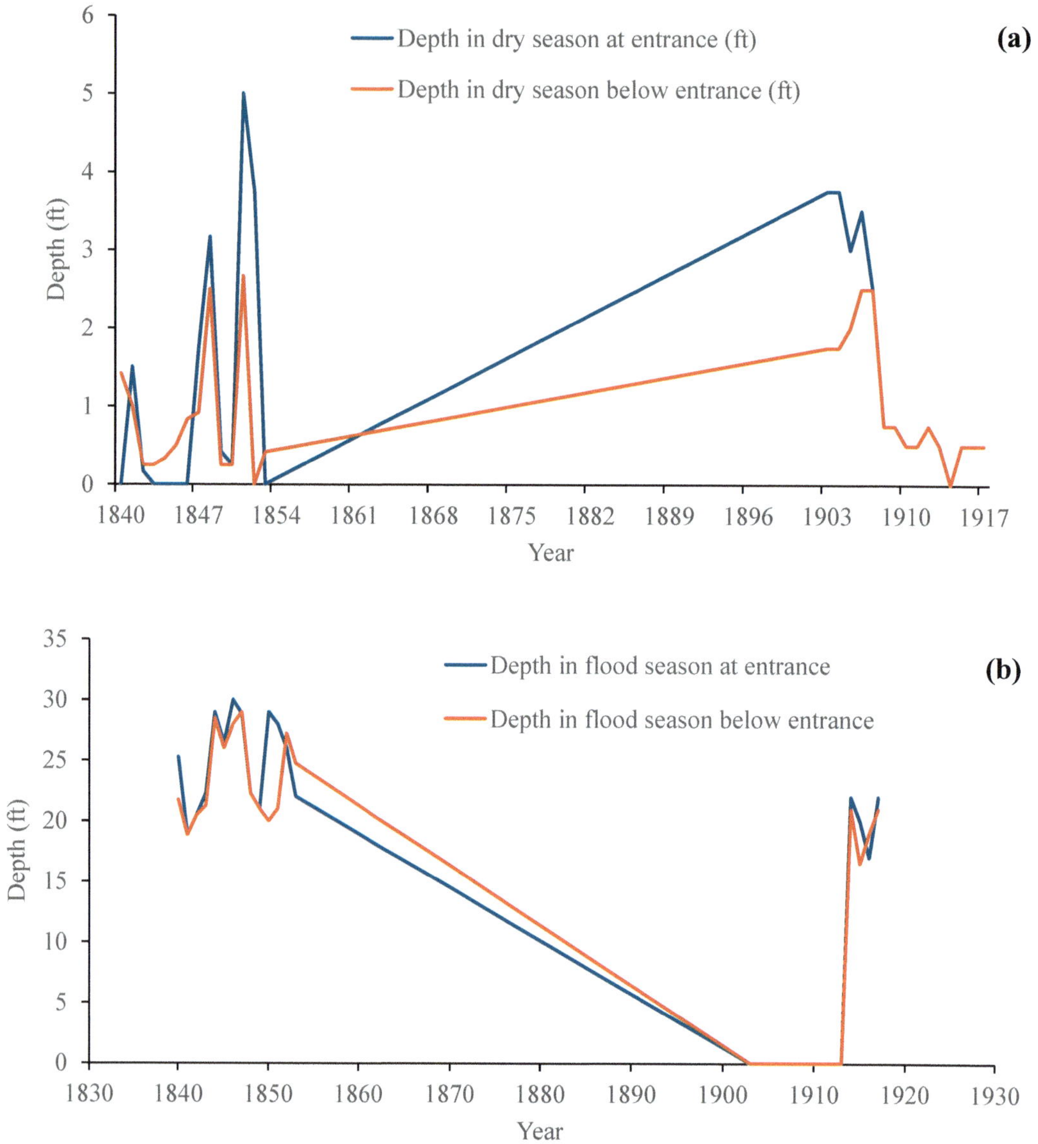

Fig. 3.26 Depths in Nadia River during the dry season and flood months (Source: 1840–53 based on Major Lang; and 1903–17 based on Reaks 1919 p. 104).

dynamic changes, the stable yet lower depth below the entrance might indicate a lower-energy environment where sediment accumulates over time, limiting the depth during the dry season. This pattern highlights the differing hydrological and sedimentary conditions across these two locations.

In light of these observations, further investigations into the historical and environmental context are necessary to understand the underlying causes of depth changes. Comparing this data with modern trends could offer valuable insights into how these patterns have evolved. Moreover, implementing sediment control measures and establishing monitoring systems could help address the challenges highlighted by this data and ensure the sustainable management of the river Jalangi.

The **flood season** depth at its entrance and below the entrance highlights significant trends over time. In the early years (1840–1850s), the depth during the flood season is relatively high at both the entrance (blue line) and below the entrance (orange line), with values fluctuating between 20 and 30 feet. Both locations display a similar pattern, indicating that the upstream and downstream sections of the river were well-connected hydraulically, with minimal obstructions to water flow. From 1854 to 1913, data are unavailable.

Between 1914 and 1917, the depths at both locations has been declined. This reduction suggests a near-complete loss of navigability and hydraulic function during the flood season. Heavy sedimentation and channel blockage along with the decline of in-flow from the feeder river are responsible for this. The consistent decline at both the entrance and below the entrance underscores the systemic nature of these changes, likely affecting the entire river system.

The implications of these trends are significant. The sharp decline in depth during the flood season over the late 19th and early twentieth centuries would have had severe ecological and socio-economic consequences. Reduced flood-season depths diminished the river's ability to recharge floodplains, support agriculture, and maintain aquatic ecosystems. For human use, these changes have restricted navigation and reduced the availability of water resources for irrigation and domestic needs during critical periods.

3.11.2 Discharge

There was a significant decline in the discharge of the river Jalangi from 1890 to 2009 (Fig. 3.27 and Table 3.8). In 1890, the discharge was a robust 100,000 cusecs, reflecting a healthy and dynamic river system. However, by 1905, this dropped by more than half to 46,336 cusecs and continued to decline steadily, reaching 31,000 cusecs by 1919. The late twentieth century saw a further sharp decline, with discharge dropping to 16,895 cusecs by 1987. While there were minor fluctuations in the following years, the discharge remained low, peaking at only 18,386.19 cusecs in 2009—an 82% reduction from 1890 levels.

This consistent decline is attributed to primarily water diversion upstream, other human activities such as deforestation, agricultural expansion. Sedimentation likely reduced the river's carrying capacity, while climate change and reduced rainfall patterns in Ganga basin (Swarnkar et al. 2021) may have further exacerbated the problem. The implications of this trend are severe, including reduced availability of water for agriculture, ecological degradation, and increased vulnerability to flooding due to channel silting. The loss of discharge also undermines the river's role in supporting livelihoods and maintaining biodiversity.

Addressing these issues requires integrated watershed management, sustainable water use, and targeted restoration of river ecosystems. Long-term monitoring of discharge and hydrological patterns is essential to implement effective solutions for the sustainable management of the Jalangi River.

3.11.3 Historical Pattern and Trend in Discharge

Figure 3.27 illustrates the average monthly discharge of the river Jalangi over three years (1915–1917), estimated using Kutter's Formula and the Float Method (Fig. 3.27 and Table 3.8). It captures seasonal fluctuations while also revealing significant year-to-year variations in discharge levels. The seasonal fluctuation aligns with the monsoon-dominated hydrology of the region, where heavy rainfall during the wet season causes high discharge, followed by a rapid reduction as rains cease.

The most striking aspect of the data is the year-to-year variation in discharge, particularly during the peak monsoon months. In **1915** discharge levels during the peak season were significantly higher than in the subsequent years. Both Kutter's Formula and the Float Method indicate peak values approaching 75,000 cusecs in August. This suggests that 1915 may have experienced a strong monsoon or favourable

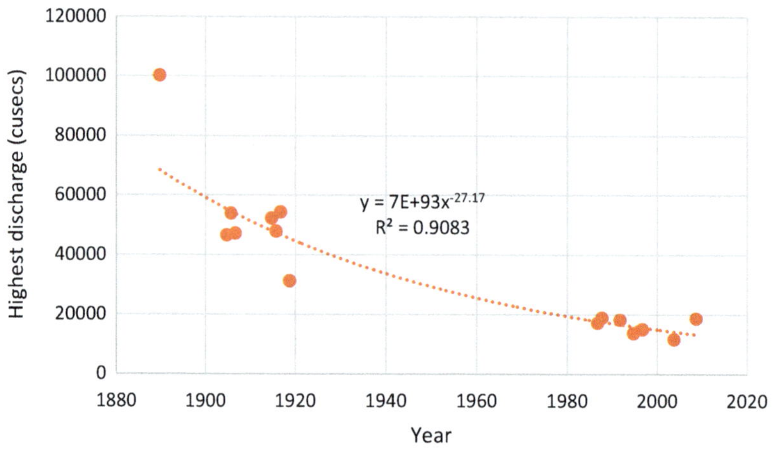

Fig. 3.27 The declining trend of the highest discharge of the river Jalangi indicates its definite fate unless a hydrogeomorphological miracle is happened in the system

Table 3.8 The declining trend in the discharge of the river Jalangi during the late 19th and early twentieth century

Year	Discharge (cusecs)	Source of information
1890	100,000	Reaks (1919)
1905	46,336	Stevension-Moor et al. (1919)
1906	53,664	Do
1907	46,946	Do
1915	52,000	Do
1916	47,640	Do
1917	54,100	Do
1919	31,000	Reaks (1919, p. 77)
1987	16,895	IWWD
1988	18,652	Do
1992	17,938.05	Do
1995	13,396.65	Do
1997	14,705.41	Do
2004	11,371.4	Do
2009	18,386.19	Do

connection with the feeder river Padma. In **1916** a noticeable decline in peak discharge is observed, with peak values closer to 63,000 cusecs. This reduction could be attributed to weaker monsoon rains or poor connection with river Padma. In **1917** the discharge levels continued to decline, with peak monsoon values dropping further to 62,400 cusecs (Fig. 3.28). The consistent downward trend over these three years (Tables 3.9 and 3.10) raises concerns about progressive changes in the river's hydraulics, possibly due to linkage issues with river Padma, increasing sediment load and higher thalweg level at the offtake of the river Jalangi & Bhairab than the water level of Padma, or climatic shifts affecting rainfall patterns. The declining trend from 1915 to 1917 is a cause for concern, as it indicates long-term degradation of the river's flow, as with the years 2021 and 2022, peak discharges reached only 56 and 45 cumecs, respectively.

3.12 Ecology and Environment

3.12.1 Ecological Water Quality Assessment

Spatial and seasonal variations in ecological risks along the Jalangi River were studied by collecting 34 water samples at ~5 km intervals from the river's source to its mouth (Islam et al. 2024a, b) and reviewing data from 119 additional samples (2012–2022). The findings reveal significant spatial and seasonal trends in water quality.

The organic pollution index (OPI) averaged 34, indicating severe contamination, with values

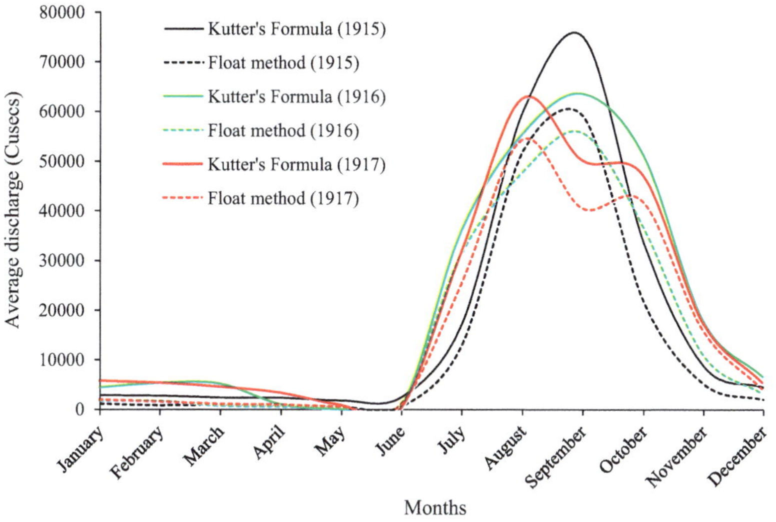

Fig. 3.28 Monthly discharge (measured using Kutter's formula & float method) pattern during the period from 1915 to 1917 (Source: Hirst 1917)

ranging from 7.17 to 588 (Fig. 3.29). OPI shows a positive trend towards downstream, with lower values in the middle section, while seasonal trends indicated a decline throughout the year. A strong positive correlation ($r = 0.99$) was observed between the organic pollution index (OIP) and chemical oxygen demand (COD), while dissolved oxygen (DO) showed a weak negative correlation ($r = -0.05$). The eutrophication index (EI) consistently exceeded the threshold of 1, highlighting severe eutrophication. EI increased with distance from the source, and seasonal patterns revealed middle-course peaks during pre-monsoon and monsoon periods, while post-monsoon values increased downstream. Strong correlations between EI, dissolved inorganic phosphate (DIP) ($r = 0.93$), and dissolved inorganic nitrate (DIN) ($r = 0.90$) underscored high oxygen demand for decomposition, leading to DO deficits for aquatic life.

Spatially, OIP ranged from 2.67 to 3.91, averaging 3.14 ± 0.29, classifying all locations as slightly polluted category (C3). Higher OIP concentrations were noted at sample points 10 and 15, while lower values were observed at points 5, 9, 24, and 34. Temporally, during pre-monsoon, OIP ranged from 2.53 to 3.18 (average: 2.8 ± 0.17), showing no clear trend. In the monsoon season, OIP varied from 2.42 to 2.95 (average: 2.69 ± 0.13), with a slight upward trend. Post-monsoon OIP ranged from 2.43 to 3.14 (average: 2.72 ± 0.15), also showing a weak upward trend. All seasons fell within the category (C3), indicating slightly polluted requires filtration and disinfection (Sargaonkar and Deshpande 2003).

3.12.2 Irrigation Water Quality Assessment

River water plays a vital role in irrigation, making its evaluation essential for sustainable crop production and irrigation development (Hoque et al. 2022). This section analyses the Jalangi's irrigation suitability using parameters such as Sodicity (SAR, %Na), Alkalinity (RSC), Permeability Index (PI), Potential Salinity (PS), and Magnesium Hazard (MH) (Fig. 3.30).

3.12.2.1 Sodicity Hazard

Sodium Adsorption Ratio (SAR) is a key indicator of soil sodium hazards and irrigation water suitability, essential for managing sodium-affected soils (Singh et al. 2023). High SAR levels in irrigation water impair soil permeability

Table 3.9 Monthly discharge (in cusecs) of the river Jalangi from 1915–1917

Year	1915				1916				1917			
Methods of estimate*	KF		OF		KF		OF		KF		OF	
	MAX	MIN	MAX	MIN	MAX	MIN	MAX	MIN	MAX	MIN	MAX	MIN
January	2910	289	1170	116	4410	437.5	1720	170.5	630	82.5	1900	83.3
February	2780	249	825	74	5300	491.8	1650	153.1	715	66.3	1510	53.8
March	2410	240	970	93	5170	512.8	680	67.5	550	54.5	1000	47.6
April	2370	227.5	710	68	1000	96	530	50.9	520	49.9	900	41.3
May	1820	184.5	520	51.5	0	0	370	36.7	500	49.6	520	41.3
June	2510	421	600	57.5	500	50.9	1040	100.3	1060	101.7	810	75.8
July	17,140	1700	12,840	1274	35,910	3562	31,260	3101	5250	5208	24,800	658.7
August	60,000	5952	52,000	5158.5	55,410	5490.6	47,640	4726	6780	672.5	54,100	877.9
September	75,000	7200	59,000	5664	63,440	6090	55,570	5335	8080	777.6	40,300	1017.6
October	33,400	3313	21,600	2142.5	51,110	5069	36,540	3625	5650	6597	41,600	813.2
November	8670	832	5100	489.5	17,610	1090.6	10,800	1036.8	2460	236.2	15,700	382.6
December	4500	446.5	2000	198.5	6440	638.8	3000	297.6	710	70.43	3900	128.5
Average discharge in cusecs for whole year	17,792		13,111		20,502		15,900		18,551		15,586	
Average discharge 1st June to 30th November	32,787		25,190		37,330		30,475		34,633		29,552	
Total discharge in millions of cubic yards for whole year	20,874.5		15,387		23,795		18,473		22,798		18,355	
Discharge in millions cubic yards from 1st June to 30th November	17,395		14,786		21,908		17,921		20,398		17,407	
Maximum discharge of any day in cusecs	84,913		64,914		68,248		65,747		69,798		71,777	

Total discharges from the river in 1915, 1916 and 1917 were 63,775 million cubic yards by KF and 48,533 million cubic yards by OF; 68,798 million cubic yards by KF and 60,626 million cubic yards by OF; and 61,556 million cubic yards by KF and 56,734 million cubic yards by OF, respectively.* Kutter's Formula (KF), Float Formula (OF). *Source* Hirst (1917)

and salinity, negatively impacting crop health and yield. In Jalangi water, SAR values ranged from 0.35 to 0.64 (mean = 0.47) with high variability (Islam et al. 2024b). Based on SAR classification, all water samples (100%) fall into the "excellent" category (SAR < 10), indicating no sodicity risk for agriculture (Islam et al. 2024b). Temporal variations (2012–2022) show SAR values ranging from 0.11 to 0.99 (mean = 0.39, CV = 33.07), while seasonal variations reveal slightly fluctuating values across pre-monsoon, monsoon, and post-monsoon seasons, all within the "excellent" category. The U.S. Salinity Laboratory (USSL) diagram confirms all water samples exhibit low sodium hazard (S1) across spatial and seasonal dimensions.

Percentage of sodium (Na), another critical parameter for sodicity hazard and irrigation water suitability, ranged from 10.16 to 20.81 (mean = 15.09, CV = 19.06). Classification

Table 3.10 Important events and discharge history of the river Jalangi

Time	Discharge	Events	Source of information
1772		"The Cossimbazar (Bhagirathi) river is almost dry from October to May; and the Jellinghy, though a stream runs in it the whole year, is in some years unnavigable during two or three months of the dry season…."	Reaks (1919)
1772		Jalangi was the only regularly navigable river and the Bhagirathi was navigable only during flood season	Reaks (1919, p. 47)
1797		The Bhagirathi and Jalangi are not navigable throughout during dry season	Colebrooke (1797, p. 47)
1825		Depth at entrance was > 2feet	Reaks (1919, p. 50)
1825		The river was closed by December during dry months from 1824 to 1847 except 1825, 1829, 1831, and 1834, the river was kept open for small boats throughout dry season	Reaks (1919, p. 86)
1831		Mr. May, superintending engineer of Nadia Rivers division proposed straightening of Jalangi by number of bend cuts and thereby increasing slope	Reaks (1919, p. 56)
1840		Depth 2.75 feet at the entrance	Reaks (1919, p. 51)
1840–1850		Jalangi & Mathabhanga had common entrance	Reaks (1919, p. 55)
1846		Depth at the entrance deepened more	Reaks (1919, p. 51)
1850		Mathabhanga got its own entrance, separated from that of Jalangi entrance	Reaks (1919, p. 46)
1850		Jalangi had two entrances from Ganga	Reaks (1919, p. 51)
1853		Large sand bars in Ganga blocked the entrance of the river Jalangi	Reaks (1919, p. 51)
1858–66		The river was main line steamer rout from Calcutta to Goalundo or Assam	Captain Brame cited in Reaks (1919, p. 87)
1874		Opening of New Bhairab River at Akriganj to fed Jalangi	Reaks (1919, p. 55)
1880		Bhairab became main entrance of Jalangi	Reaks (1919, p. 77)
1890	100,000 cusecs	Maximum discharge at Panditpur	Reaks (1919, p. 77)
1891–96		The river was again main line steamer route from Calcutta to Goalundo or Assam	Captain Brame, cited in Reaks (1919, p. 87)
1902–07		During dry months, depth ranged from 1.75ft to 2.5ft	Reaks (1919, p. 86)
1907		Navigable for small crafts. The upper reach of the Jalangi, above Karimpur, is believed to have historically branched off from the Ganges river near Karimpur. However, this channel has become completely silted over a stretch of approximately eight miles. The riverbed consists primarily of a chain of disconnected pools	Reaks (1919, p. 88) Inglis et al. (1907)
1913		Six year back (1919–6), a neck cut-off of a loop below Patkabari created Sardanga oxbow lake, Tehatta-I	Reaks (1919, p. 56)
1913–17		Entrance depth at Akhriganj -3.5 to 5 ft	Reaks (1919, p. 87)
1914	85,000 cusecs	Maximum discharge at Akhriganj	Reaks (1919, p. 77)
1915	78,000 cusecs	Maximum discharge at Akhriganj	Reaks (1919, p. 77)
1915	84,900 cusecs	Maximum discharge at Panditpur	Reaks (1919, p. 77)

(continued)

Table 3.10 (continued)

Time	Discharge	Events	Source of information
1915		The steamer transport services by Indian steam companies had been abandoned & pilots withdrawn	Reaks (1919, p. 87)
1916	79,000 cusecs	Maximum discharge at Akhriganj	Reaks (1919, p. 77)
1916–17		13 important shoals with a length of 21,485 feet were dredged	Reaks (1919, p. 77)
1917		The mean slope from Akhriganj to Swarupganj is 2.7 inches/mile	Reaks (1919, p. 77)
1919		Jalangi was opened to the Ganga at Bausmari and Mathabhanga was taking off from the Jalangi entrance	Reaks (1919, p. 46)
1919	31,000 cusecs	Average rainy season discharge at Panditpur	Reaks (1919, p. 77)

reveals 91.18% of water samples as "excellent" (<20) and 8.82% as "good" (20–40), indicating suitability for irrigation. Temporal % Na variations (2012–2022) ranged from 5.56 to 30.30 (mean = 14.75, CV = 29.82), while seasonal variations showed higher % Na during the monsoon. Across pre-monsoon, monsoon, and post-monsoon, 92.31%, 80%, and 90% of water samples were classified as "excellent," respectively. Wilcox diagram analysis further confirms that all Jalangi River water samples are "excellent" for irrigation across spatial and temporal scales.

3.12.2.2 Alkalinity Hazard

Residual sodium carbonate (RSC) is a key parameter for assessing water alkalinity hazards that can hinder crop growth by restricting root water supply. RSC values in the study ranged from −3.54 to 36.39 meq/l (mean = 11.47 meq/l) with high spatial variability (CV = 54.40). Classification based on irrigation suitability shows 94.12% of samples as unsuitable (RSC > 2.5), while only 2.94% were good (RSC < 1.25) or doubtful (1.25 < RSC < 2.5). Temporal analysis (2012–2022) revealed RSC values from 1.45 to 14.82 meq/l (mean = 9.01 meq/l, CV = 32.07). Seasonal variation showed the highest RSC in pre-monsoon (mean = 10.77 meq/l, CV = 17.75) and the lowest in monsoon (mean = 6.88 meq/l, CV = 28.87). Post-monsoon values averaged 9.43 meq/l (CV = 33.26). Alkalinity hazards were highest in pre-monsoon and lowest in monsoon, but irrigation unsuitability persisted across all seasons, with only 2.5% of post-monsoon water classified as doubtful.

3.12.2.3 Permeability Hazards

Permeability Index (PI) is a key parameter for assessing irrigation water quality. This study calculated PI values to evaluate the Jalangi River's suitability for irrigation. PI values ranged from 27.81% to 88.51% (mean: 62.12%) with low variability. They are classified as good (<80), moderate (80–100), and unsuitable (>100). Results showed 91.18% of samples were good and 8.82% moderate.

From 2012 to 2022, PI values ranged from 55.19 to 104.46 (mean: 71.46, CV: 15.54). Seasonal averages were 65.95 (pre-monsoon), 79.30 (monsoon), and 68.98 (post-monsoon) with higher permeability hazard during monsoon. Categorical analysis revealed 94.87%, 47.5%, and 82.5% of water samples were good for irrigation in pre-monsoon, monsoon, and post-monsoon seasons, respectively, with only 2.5% classified as unsuitable in pre-monsoon. Overall, water posed minimal permeability hazards in pre- and post-monsoon seasons.

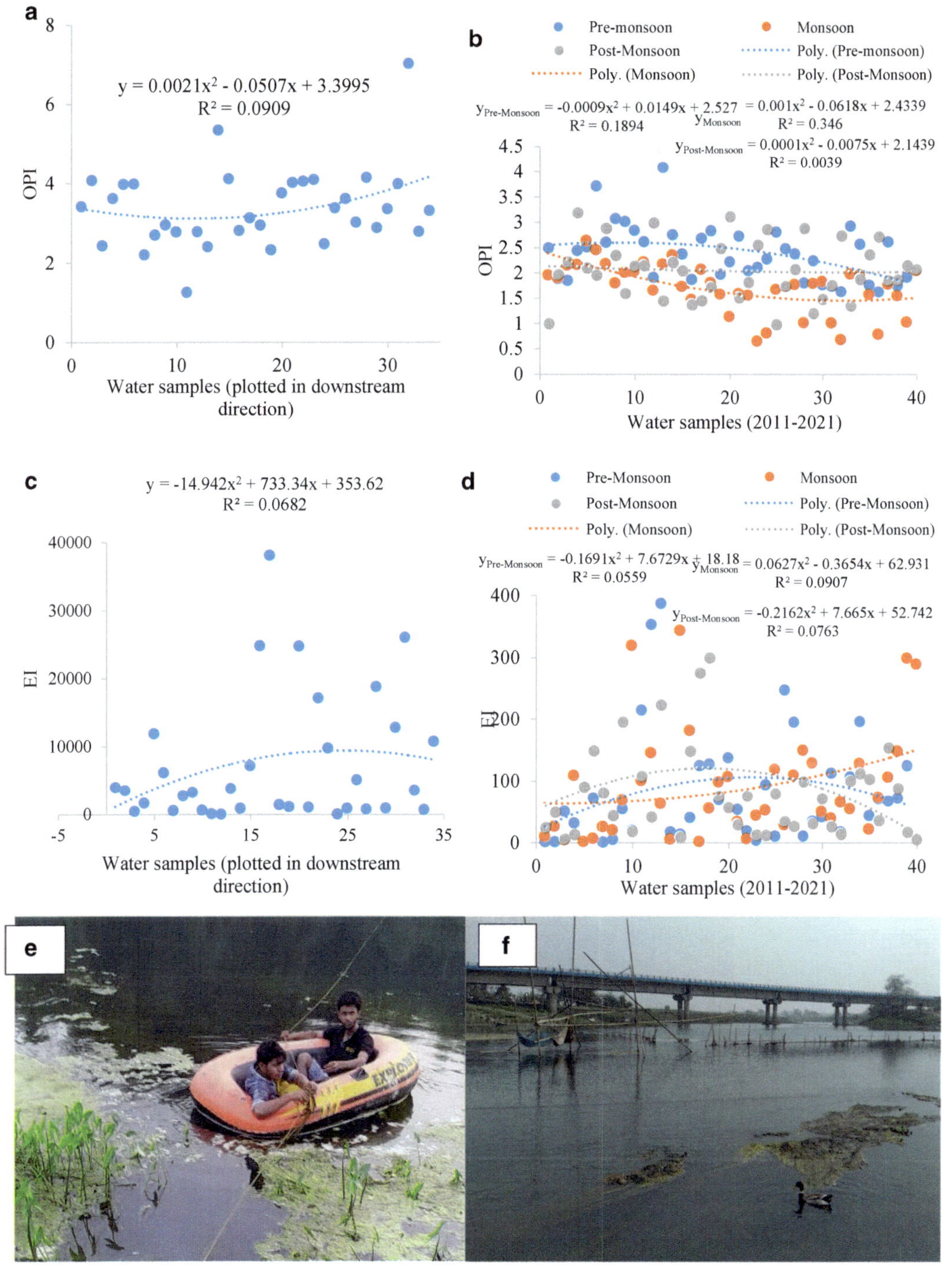

Fig. 3.29 Spatial and temporal variations in ecological risk **a** Spatial variation of OPI, **b** Seasonal variation of OPI **c** Spatial variation of EI, **d** Seasonal variation of EI, **e** & **f** Field photographs showing the nature of water pollution and eutrophication (Source: Islam et al. 2024a).

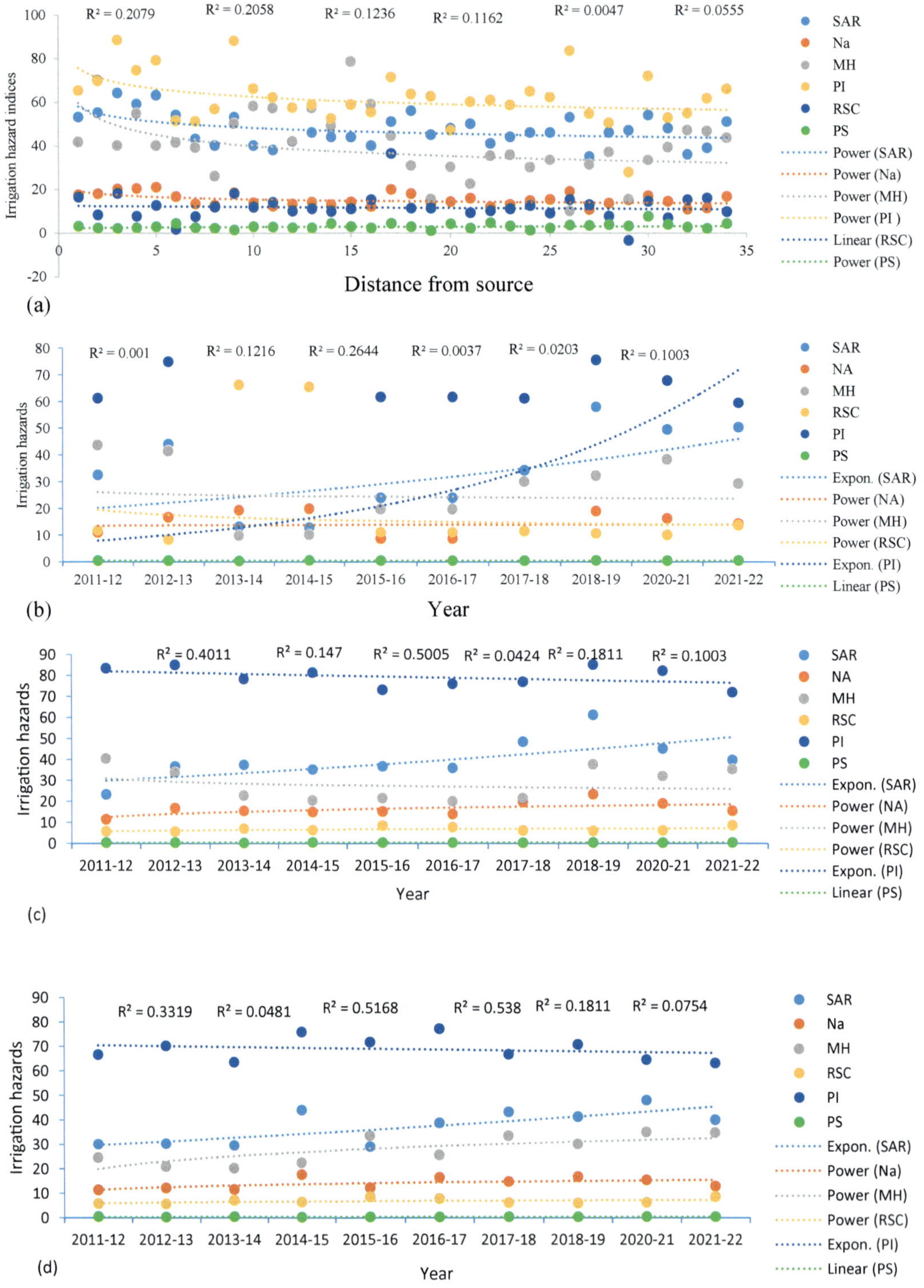

Fig. 3.30 Spatio-temporal variation of the irrigation hazards **a** variation from the source to the mouth of the Jalangi River, **b** Pre-monsoon variation, **c** Monsoon variation, **d** Post-monsoon variation (Source: Islam et al. 2024b).

3.12.2.4 Salinity Hazard

Potential salinity (PS) serves as a key indicator of irrigation water quality. PS values in the study ranged from 1 to 7.58, with a mean of 2.97 and high variability (CV = 42.07). Of the water samples, 58.82% were classified as excellent to good, 38.24% as good to injurious, and 2.94% as injurious to unsatisfactory (Islam et al. 2024b). Temporal analysis (2012–2022) showed PS values ranging from 0.15 to 0.73 meq/L (mean = 0.41, CV = 30.12). Seasonal PS variations included pre-monsoon (0.25–0.61 meq/L, mean = 0.46, CV = 18.09), monsoon (0.15–0.73 meq/L, mean = 0.38, CV = 41.10), and post-monsoon (0.22–0.60 meq/L, mean = 0.39, CV = 26.74).

Pre-monsoon had the highest salinity hazard, followed by post-monsoon and monsoon. Despite these variations, PS values categorized the water as excellent to good and free from salinity hazards for agriculture year-round. The US Salinity Laboratory (USSL) diagram classified most water samples as medium salinity hazard (C2), though monsoon samples fell between low (C1) and medium salinity (C2). These results suggest that Jalangi River water poses a medium salinity hazard for irrigation, with lower risks during the monsoon season.

3.12.2.5 Magnesium Hazard

Magnesium hazard (MH) is an important water quality parameter for evaluating irrigation water suitability. High magnesium levels in irrigation water can harm soil structure, reduce infiltration rates, and increase alkalinity due to interactions with clay particles. An MH value below 50 indicates safe water, while values above 50 are unsafe.

Jalangi River's MH values ranged from 10.16 to 78.66 (mean: 40.98, CV: 36.29), with 76.47% of samples deemed safe and 23.53% unsafe. Temporal analysis (2012–2022) showed MH values between 6.90 and 60.39 (mean: 28.85, CV: 34.77), varying across seasons: pre-monsoon (6.90–47.95, mean: 29.45, CV: 39.12), monsoon (16.39–48.70, mean: 29.57, CV: 32.41), and post-monsoon (14.15–60.39, mean: 27.54, CV: 32.61). The highest MH occurred during the monsoon, followed by pre-monsoon and post-monsoon seasons.

Despite some seasonal variations, the results show water is generally suitable for irrigation, with only 5% of post-monsoon samples classified as unsuitable. Overall, water in the study area is largely free from magnesium hazards across all seasons.

3.12.3 Drinking Water Quality and Human Health Risks Analysis

3.12.3.1 Drinking Water Quality Through WQI

The water quality was evaluated for drinking purposes using the weighted arithmetic Water Quality Index (WQI) method (Brown et al. 1972), yielding calculated values ranging from 6.23 to 140.83, with a mean of 92.29 (Fig. 3.31). These values span from excellent to unsuitable for drinking across different spatial locations. Categorical classification of water quality indicated that 32.35% were unsuitable, 47.06% very poor, 14.71% poor, 2.94% good, and 2.94% excellent for drinking. Furthermore, temporal assessment revealed varying quality: from 55.39 to 130.78 with a mean of 84.57 (poor to unsuitable) in pre-monsoon, 53.05 to 99.44 with a mean of 74.30 (poor to very poor) in monsoon, and 49.65 to 91.56 with a mean of 71.15 (good to very poor) in post-monsoon. Categorical analysis showed that in pre-monsoon, 28.21% were poor, 51.28% very poor, and 20.51% unsuitable; in monsoon, 60% poor and 40% very poor; and in post-monsoon, 2.5% good, 62.5% poor, and 35% very poor for drinking. These findings suggest the water quality of the Jalangi River is not suitable for drinking. Spatially, water quality worsened downstream, indicating poorer quality in the upward direction. Additionally, WQI values peaked during the pre-monsoon season compared to monsoon and post-monsoon seasons (Hoque et al. 2023).

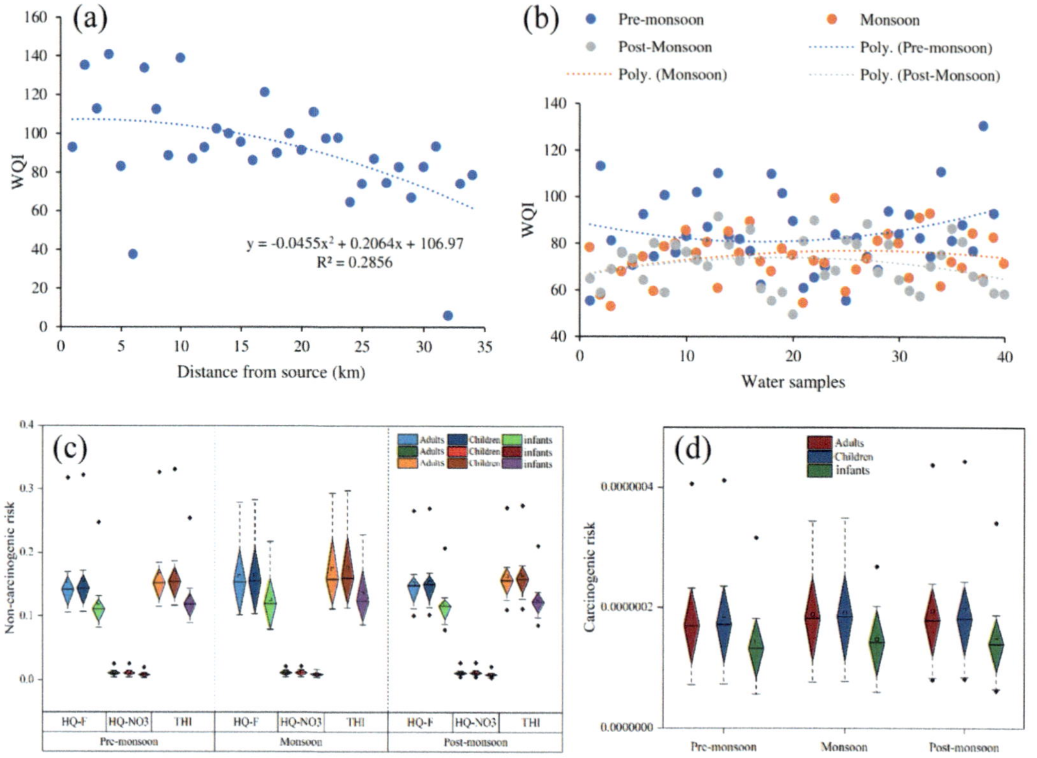

Fig. 3.31 Drinking water quality and human health risk analysis, (**a**) spatial variation and declining trend of WQI with distance downstream, (**b**) Seasonal variation of WQI in pre-monsoon, monsoon and post-monsoon, (**c**) Box plot showing the non-carcinogenic risk of adults, children and infants in pre-monsoon, monsoon and post-monsoon seasons, and (**d**) box plot illustrating the seasonal variation in carcinogenic risk associated with water consumption for adults, children, and infants (Source: Hoque et al. 2023).

3.12.3.2 Non-carcinogenic Risk Assessment

This section illuminates non-carcinogenic health risks from fluoride (F^-) and nitrate ($NO-_3$) contamination, a significant issue in rural areas. Sunitha et al. (2021) reported elevated F^- and $NO-_3$ levels in South Indian districts' drinking water, leading to illnesses. Using the USEPA methodology, risks for adults, children, and infants were assessed via Hazard Quotient (HQ) and Total Hazard Index (THI).

Jalangi water analysis revealed that 32.35% of F-levels exceeded permissible limits, while NO^-_3 remained within safe levels. HQ values for F-ranged from 0.03 to 1.95, with 32.35% exceeding safe limits, indicating potential health risks. For $NO-_3$, HQ values were significantly lower, posing minimal risk. THI values, though generally within acceptable limits, exceeded safe thresholds in 32.35% of samples due to F-contamination (Hoque et al. 2023).

The study also examined seasonal variations in health risks (2012–2022) of Jalangi water, downstream of the Dwijendra Setu on NH-12 at Krishnagar. F-risks were highest during the monsoon and lowest post-monsoon, while NO^-_3 risks peaked post-monsoon. Mean HQ and THI values indicated higher risks from F-compared to $NO-_3$, though all seasonal THI values remained below 1, suggesting no overall non-carcinogenic risk. These findings highlight the need for continuous monitoring and mitigation, particularly in rural areas.

3.12.3.3 Carcinogenic Risk (CR)

This section evaluates the carcinogenic risk of nitrate (NO^-_3) using the methodology recommended by the U.S. Environmental Protection

Agency (USEPA) for adults, children, and infants. The risk values for adults ranged from $3.1 \times 10-^{9}$ to 2.05×10^{-6}, for children from $3.1 \times 10-^{9}$ to 2.08×10^{-6}, and for infants from $2.4 \times 10-^{9}$ to 1.6×10^{-6}. Only 5.88% of adult and child samples exceeded the permissible carcinogenic risk (CR) limit of $1 \times 10-^{6}$, while all infant samples stayed within safe levels. This indicates minimal carcinogenic risk in the water samples.

A seasonal assessment of carcinogenic risks from 2012 to 2022 at the Jalangi station (downstream of the Krishna River) showed CR values for adults, children, and infants ranging from 7.18×10^{-8} to $4.38 \times 10-^{7}$, with averages around 1.88×10^{-7} for adults, $1.92 \times 10-^{7}$ for children, and 1.47×10^{-7} for infants. Seasonal variations showed higher risks during the monsoon, but all values remained below the safe threshold of $1 \times 10-^{6}$, confirming no carcinogenic health risk from the river water (Hoque et al. 2023).

3.12.4 Linking River Decay, Pollution, and Water Quality

Flowing rivers have a profound impact on ecosystems, ecology, and human life. These lotic systems provide essential resources such as nutrient rich silt for agricultural land, clean drinking water, fish, energy, irrigation, industry, and biodiversity benefits, while also supporting cultural and aesthetic values. Anthropogenic activities (Fig. 3.32) along with natural processes have significantly altered water quality and channel geomorphology of the river Jalangi. Water quality parameters like BOD, COD, and DO highlight the ecological risks faced by aquatic ecosystems, which can impair the natural functioning of aquatic life. The concentration of BOD and COD is inversely related to the concentration of DO in the water (Dobbins 1964; Li et al. 2018), meaning that higher BOD and COD values lead to lower dissolved oxygen levels. BOD represents the amount of oxygen needed by microorganisms to break down organic waste (Jouanneau et al. 2014). Higher BOD concentrations (1.0 to 5.7 mg/l) of Jalangi (Das 2023b) limit microbial respiration and metabolism, leading to the death of aquatic plants and fish, as well as disruptions to ecosystems (Deswati et al. 2022). COD measures the amount of oxygen that can be consumed through reactions in a given solution (Kosseva 2012), with higher COD concentrations indicating more polluted water (Li and Liu 2018). Thus, elevated COD levels of Jalangi water (Das 2024b) correspond to a greater presence of oxidizable organic material, reducing DO levels, which in turn promotes anaerobic conditions harmful to aquatic life (Liu et al. 2021). During field investigations of the Jalangi River, significant eutrophication and frequent fish kills were observed.

To dilute pollutants, rivers require a minimum flow. Research indicates that algal blooms rarely occur when flow velocities exceed 0.1–0.2 m/s (Escartín and Aubrey 1995), with minimal flow velocity for good hydrological connectivity set at 0.1 to 0.2 m/s (Yin et al. 2018). In the case of the Jalangi River, inadequate discharge is crucial in determining its physical habitat and the taxonomical and functional structure of its community. The Jalangi River faces similar discharge issues as other distributaries of the Ganga River. The river has struggled with discharge problems for a long time (Das 2015). One of the main contributors to the river's decline is the closure of its off-take from the Padma River, which feeds into the Jalangi. This problem is mainly driven by the eastward tilting of the Bengal basin, which caused the disconnection of South Bengal's rivers from the Ganga-Padma system (Allison et al. 2003; Goodbred and Kuehl 2000; Sarkar et al. 2021). The off-take of the Jalangi River has been detached and become silt-blocked due to this movement. Consequently, river discharge is primarily controlled by base flow (Das 2015). Therefore, low river flows contribute to natural variations in water quality, such as increased temperature and reduced dissolved oxygen levels from sustained base flows (Nilsson and Renöfält 2008). Native species may survive in such conditions, especially if there is

Fig. 3.32 Idol immersion and consequence at Kadamtala Ghat, Krishnagar

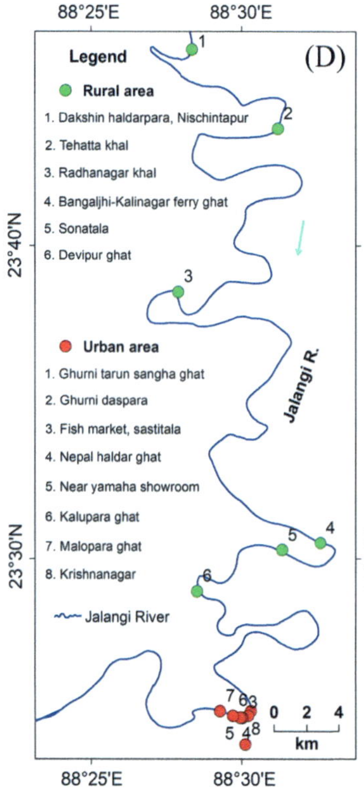

Fig. 3.33 Nallah (canal) untreated discharge (red locations are urban effluent points mainly from Krishnagar Municipality and green locations denote the rural ghats (bathing place) and effluent points

considerable habitat variability. Additionally, growing human activities, including channel narrowing, flow blockages, and the mixing of urban effluents (Fig. 3.33), have worsened these issues.

3.13 Riverine Society, Economy and Culture

3.13.1 Agriculture

The Jalangi River basin, located in the deltaic region of West Bengal, is a highly fertile agricultural belt that supports the cultivation of a wide variety of crops, including rice, jute, pulses, oilseeds, vegetables, fruits, flowers and more (Fig. 3.34). The rich alluvial soil and favourable climatic conditions make this area a vital hub for agricultural productivity. However, water scarcity, particularly in the upper reaches of the river, poses a significant challenge to sustainable farming practices. While the water quality of the Jalangi River is generally suitable for irrigation, concerns about alkalinity, as well as potential carcinogenic and non-carcinogenic hazards, highlight the need for careful water

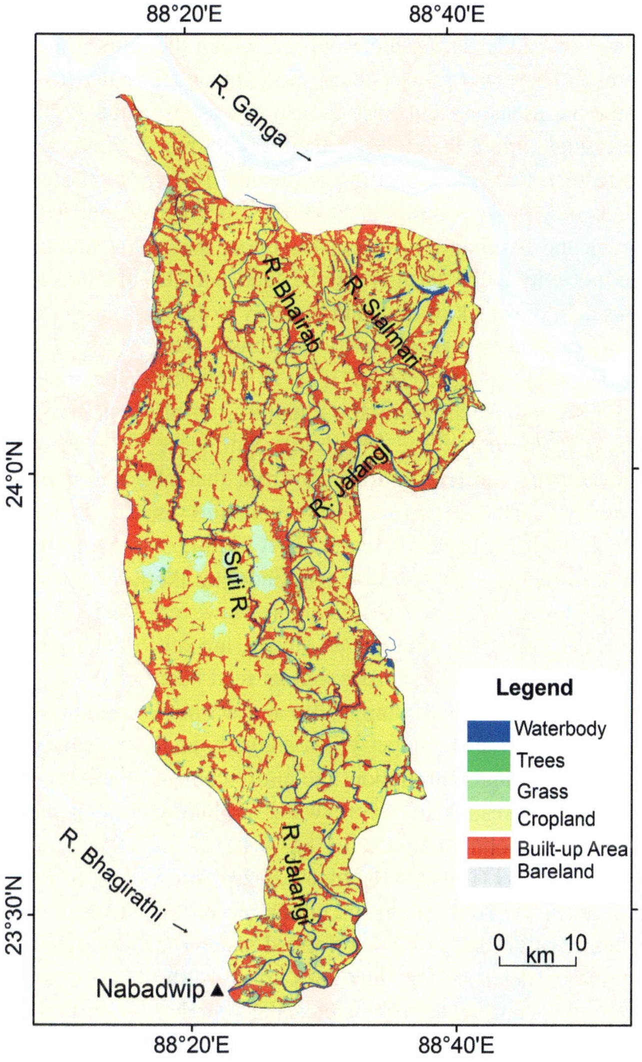

Fig. 3.34 Land use and cover of the Jalangi River basin

management and monitoring to ensure the long-term health of both crops and farmers (Islam et al. 2024a, b). Addressing these issues is crucial for maintaining the region's agricultural output and livelihoods. Yet the volume of water in the river, rather than its quality, is the primary concern.

The stretch of the river, divided into two reaches—the first stretch from the closed off-take at Gopalpur-Char Madhubona to Moktarpur, and the second stretch from Char Moktarpur to the confluence at Swarupganj—exerts different effects on agriculture and irrigation along its banks. From the off-take to the Bhairab confluence at Char Moktarpur, the river is virtually lifeless. It has transformed into stagnant swamps in some areas and almost flat terrain in others. From 2.5 km upstream to 3 km downstream at Karimpur, the river resembles a swamp or marsh. At the confluence of the Sialmari River at Kupila, opposite Jayrampur, the river has become another swamp. These depressed areas are utilized for rotting and washing jute fibres during the rainy season. Parts of the riverbed have been divided into personal ponds in this stretch. Apart from these depressions, the total length (45.5 km) of this stretch has been converted into cultivable land for villagers along the banks. They cultivate Ravi crops during the winter and paddy during the monsoon within the riverbed. They also grow jute on the banks during the pre-monsoon and early-monsoon seasons. Interestingly, more than 60% of them do not recognize that they are cultivating within the riverbed, as they possess deeds or land titles for their plots in the riverbed. However, this upper reach does not enrich the fertility of the lands on its banks due to its lack of flow and inability to carry silt. Consequently, productivity is lower than that of the lower reach (Das 2013). The impacts of the second reach, from the Bhairab confluence at Moktarpur to the Bhagirathi confluence at Swarupganj, differ significantly from that of the upper reach. This lower reach maintains its flow through the Bhairab, particularly during a few days of the monsoon months, rendering the riverbed unsuitable for crop cultivation. However, the banks, especially the convex slopes, are utilized for cultivating Ravi crops and as seed beds for Boro paddy. Jute decomposition and washing occur primarily in the upper reach annually and, if rainfall is scarce, throughout the entire course. This practice results in severe water pollution, causing fish population decline as observed in September 2009 (Das 2013). River Lift Irrigation (RLI) from the river Jalangi is almost absent in the upper reach due to its deteriorated state. The severity of deterioration decreases towards the Bhagirathi confluence with increasing number of RLI, demonstrating a strong negative correlation between deterioration degree and the number of RLI systems. Conversely, there is a positive correlation between the number of River Lifting Irrigation systems and rice yield, indicating a strong negative relationship between deterioration degree and rice yield (Das 2013).

3.13.2 Fishing

The impact of changes, particularly the deterioration of the river Jalangi, is the most notable among those whose lives revolve around the river, particularly fishermen. The decline in per capita fish catch is attributed to various factors, as identified by the fishermen themselves. The detachment of the Jalangi from the river Padma and the subsequent absence of silt-laden flow is cited as the primary cause by 36.6% of fishermen (Das 2013). Additionally, 15.6% attribute the decrease in fish catch to the reduced volume of the river, while 15.1% point to water pollution. A joint factor cited by 21.4% of fishermen is the detachment of the river Jalangi from the Padma and the concurrent decrease in fish population and river volume. A small percentage (3.5%) opines that the increase in the number of fishermen has sharply reduced their per capita catch. Furthermore, 7.4% of fishermen attribute the decrease in per capita fish catch to a combination of all the aforementioned causes.

The alarming decrease in fish population in the river Jalangi has made it nearly impossible to sustain a livelihood solely through fishing. Consequently, many fishermen have either taken up additional occupations alongside fishing or have switched to alternative livelihoods offering higher income. Despite some nostalgia expressed by older respondents, it is evident that per capita fish consumption and catch have significantly decreased compared to 30 years ago (Das 2013). Factors contributing to this decline include the multi-fold reduction in fish catch from the river Jalangi due to river deterioration, the introduction and popularity of substitute protein sources such as poultry, and the impact of population growth on reducing per capita fish consumption. A field survey conducted in Halderpara village, Panditpur, C.D. Block Krishnagar-II, revealed that fishermen's families are now purchasing fish from vendors due to the scarcity of fish in the river Jalangi. The dwindling fish population in the river Jalangi and the subsequent decline in catch are also evident in local fish markets, where the quantity of local fish and those from the river Jalangi has sharply decreased compared to 30 years ago.

3.13.3 Ferry Service

The Jalangi River, a distributary of the Ganges in West Bengal, has historically been a key route for navigation and ferry services, supporting the transportation needs of local communities. Ferries have provided an affordable and efficient means of crossing the river, fostering trade, cultural exchange, and daily commuting. However, the health of the river is increasingly threatened by unsustainable practices tied to these services (Fig. 3.35).

The Jalangi River, a crucial National Waterway No. 47 (Ministry of Jal Shakti 2023) in West Bengal, is heavily relied upon for ferry services, yet the methods employed for these services

Fig. 3.35 River-road crossing on the river Jalangi. **(a)** The permanent brick road in the river bed at Karimpur-Bakshipur ferry-ghat. The bamboo bridge on which the man is stepping towards Karimpur leaves a very narrow passage for the water pass. **(b)** The raised bamboo bridge on the right side is used during the monsoon season. (Source: Das & Bhattacharya 2020).

have led to significant environmental concerns. Across various stretches, practices like constructing earthen dams and narrowing channels with putting soil and bricks from both banks, have become common to facilitate ferry services and vehicular crossings (Table 3.11). Locations such as Char Moktarpur and Karimpur demonstrate how these temporary modifications, such as tying boats together and creating bamboo bridges, have altered the river's natural flow (Das 2013; Das and Bhattacharya 2020). These interventions have severe negative impacts on the river's health. Earthen dams, constructed to shorten ferry routes, disrupt the river's free flow, causing accelerated bank erosion and altering its course. The repeated introduction of large volumes of soil—estimated at 15,000 cubic feet annually at each ferry ghat—leads to sediment charge during the rainy season. The Jalangi, with insufficient flow energy, cannot transport all the introduced soil, resulting in cumulative sediment buildup on the riverbed and banks. This not only degrades the aquatic ecosystem but also increases the risk of flooding and disrupts the livelihoods dependent on the river. The construction of bandhs for fishing and ferry services across the river—activities that adversely affect the river's health—has long been brought to the attention of government authorities. However, official negligence has persisted since 1948, as evidenced by newspaper reports from that time (Fig. 3.S3).

At lower reaches, particularly Swarupganj-Mayapur ferry operations often contribute to pollution, with oil spills, discharge of untreated waste, and plastic litter becoming commonplace. Frequent dumping of non-biodegradable materials and the lack of proper docking facilities exacerbate the problem, leading to habitat degradation and water quality deterioration. Such practices disrupt aquatic ecosystems and harm the river's overall health, undermining its ecological and cultural significance. Sustainable measures, including stricter regulation of ferry operations and awareness campaigns, are essential to balance navigation needs with the river's preservation. Moreover, the use of bricks to permanently reduce water passages exacerbates the damage by creating artificial barriers that hinder the natural movement of water and aquatic species (Das 2024a). These practices, while aimed at facilitating transportation, prioritize convenience over sustainability, causing long-term harm to the river's ecological balance. Sustainable alternatives, such as modernized ferry systems, eco-friendly infrastructure, and stricter enforcement of environmental guidelines, are essential to mitigate these impacts and preserve the Jalangi River for future generations.

3.13.4 The Fragile Bond Between Man and Water in the River's Decline

3.13.4.1 Boatmen and the River

This is the era of the Anthropocene. Several human activities have degraded the river both its magnitude and quality. With the transformation of the Jalangi River ecosystem from lotic to lentic, both farmers and fishermen, whose livelihoods are intricately tied to the river, have altered their interactions with it. When the river surged with full force, no one dared to challenge its power by placing jute in its turbulent waters. In stagnant river water, farmers now dare to heap their jute during the monsoon months for retting purposes, leading to a decline in water quality. Consequently, the riverbed accumulates silt and organic deposits, resulting in shallower and poorer waters.

The Jalangi River, a vital lifeline for local communities, illustrates a complex and often damaging man-river interaction, particularly in its upper reaches. The government leases ferry ghats at a nominal revenue (Table 3.12) and, once leased, pays little attention to monitoring river conditions or ensuring adequate services for the people. Earthen dams across the river, built to reduce the length of river crossings, are a common feature (Das 2013; Das and Bhattacharya 2020). These dams are constructed using soil and bricks from both banks, narrowing the river's channel and significantly altering its natural flow. Boatmen tie 2–3 boats together at these narrow stretches to facilitate ferry services, even enabling heavy vehicles

Table 3.11 Location of ferry services and state of river and means of river crossing at Ferry Ghats

Reach		Length (km)	Sl. No.	Location of ferry Ghat with earthen Bāndh	Comments
From	To				
Offtake at Charmadhubona	Sialmari Confluence at Kupila	9.10		Nil	The reach is erased out and no need to make any Bāndh to cross the river
Sialmari Confluence at Kupila	Bhairab Confluence at Moktarpur	41.9	1	Jayrampur-Kupila ferry ghat	Bamboo bridge
			2	Durlabhpur-Rajapur Ferry Ghat	Earthen Bāndh & Bamboo bridge
			3	Kuchaidanga Ferry Ghat	Boat/Bamboo bridge
			4	Karimpur- Pattabuka Ferry Ghat	Boat/Bamboo bridge
			5	Karimpur-Boxipur Ferry Ghat	Channel width narrowing by earthen Bāndh & bridging the gap by boats tightened together
			6	Karimpur Shisa Kadamtala Ferry Ghat	Boat/ Bamboo bridge
			7	Baishnabpara Ferry Ghat	Bamboo bridge
			8	Sahebpara Ferry Ghat (Simultola)	Bamboo bridge
			9	Dogachi Kushaberia (Ferry) Ghat	Earthen Bāndh & Bamboo bridge
			10	Char Moktarpur Ghat	Bamboo bridge
Bhairab Confluence at Moktarpur	Suti Confluence at Bali-Tungi	42.7	11	Bhelanagar Ferry Ghat	Channel width Narrowing by earthen Bāndh & bridging the gap by boats tightened together
			12	Amtala-Fazilnagar Ferry Ghat	Wide bamboo bridge on the pier of sand-bags with the intermittent passage of water flow (Fig. 3.35b)
			13	Bonchadanga Ferry Ghat	Do
			14	Gopalnagar-Paschim Char Brindabanpur Ferry Ghat	Bamboo bridge
			15	Paranpur Ferry Ghat	Wide bamboo bridge on the pier of sand-bags with the intermittent passage of water flow
			16	Char Jagaipur ferry ghat	Wide bamboo bridge on the pier of sand-bags with the intermittent passage of water flow. A brick kiln on the right bank is near the ferry ghat
			17	Kanainagar ferry ghat	Wide bamboo bridge on the pier of sand-bags with the intermittent passage of water flow

(continued)

Table 3.11 (continued)

Reach		Length (km)	Sl. No.	Location of ferry Ghat with earthen Bāndh	Comments
From	To				
			18	Kanainagar 2nd ferry ghat	Earthen Bāndh & Bamboo bridge
			19	Uttar Char Chandpur Ferry Ghat	Wide bamboo bridge on the pier of sand-bags with the intermittent passage of water flow
			20	Patikabari Ferry Ghat downstream to Bridge	Wide bamboo bridge on the pier of sand-bags with the intermittent passage of water flow
			21	Raninagar-Patikabari Ferry Ghat	Bamboo bridge
			22	Raninagar-Tungi Ferry Ghat	Wide bamboo bridge on the pier of sand-bags with the intermittent passage of water flow
Suti Confluence	Kalma Khal Confluence at Radhanagar	55.2	23	Gopinathpur-Bali Ferry Ghat	Bamboo bridge
			24	Gopinathpur Ferry Ghat	Bamboo bridge
			25	Raghunathpur-Iswarchandrapur Ferry Ghat	Bamboo bridge
			26	Krishnachandrapur ferry ghat	Bamboo bridge
			27	Tehatta ferry ghat	Earthen Bāndh & Bamboo bridge
			28	Hanspukuria Ferry ghat	Bamboo bridge
			29	Kulgachi-Chanderghat Ferry	Bamboo bridge
			30	Nimtala-Chanderghat ferry	Do
			31	Putimari-Shibpur Ghat	Do
Kalma Khal Confluence at Radhanagar	Bhagirathi-Jalangi Confluence	84.1	32	Birpur ghat ferry	Wide bamboo bridge
			33	Bara Abdulia Ferry ghat	Bamboo bridge
			34	Gopinathpur Teghari Ferry Ghat	Narrowing of the channel by Earth filling and bamboo bridge
			35	Brittihuda Ferry Ghat	Narrowing of the channel by Earth filling and bamboo bridge
			36	Gokhurapota ferry ghat	Do
			37	Sonatala Ferry Ghat	Do

such as trucks and lorries to cross. While practical for transportation, these practices are taking a severe toll on the river's health. This cyclical process of annual soil (and bricks) addition and deposition in the river bed leads to cumulative sediment buildup, causing severe degradation of the river's health and altering its natural course. These practices highlight the unsustainable interaction between human activities and the river ecosystem. The reliance on earthen

Table 3.12 Amount of revenue collection from some of Ferry Ghats by Nadia district authority

Sl. No.	Ferry ghat	River	C. D. Block	Base price of Auction (Rs)	Comments (Memo No and date)
1	Madhpur Ferry Ghat	Bhairab	Karimpur–II	150,205	835/NZP Dt. 01.04.2024
2	Iswarchandrapur Ferry Ghat	Jalangi	Tehatta–I	226,600	838/NZP Dt. 05.03.2024
3	Gobra (Basatpur) Ferry Ghat	Gobra *BeelBeel*	Kaliganj	15,840	825/NZP Dt. 05.03.2024
4	Hazrapur-Byaspur ferry ghat	Churni	Ranaghat–II	49,500	824/NZP Dt. 05.03.2024
5	Uttar Panditpur	Jalangi	Krishnagar–II	25,740	821/NZP Dt. 05.03.2024
6	Taldaha Ferry ghat	Churni	Krishnagan	63,800	823/NZP Dt. 05.03.2024
7	Nidaya Ferry ghat	Bhagirathi	Nabadwip	90,200	836/NZP Dt. 05.03.2024
8	Gobindapur Ferry ghat	Churni	Krishnaganj	22,660	820/NZP Dt. 05.03.2024

dams and the recurring introduction of soil not only harm the river's flow dynamics but also compromise its ecological balance. Sustainable alternatives, such as regulated ferry operations, permanent infrastructure for crossings, and sediment management, are urgently needed to mitigate these impacts and preserve the Jalangi River for future generations.

3.13.4.2 Fishermen and the River

The deterioration of the Jalangi River and changes in its course have had a profound impact on the livelihoods of people who depend on it directly, particularly fishermen. Field surveys reveal that many fishermen have been forced to abandon their ancestral profession due to declining fish stocks and poor river health. The once-thriving fishing communities are now facing dwindling opportunities, leading to significant occupational shifts. Most fishermen, unable to sustain their traditional livelihood, have turned to alternative occupations such as rickshaw pulling or working as day labourers (Das 2015). In many cases, they have migrated to other states like Kerala and Maharashtra to find work, often taking up roles as waiters in hotels or construction workers. Similarly, there has been a notable increase in the percentage of individuals engaged in handloom weaving, as the younger generation moves away from fishing. The man-river interaction in this context highlights the dire consequences of environmental degradation on human occupation. The loss of livelihoods tied to the river not only disrupts traditional knowledge systems but also forces communities to adapt in ways that often lead to economic insecurity and cultural displacement.

3.13.4.3 The Functional Bond Related to Engineering Works

The emotional and functional bond between humans and rivers, as seen with the Jalangi River, has eroded over time, mirroring the river's own decline. Once a lifeline for communities, the Jalangi not only nurtured livelihoods through navigation, fishing, and irrigation but also held spiritual significance, inspiring reverence in rituals and literature. However, as the river's flow weakened and its capacity to support life diminished, so did the emotional ties of the people who depended on it. Construction projects, like bridges with broad dais at piers bases, further compromise the river's natural geomorphology, obstructing flow and aquatic ecosystems (Das 2019; Das and Bhattacharya 2020; Das et al. 2024a, b). The Jalangi's transformation from a vibrant distributary of the Padma to a stagnant, hyacinth-choked watercourse exemplifies how neglect and exploitation of rivers can fracture the deep-seated connection between man and nature. This shift underscores the urgency of preserving river health, not just for ecological balance but for maintaining the cultural and emotional ties that define communities.

3.13.4.4 The River Dynamics and Debate on the Birthplace of Sri Chaitanya

The Jalangi River, intertwined with the lives and culture of those residing along its banks, has played a pivotal role in shaping the historical and spiritual narratives of the region. One of the most significant debates rooted in the changing course of the Bhagirathi-Jalangi confluence concerns the birthplace of the revered saint and spiritual leader, Sri Chaitanya (Chaudhuri 2004). This debate—whether his birthplace lies in Nabadwip or Mayapur—emerged due to the historical channel migration and the shifting course of the Bhagirathi River from the west to the east of Nabadwip. The dynamic nature of the river system reflects the ever-changing relationship between humans and rivers, where natural processes such as channel migration have deeply influenced religious geography and human settlements. The question of Sri Chaitanya's birthplace is not merely an academic or archaeological inquiry but also a spiritual and cultural concern for millions of his followers. The shifting rivers have redefined the sacred landscape, transforming how people perceive and engage with this historically significant region. This interaction between man and river continues to shape the narrative of faith, identity, and the living heritage of the Bhagirathi-Jalangi system (Das 2023a).

3.13.4.5 Degraded River Intertwined with Culture and Emotion

Historically, the Jalangi served as a physical and spiritual source of livelihood, sustenance, and cultural identity for the villagers. The fishermen of Panditpur, for example, once worshipped the river as a goddess, symbolizing reverence for the river's role in their daily survival (Das 2013). This deep, almost familial connection to the river is evident in their practices, such as their worship rituals and the yearly Chhat Puja or Pous Parvana ceremonies, where the river served not only as a physical resource but also as a sacred entity.

However, the river's decaying condition due to natural changes and human interventions has resulted in a profound transformation of this relationship. The villagers' refusal to worship Jalangi today marks a sharp contrast to their past devotion. As the river no longer supports their livelihoods—no longer providing fish or irrigating their fields—the emotional connection has weakened. This abandonment of religious/cultural rituals signifies a loss of faith, not just in the river itself, but in the continuity of the way of life that once thrived along its banks. For the women of Karimpur, who once visited the river during Chhat Puja, and the Bangalee Hindu wives who performed sacred rituals like washing rice for Pous Parvana, the river's changed state has caused a rupture in tradition. With the river's polluted shallow waters and reduced flow, it has become a symbol of absence rather than presence. The once sacred river is now avoided, and rituals that once honored the river are no longer observed. This shift is a testament to the power of rivers not just as physical resources, but as cultural and emotional anchors in communities.

The poet Jibanananda Das's poem, আবার আসিব ফিরে (Ābāra Āsiba Phirē), beautifully captures the emotional and nostalgic longing for the river. His words reflect a deep yearning for the return to a time when the Jalangi was full of life, nurturing both the land and its people. The poem evokes a sense of return to a past that can never truly be reclaimed. In the literary works that discuss the Jalangi, the river is portrayed not only as a physical entity but as a symbol of hope, fear, and an emotional bond with the river and nature.

The contrast between the past and present relationships with the river is stark. Where the river once provided sustenance, joy, and spiritual fulfilment, it now represents loss, decay, and a changing way of life. This profound shift in the human-river emotional connection is not just a local phenomenon but a reflection of broader environmental and cultural shifts that many riverine communities around the world experience as their natural resources diminish.

3.13.5 The Eager Want for Bygone Days of the River Woven with Society

The Jalangi and Churni rivers are currently undergoing rapid deterioration, their once lively flow are now diminished to stagnation and decay. Their plight mirrors a deeper fracture between society and its rivers, driven by governmental negligence and the apathy of the very people who once revered these water bodies.

Amid this crisis, a growing movement is rising to reclaim the glory of bygone days. Sensible individuals and dedicated NGOs are striving to revive the spirit of the rivers, countering indifference with action. Through a blend of cultural, educational, and activist initiatives, they aim to rekindle the connection between society and its waterways. Festivals celebrating river-centric songs, dramas, and poetry have become platforms for spreading awareness. River walks and cycle rallies emphasize the importance of active engagement while drawing competitions and cultural rituals foster a sense of responsibility among the younger generation (https://www.facebook.com/share/g/19pCfwR6Yo/). These efforts seek not only to restore the rivers' physical health but also to revive their place in the collective consciousness of society.

One remarkable success story emerges from the Jalangi River, where the activist group 'Save Jalangi' (https://www.instagram.com/savejalangi/) brought the negligence of construction agencies to light (Fig. 3.36). The debris left behind by construction projects had severely impeded the river's flow, further endangering its fragile ecosystem. Persistent advocacy led to a critical intervention in June 2024, when KCC Bildcon Pvt. Limited, under pressure from the National Highway Authority of India (NHAI), cleared earth and concrete slabs from the riverbed and banks (Das et al. 2024b). This act of

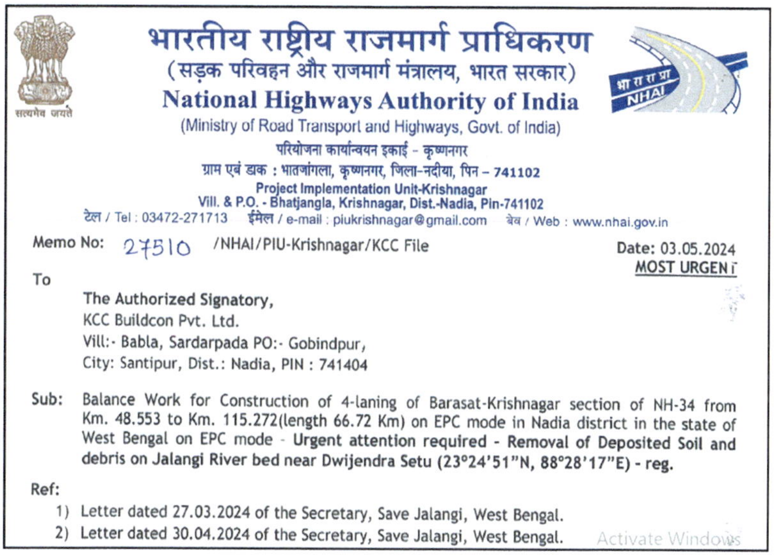

सत्यमेव जयते

भारतीय राष्ट्रीय राजमार्ग प्राधिकरण
(सड़क परिवहन और राजमार्ग मंत्रालय, भारत सरकार)
National Highways Authority of India
(Ministry of Road Transport and Highways, Govt. of India)
परियोजना कार्यान्वयन इकाई - कृष्णनगर
ग्राम एवं डाक : भातजांगला, कृष्णनगर, जिला-नदीया, पिन - 741102
Project Implementation Unit-Krishnagar
Vill. & P.O. - Bhatjangla, Krishnagar, Dist.-Nadia, Pin-741102
टेल / Tel : 03472-271713 ईमेल / e-mail : piukrishnagar@gmail.com वेब / Web : www.nhai.gov.in

Memo No: 27510 /NHAI/PIU-Krishnagar/KCC File Date: 03.05.2024
MOST URGENT

To
The Authorized Signatory,
KCC Buildcon Pvt. Ltd.
Vill:- Babla, Sardarpada PO:- Gobindpur,
City: Santipur, Dist.: Nadia, PIN : 741404

Sub: Balance Work for Construction of 4-laning of Barasat-Krishnagar section of NH-34 from Km. 48.553 to Km. 115.272(length 66.72 Km) on EPC mode in Nadia district in the state of West Bengal on EPC mode - **Urgent attention required - Removal of Deposited Soil and debris on Jalangi River bed near Dwijendra Setu (23°24'51"N, 88°28'17"E) - reg.**

Ref:
1) Letter dated 27.03.2024 of the Secretary, Save Jalangi, West Bengal.
2) Letter dated 30.04.2024 of the Secretary, Save Jalangi, West Bengal.

Fig. 3.36 Repeated petitions by Save Jalangi, West Bengal made it possible to remove soil and debris (deposited during the construction of the second highway bridge) from the river bed by the National Highways Authority of India's authorized agency restoration, though a small step, is a testament to the power of collective will and the impact of determined activism.

The struggle to rejuvenate the rivers of Nadia is not just an environmental battle; it is a cultural and existential one. These rivers, woven into the tapestry of society, carry the memories of a time when life and water were inseparably linked. The eager want for those bygone days reflects a deeper yearning—a hope to reconnect with a way of life where nature was respected and celebrated. The challenge lies not only in reversing the environmental damage but also in overcoming societal apathy and institutional inertia. It is a call to action for all to honour these rivers, not just as geographical entities, but as living legacies vital to the transformation of society.

3.14 Proposed Plan for the Dead Reach of the River and Palaeo Channels

The upper 30 km stretch of the river Jalangi (Fig. 3.37), covers an area of approximately 3 km^2 (735 acres), has fallen into a state of environmental degradation over the decades. Once a thriving waterway supporting agriculture, irrigation, and livelihoods, it now suffers from siltation, water hyacinth cover, reduced water flow, and neglect. However, this decayed condition also offers an opportunity for a transformative plan that can restore the river and empower the surrounding communities. By adopting a sustainable, multi-pronged approach, the proposed plan aims to rejuvenate the river while fostering economic growth through agriculture, fish farming, duck farming, vegetable and fruit cultivation, and rural tourism.

A central component of the plan is the utilization of the river's water to support agriculture and irrigation. Revitalizing the river would create a dependable source of water, enabling farmers to grow crops more efficiently and boosting agricultural productivity. In addition, the plan includes fish farming and duck farming, which can be developed in and around the river. These ventures are expected to generate significant income for local residents while maintaining ecological balance, making them a sustainable addition to the region's economy. Vegetable and fruit cultivation on the fertile

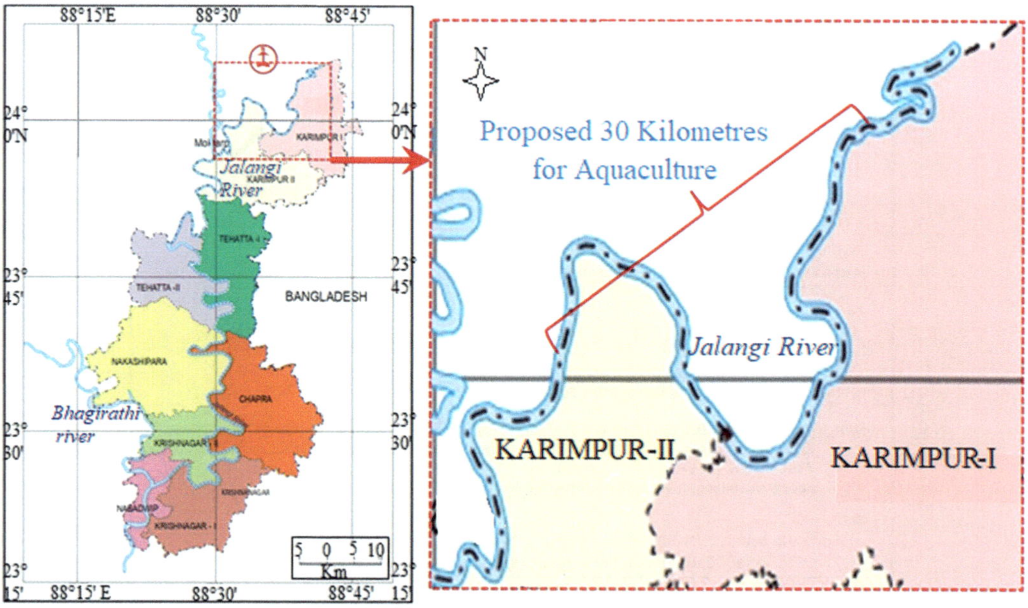

Fig. 3.37 The proposed 30 km dead stretch of the Jalangi River near Karimpur has the potential to transform from a hyacinth-covered, filthy, mosquito-breeding area into a valuable source of resources

banks of the Jalangi is another critical element of the plan. By promoting diverse agricultural practices, this initiative will ensure food security and offer farmers additional revenue streams. Furthermore, the project seeks to tap into the potential of rural tourism. The serene beauty of the river and its surroundings, combined with eco-friendly tourism activities such as boating, cultural events, and farm stays, can attract visitors and generate supplementary income for the community.

A unique feature of this plan is its emphasis on community engagement and incentives. Local residents will be involved at every stage, from planning to implementation, ensuring a sense of ownership and responsibility. Incentives tied to production and participation will motivate communities to maintain the river and contribute to the project's success. This participatory model is designed to avoid the pitfalls of previous government initiatives (e.g. Nawpara Beel), which often failed due to a lack of local involvement with accountability.

By integrating environmental restoration with economic development, the proposed plan offers a promising future for the upper Jalangi region near Karimpur. The combination of diverse income streams, sustainable practices, and rural tourism can transform the river and its surroundings into a hub of prosperity. With careful planning and execution, this initiative has the potential to restore the Jalangi as a lifeline for the community and set an example for sustainable development in similar regions.

A key focus of the proposed plan is fish farming, a sector that has historically underperformed in similar water bodies in the region. Data from five *Beels* (wetlands) in CD Block Krishnagar - II (Das 2017), West Bengal reveal a stark gap between expected and actual incomes. The Department of Fishery, Government of West Bengal, set an expected income of Rs. 180,000 per acre per annum in 2013. For 551.25 acres of Beels, this translates to an expected total income of Rs. 9.92 crore per year. However, the actual income was only Rs. 30.23 lakh—a mere 0.045% of the target. Certain Beels, such as Hansadanga, Nawpara, and Bara Beels, generated almost no returns, while Satitala Beel achieved 38.43% of its expected income, yielding Rs. 30 lakh. This significant disparity highlights inefficiencies in fish farming practices, resource management, and operational challenges, despite the Beels' ample size and potential. For the Jalangi River's 735-acre stretch, a sustainable and integrated approach is essential to unlock its potential (Table 3.13).

Based on the Department of Fishery's benchmark of Rs. 180,000 per acre annually, the total expected income from the 735-acre river course could reach Rs. 13.23 crore per year. However, considering the actual performance of the five Beels, which yielded only 3.05% of the expected income, the realistic revenue from fish farming alone might initially be around Rs. 40.36 lakh annually unless significant improvements in management are implemented.

Comprehensive Utilization Plan: To maximize the river's economic and ecological utility, the following initiatives are proposed-

Integrated Fish Farming: Adoption of advanced aquaculture practices, better resource allocation, and local capacity building to improve fish yield and income.

Duck Farming: Utilizing wetland areas for duck rearing, generating supplementary income for local farmers.

Vegetable & Fruit Cultivation: Promoting agricultural activities along the riverbanks using

Table 3.13 Showing area, perimeter, and average depth of five beels of CD Block Krishnagar -II with immense potential for aquaculture

Sl. No.	Name of the *Beel*	Area (Acre)	Perimeter (m)	Average depth (m)
1	Hansadanga *Beel*	178.08	9462.19	1.92
2	Nawpara *Beel*	240.89	9855.75	4.19
3	Satitala *Beel*	43.38	3420.31	1.48
4	Bara *Beel*	87.46	5180.83	1.44
5	Rukunpur *Beel*	1.45	301.68	–

organic practices will support food security and diversify incomes.

Rural Tourism: Developing eco-tourism activities, including boating, birdwatching, and cultural tours, to attract visitors and create jobs for local communities.

Community Engagement & Training: Empowering local stakeholders through skill development programs and cooperative management structures to ensure long-term sustainability.

By addressing the challenges observed in similar wetlands and adopting innovative and community-centred strategies, the Jalangi River's decayed stretch can be transformed into a vibrant hub of economic and ecological activity, ensuring sustainable livelihoods and environmental conservation.

3.15 River Linking and Diversion of Water from the Jalangi Basin

The proposed river linking and water diversion project planned by the West Bengal Government aims to rejuvenate the ecology of the Indian Sundarbans by addressing salinity issues in its creeks and lands (Halder 2025). The project is supposed to be funded by the Asian Development Bank (https://bartamanpatrika.com/detailNews.php?cID=12&nID=473002&P=1&nPID=20230720) and/ or National Mission for Clean Ganga (NMCG). The plan involves channelling fresh water through the Ichhamati River by linking it with the Bhairab River via the Jalangi River, Kalinga Khal, and other paleochannels. The Mathabhanga and Ichhamati rivers form critical links in this ambitious project. If successful, this initiative could substantially restore the region's ecological balance, benefiting biodiversity and local communities reliant on the Sundarbans' unique environment. Furthermore, dredging the Bhairab River's channel, particularly at its offtake, would enhance water flow from the Padma, revitalizing the riverine ecology along its course. However, while the objectives are promising, the project is fraught with challenges.

One major concern is the sustainability of the narrow and highly curved bypass channel of the Padma, proposed to feed the Bhairab River. The precarious nature of these channels raises doubts about their long-term viability, especially since the main flow of Padma could shift to the broader channels within Bangladesh. In that situation, any effort could lead to the resurgence of Indo-Bangladesh water crisis issues, potentially hindering the project's success. The project's success also hinges on maintaining a critical water flow through the linked rivers. However, the negligible slope (0.000032) of the river courses may inhibit sustained flow over such a long distance, undermining the project's goals (Fig. 3.38). There is also a risk of unintended consequences for the Jalangi River's lower reach, which could be deprived of adequate water flow, leading to social, economic, and ecological challenges for the communities and ecosystems reliant on it.

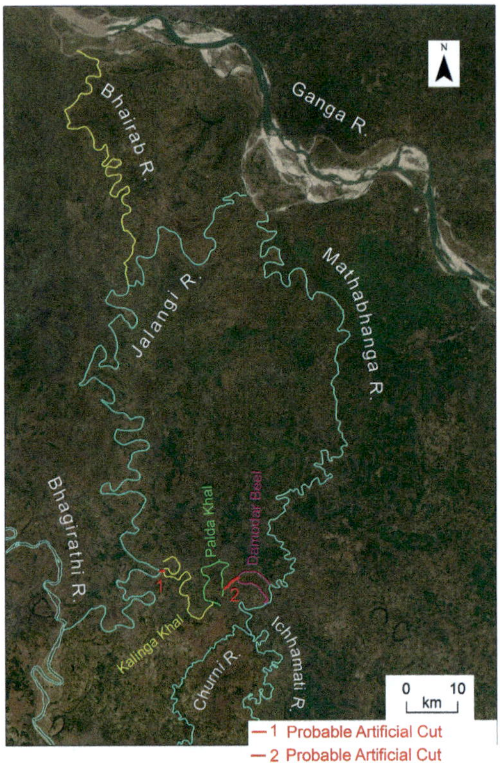

Fig. 3.38 Plan for proposed river linking and diversion of water from Jalangi basin to the Indian Sundarban through Bhairab-Jalangi-Kalinga-Mathabhanga-Ichhamati Rivers.

Another pressing issue is whether sufficient water will ultimately reach the Sundarbans. The authorities at River Research Institute (RRI) estimate that after excavation of Bhairab River, 800 cumecs of water could be available during 2–3 monsoon months, with 500 cumecs directed toward the Bhagirathi River at Swarupganj and the remaining 300 cumecs diverted toward the Ichhamati. However, the historical discharge trends of the Jalangi River cast doubt on this projection. While the river recorded a peak discharge of 800 cumecs in 2003, the volume significantly declined in subsequent years, with only 180 cumecs in 2023 and as low as 56 and 45 cumecs in 2021 and 2022, respectively. This dwindling water flow raises questions about whether the project will achieve its intended benefits or exacerbate existing problems. As presented in District Ganga Committee meeting on Jalangi River Issues, dated 12/09/2024, RRI estimated storage of 31.05×10^6 m^3 of water from several points of check dam within the basin to ensure a flow of 110 cusecs for four months. From Chand beel and Bagakhali beel, an additional flow of 450 cusecs for four months of lean period has been estimated.

Another critical issue that may arise from diverting water from the Jalangi Basin is the potential disruption of the hydrogeomorphic role back swamps and wetlands play in sustaining the deltaic ecosystem. These natural features are essential for delta building and maintaining soil fertility. During floods, backswamps and wetlands act as natural reservoirs, accommodating silt-laden water and gradually depositing nutrient-rich sediments across the landscape. This process not only replenishes the fertility of agricultural lands but also strengthens the delta structure by adding new layers of sediment. Furthermore, wetlands and backswamps play a crucial role in groundwater recharge by allowing water to percolate into underground aquifers. This helps maintain the water table, ensuring a reliable groundwater source for irrigation and domestic use. During lean months, these wetlands act as a source of base flow for rivers, sustaining their flow when monsoon-fed channels dry up (Majumder 1941). Diverting water from the Jalangi Basin and draining these swamps and wetlands through cut canals could severely compromise these natural processes and groundwater reserves (Jiang et al. 2022).

By disrupting the hydrological balance of the Jalangi Basin, this project risks diminishing groundwater availability, reducing agricultural productivity, and increasing the region's vulnerability to floods and droughts. The loss of back swamps as silt reservoirs could also lead to greater sediment load in the main river channels, causing siltation, reduced flow capacity, and an increased likelihood of riverbank erosion. If these critical wetland systems are degraded or drained, the Jalangi Basin may face long-term ecological and socio-economic challenges, threatening the region's and its communities' sustainability.

In conclusion, while the proposed river linking and water diversion initiative holds the potential to rejuvenate the Indian Sundarbans and its riverine ecosystems, it is not without significant risks. The challenges of maintaining adequate flow, sustaining the narrow feeder channel of Padma, and ensuring water reach the target areas highlight the need for careful planning, rigorous assessments, and adaptive management strategies. The success of this ambitious project depends on addressing these concerns comprehensively to avoid inadvertently creating more harm than good (Table 3.14).

Table 3.14 Showing probable length of paths for proposed river linking and water diversion from the Jalangi basin to the river Ichhamati

Paths	Length (km) estimated by present authors	Length (km) estimated by the proposed project
Bhairab	78	78*
Jalangi	135	100
Kalinga	30	52
Other	22	10
Mathabhanga	15	15*
Total	280	255

*The asterisk next to the river length values in the table indicates that these lengths are excluded from the RRI estimation, resulting in a significantly reduced length

3.16 Conclusion

The transformation of the Jalangi River from a vibrant, natural watercourse to a geomorphologically altered and deteriorated system reflects the complex interplay between natural processes and human interventions. This transition to transformation highlights critical changes across both channel and extra-channel geomorphology, driven by physical modifications and developmental pressures. Human activities such as channel dredging, barrier constructions, division of the river into ponds, encroachments, and infrastructural developments (like road and rail crossings and construction of buildings) have fundamentally altered the river's geomorphic features. These anthropogenic influences have accelerated channel degradation, sediment deposition, and the loss of natural hydraulic efficiency over time.

The Jalangi River, which once demonstrated natural geomorphic dynamism through processes like meandering and sediment redistribution, is now constrained by extensive anthropogenic alterations. Channel migration, a fundamental fluvial process, has been heavily restricted due to human interventions such as upstream water withdrawal from the feeder river Ganga, embankments, encroachments, and infrastructure projects. As seen near Dwijendra Setu in Krishnanagar, such modifications have disrupted the natural geomorphic equilibrium, leading to channel narrowing & shallowing, soil deposition, and further geomorphic and ecological deterioration of the river. These changes also extend beyond the channel itself, impacting the river's floodplain and connected landforms. The resultant disruptions have not only altered flow patterns but also impacted the surrounding landscape, reducing the capacity for natural channel adjustment.

The changes in the geomorphology of the Jalangi River have led to significant alterations in its hydraulics, reducing flow efficiency and causing irregular water discharge patterns. These disruptions exacerbate sediment deposition within the channel and spillage onto the adjacent floodplains, especially during the monsoon season. The river's capacity to transport sediment has diminished, leading to the buildup of sandbars and shoals, further fragmenting the channel. Additionally, human-led projects like embankments and barriers have disturbed the river's natural connectivity, causing stagnant water pools and localized flooding, particularly during peak discharge.

The geomorphic and hydraulic transformations of the Jalangi River have had cascading effects on its ecology and environment. Reduced water flow, sedimentation, and pollution have significantly deteriorated the riverine ecosystem, threatening aquatic biodiversity. The habitat for various fish species, which are integral to local livelihoods, has been shrinking, leading to a decline in fish catch. Similarly, the surrounding riparian vegetation and wetland ecosystems have been compromised, reducing their capacity to support biodiversity and regulate natural processes like nutrient cycling. This environmental decline also impacts the broader ecosystem services provided by the Jalangi River, such as groundwater recharge, water purification, and flood regulation. The decline in ecological health further diminishes the river's ability to act as a socio-environmental lifeline for communities that depend on it for fishing, agriculture, and domestic needs.

The transformation of the Jalangi River underscores the critical relationship between humans and rivers in the Anthropocene era. While human activities—ranging from infrastructure projects to agricultural practices—have reshaped the river's dynamics, this relationship remains paradoxical. Local communities heavily depend on the river for livelihoods, irrigation, fishing, and daily activities, yet these very human pressures have accelerated its deterioration. For instance, fishing practices and encroachments have fragmented the river, restricting its flow and natural geomorphic processes. The decline of the Jalangi River highlights the urgent need to reconcile human needs with ecological preservation.

A significant finding of this analysis is the lack of comprehensive monitoring of the river and regular interval data regarding sediment displacement, geomorphic processes, and long-term hydraulic changes. To address this, the study emphasizes the need for intensive, systematic research that includes monitoring earth material transport, geomorphic evolution, and hydrodynamics. Data-driven studies will enable the development of predictive models and sustainable management strategies to mitigate further degradation of the Jalangi River.

The ongoing deterioration of the Jalangi River raises critical concerns for future construction projects within the basin area. Infrastructure developments, such as bridges, embankments, and roads, need to incorporate research findings to minimize adverse impacts like significant channel modification, soil deposition and altered flow patterns. Integrating anthropogeomorphic factors into infrastructure planning will ensure that development activities align with environmental sustainability goals. A proactive approach—one that considers geomorphic sensitivity, sediment dynamics, and ecological needs—can help mitigate human impacts while fostering sustainable resource management.

The Jalangi River's transformation serves as a cautionary tale for other rivers in the region, such as the Churni River, which has similarly experienced reduced flow following upstream interventions like the Farakka Barrage. While the Churni remains partially connected to its feeder river Padma, during monsoon months only, its diminished flow highlights the potential for a fate similar to that of the Jalangi River. These cases reflect broader challenges faced by rivers subjected to human pressures, emphasizing the need for integrated water management approaches and cross-regional cooperation to restore river systems.

Finally, the transition and transformation of the Jalangi River embody the delicate balance between human intervention and natural processes. While anthropogenic activities have reshaped landforms, altered channel dynamics, and disrupted ecological systems, they also offer a pathway for restoration and sustainability. Through robust research, sustainable policies, and community-driven initiatives, it is possible to mitigate the adverse impacts on the Jalangi River and promote its ecological resilience. Addressing Jalangi's challenges requires a holistic, multi-disciplinary approach that prioritizes ecological restoration, geomorphic stability, and sustainable development through community engagement fostered by incentives as well as liabilities. By acknowledging the river's vital role in human and environmental systems, we can work toward a future where the Jalangi River not only survives but thrives, ensuring its legacy as a vital watercourse for generations to come.

Unlike the Jalangi River, the Churni River isn't entirely disconnected from its feeder river Padma. However, following the Farakka episode, the Churni's flow has diminished significantly. This decline raises concerns that it might eventually suffer a fate similar to the Jalangi, which is illuminated in detail in the next Chapter. Welcome to Chap. 4.

References

Allison MA (1998) Historical changes in the Ganges-Brahmaputra delta front. J Coast Res 1269–1275

Allison MA, Khan SR, Goodbred SL Jr, Kuehl SA (2003) Stratigraphic evolution of the late Holocene Ganges-Brahmaputra lower delta plain. Sed Geol 155(3–4):317–342

Bagchi K (1978) Diagnostic Survey Of Deltaic West Bengal, A Research And Development Project, Department Of Geography, Calcutta University, P- 17

Barlow R (1887) The diary of William Hedges, Esq. During his agency in Bengal, as well as his voyage out and return over land I(LXXIV):1681–1687. P. 38–39

Bastia F, Equeenuddin SM (2016) Spatio-temporal variation of water flow and sediment discharge in the Mahanadi River, India. Glob Planet Change 144:51–66. https://doi.org/10.1016/j.gloplacha.2016.07.004

Bhattacharyya A (1974) Historical geography of ancient and early medieval Bengal. Thesis submitted for the Ph.D. Degree of the Calcutta University. P. 39–44

Biswas KR (2001) Rivers of Bengal. Vol-II, Govt. of West Bengal, pp 1–15

Biswas M (2024) Khāla-bilēra ākhyāna (aṣṭama parba). Samanbaẏa. Dbitīẏa barṣa. Caturtha saṅkhyā. https://www.samanwoy23.com/issues/2nd-Year-4th-Issue-15-April-2024/details/?details=356

Bose NK (1972) The Bhagirathi - Hooghly Basin- A Few Remarks, Proceedings of the inter disciplinary symposium Edtd by K. G. Bagchi, p- viii, xii

Brockington J (2022) The Sanskrit epics. Wiley Blackwell companion to Hinduism, pp 81–91
Captain Lang (1848) Report to Government, dated 14th July 1848. In Hunter (1875). Statistical account of Bengal. The Presidency division. Comprising the districts of 24 Parganás, Nadiyá and Jessor. Under the Jurisdiction of the Lieft. Government of Bengal, vol II, p 19
Captain Colebrooke RH (1797) In a memoir on the course of the Ganges (1797), writes thus:—*The Bhagirathi and Jalangl are not navigable throughout during the dry season. There have been instances of all these rivers continuing open in their turn during the dry season. The Jalangi used formerly to be navigable during the whole or greater part of the year. The Bhagirathi was navigable in the dry season of 1796. The Mathabhanga, when surveyed in 1795, navigable throughout in the dry season for boats of a moderate burden. This year (1797), however, I was informed that the passage was no longer practicable for boats proceeding to Calcutta. Experience has shown that none of these rivers are to be depended on
Chaudhuri J (2004) Srichaitanyadeva O Samakalin Nabadwip. Nabadwip Puratattva Parishad, Nabadwip, p-8
Chatterjee R, Samadder S, Mondal D, Adhikari K (2020) Analysis of spatio-temporal trend in groundwater elevation data from arsenic affected alluvial aquifers–Case study from Murshidabad district, West Bengal, Eastern India. J Earth Syst Sci 129:1–15. https://doi.org/10.1007/s12040-020-01489-8
Colebrooke RH (1803) On the course of the Ganges, Asiatic researches, or transactions of the society, instituted in Bengal for inquiring into the history and antiquities, the arts, sciences, and literature of Asia, vol 7. London. Through Bengal, pp 1–31
Das B (1535) Chaitanya Bhagavata. Nadia, 3rd ed., pp. 474–75 (in Bengali) https://archive.org/details/ChaitanyaBhagavat/mode/2up
Das BC (2013) Changes and deterioration of the course of river Jalangi and its impact on the people living on its banks Nadia West Bengal. Ph.D. Thesis. University of Calcutta, pp 47–50
Das BC (2014). Impact of In-Bed and On-Bank Soil Cutting By Brick Fields on Moribund Deltaic Rivers: A Study of Nadia River in West Bengal. The NEHU Journal, India. ISSN. 0972 - 8406, Vol. XII, No. 2, July - December 2014, Pp. 101–111.
Das BC (2015) Socio-economic impact of a decaying river on fishermen: a case study of taranipur village, West Bengal. Int J Res Manag Sci Technol, India 3(4):141–149. ISSN:2321-3264
Das S (2017) Evolution of drainage and morphology of upper Bhagirathi Ganga interfluve region of West Bengal with special reference to Palaeochannels. Ph.D. Thesys. The University of Calcutta
Das BC (2019) A study on impact of bridge construction on channel dynamics, West Bengal, India. Sci J KFU (Humanit Manage Sci). Saudi Arabia 20(1):265–279. https://services.kfu.edu.sa/scientificjournal/Handlers/FileHandler.ashx?file=h20114.pdf&Folder=UploadFiles. ISSN-1319-6944
Das BC (2023a) In Search of Birthplace of Sri Chaitanya: Part I. A Study on Hagiographies Up to 17th Century. Studia Universitatis Babeş-Bolyai Geographia 68(1) 63–91. https://doi.org/10.24193/subbgeogr.2023.1.05
Das GK (2023b) Water quality index of Churni and Jalangi rivers, West Bengal, India. Int Res J Environ Sci 12(1):14–18
Das BC (2024a) Exploring the past to discern the present: the role of historico-geographical study in differentiating natural river channels from artificial canals. River 1–18. https://doi.org/10.1002/rvr2.73
Das GK (2024b) Water Quality Determinants of Mathabhanga, Churni and Jalangi Rivers. In River Systems of West Bengal: Water and Environments (pp. 129–140). Cham: Springer Nature Switzerland
Das BC, Bhattacharya S (2020) The Jalangi: A Story of Killing of a Dying River. In Das et al. (2020) eds. Anthropogeomorphology of Bhagirathi-Hooghly River System in India. Taylor & Francis Group. ISBN (eBook) 9781003032373. https://doi.org/10.1201/9781003032373
Das BC, Das D (2020) The Anjana: a journey from river to canal. In: Das et al (eds) (2020) Anthropogeomorphology of Bhagirathi-Hooghly river system in India. Taylor and Francis Group. https://doi.org/10.1201/9781003032373. ISBN (eBook) 9781003032373
Das BC, Islam A, Biswas B (2020) Morphometry as tool to trace out the genealogy of oxbow lake. Environ Earth Sci 79:137, 1866–6299 (web). https://doi.org/10.1007/s12665-020-8854-3. ISSN: 1866-6280 (print)
Das BC, Das S, Sarkar B, Das S, Adhikari R, Saha D et al (2024a) Geospatial assessment of water quality of a dying tropical river and its environmental implications. In: Spatial modeling of environmental pollution and ecological risk. Woodhead Publishing, pp 409–421
Das BC, Islam A, Das S, Adhikary R, Sarkar R (2024b) Bridge construction impacts the adjacent channel morphology and turbidity: a study on a dying distributary of river padma. Earth Sci, Syst Soc 4:10099. https://doi.org/10.3389/esss.2024.10099
Das BC, Islam A Sarkar B (2022) Drainage Basin Shape Indices to Understanding Channel Hydraulics. Water Resour Manage 36, 2523–2547 https://doi.org/10.1007/s11269-022-03121-4
Deswati D, Safni S, Khairiyah K, Yani E, Yusuf Y, Pardi H (2022) Biofloc technology: water quality (pH, temperature, DO, COD, BOD) in a flood & drain aquaponic system. Int J Environ Anal Chem 102(18):6835–6844
Dev S (2004) Kathamukh. In Bengali in Chaudhuri, Jajneswar (2004) Sri Chaitanyadeva O Samakalin Nabadwip, Nabadwip Puratattva Parishad, Nabadwip, p-a
Dobbins WE (1964) BOD and oxygen relationships in streams. J Sanit Eng Div 90(3):53–78

DSR (2022) District survey report of Nadia district. (For mining of minor minerals) As per Notification No. S.O.141 (E) New Delhi dated 15th of January 2016, S.O.3611 (E) New Delhi dated 25th of July 2018 and enforcement and monitoring guidelines for sand mining (EMGSM) January 2020, Issued by ministry of environment, forest and climate change (MoEF&CC) July, 2022. Department of industry, commerce and enterprises, Government of West Bengal

Eaton RM (1993) The rise of Islam and the Bengal frontier, 1204–1760, vol 17. University of California Press, p 63. ISBN 978-0-520-20507-9

Escartın J, Aubrey DG (1995) Flow structure and dispersion within algal mats. Estuarine, Coastal and Shelf Science 40(4):451–472

Fergusson J (1863) On recent changes in the delta of the ganges. Q J Geol Soc Geol Soc London XIX:321–354

Gibb HAR (1971) Travels of Ibn Battuta, AD 1325–1354. Vol. III. ISBN: 0521010330. The Hakluyt Society. p 267

Goodbred Jr, SL, Kuehl SA (2000) The significance of large sediment supply, active tectonism, and eustasy on margin sequence development: late Quaternary stratigraphy and evolution of the Ganges–Brahmaputra delta. Sediment Geol 133(3–4), 227–248

Goodbred SL Jr, Kuehl SA, Steckler MS, Sarker MH (2003) Controls on facies distribution and stratigraphic preservation in the Ganges-Brahmaputra delta sequence. Sed Geol 155(3–4):301–316

Greb SF, DiMichele WA, Gastaldo RA (2006) Evolution and importance of wetlands in Earth history. In SF Greb, WA DiMichele (Eds) Boulder, CO: Geological Society of America. https://doi.org/10.1130/2006.2399(01)

Halder S (2025) Anandabazar Patrika. 19th June 2025. https://epaper.anandabazar.com/calcutta-north/15/2025-06-19/page-1.html

Hey RD (1976) Geometry of river meanders. Nature 262:482–484

Hickin EJ, Nanson GC (1975) The character of channel migration on the Beatton River, north-east British Columbia, Canada. Bull Geol Soc Am 86:487–94

Hirst FC (1916) Report on the Nadia Rivers 1915. The Bengal Secretariat Book Depot, Calcutta, p 1916

Hirst FC (1917) Note on the geological aspect of the changes that have taken place in the rivers of Bengal. Received with director's letter no. 1764, dated the 31st July 1917. In: Stevenson-Moor et al. (1919) Report on the Hooghly River and its headwaters. Reprinted by Biswas KR. Rivers of Bengal, vol II. Gazetteer Department, Government of West Bengal. Kolkata, pp 105–107

Hickin EJ, Nanson GC (1984) Lateral migration rates of river bends. J Hydraul Eng 110:1557–1567

Hoque M, Islam A, Ghosh S (2022) Environmental flow in the context of dams and development with special reference to the Damodar Valley Project, India: a review. Sustain Water Resour Manage 8(3):1–27. https://doi.org/10.1007/s40899-022-00646-9

Hoque MM, Islam A, Islam ARMT, Das BC, Pal SC, Arabameri A, Khan R (2023) Spatio-temporal assessment of water quality of a tropical decaying river in India for drinking purposes and human health risk characterization. Environ Sci Pollut Res 30(45):101653–101668

Hunter WW (1877) A statistical account of Bengal, vol II. Trubner & Co., London, p 19

Inglis WA (1909) The canals and flood banks of Bengal. Calcutta. The Bengal Secretariat Press. Cited from Biswas KR (2002) Edited. Rivers of Bengal. Vol. V. Part II. Kolkata. West Bengal Dist gazetteers. Higher Education Department. Government of West Bengal. P-48

Inglis WA, Clarkson FC, Wheeler H (1907) Report of the drainage committee, Bengal. (Presidency Division). The Bengal Secretariate Press, Calcutta, p 28

Islam A, Das BC, Mahammad S et al (2024a) Assessing river water quality for ecological risk in the context of a decaying river in India. Environ Sci Pollut Res. https://doi.org/10.1007/s11356-024-33684-1

Islam A, Hoque MM, Ghosh S, Sarkar B, Mahammad S, Das BC et al (2024) Hydro-chemical characterization and irrigation suitability assessment of a tropical decaying river in India. Sci Rep 14(1):20096

Islam A, Das BC, Kumar N, Barman SD (2020) Assessing meander belt width of Bhagirathi-Jalangi river system in lower Ganga delta, India. Eur J Geogr 11(1)

Islam SA (1993) Meherpurer Itihas. (In Bengali). Meherun Islam. Boisagar. Meherpur. Bangladesh, pp 50–53

Islam A, Hoque MM, Ghosh S, Sarkar B, Mahammad S, Das BC et al (2024b) Hydro-chemical characterization and irrigation suitability assessment of a tropical decaying river in India. Sci Rep 14:20096. https://doi.org/10.1038/s41598-024-70851-3

Jiang X, Ma R, Ma T, Sun Z (2022) Modeling the effects of water diversion projects on surface water and groundwater interactions in the central Yangtze River basin. Sci Total Environ 830:154606.

Jouanneau S, Recoules L, Durand MJ, Boukabache A, Picot V, Primault Y, Lakel A, Sengelen M, Barillan B, Thousand G (2014) Methods for assessing biochemical oxygen demand (BOD): a review. Water Res 49:62–82

Kosseva MR (2012) Sources, characterization, and composition of food industry wastes. Food Ind Wastes 37–60. https://doi.org/10.1016/B978-0-12-391921-2.00003-2

Knighton D (1998) Fluvial forms and processes: a new Perspective. Routledge, Oxford UK

Lang J (1848) Report on the Nuddea Rivers and the advantages derived from the measures annually adopted for facilitating their navigation. No. 89, dated Kishnaghur, the 14th July 1848. From Captain John Lang, 36th regiment, Bengal N. I, officiating superintendent, Nuddea Rivers. To the officiating superintending engineer, lower provinces, in Nadia Rivers (From 1848 to1926), p 72

Lang J (1854) Memorandum on the Ganges and the entrance of the Nuddea Rivers. Bengal secretariate Book Depot Calcutta, p 109

Lee S ed (1896) Rennell, James. Dictionary of National Biography. Vol. 48. London: Smith, Elder & Co

Li D, Liu S (2018) Water quality detection for lakes. Water quality Monitoring Manage 221–231. https://doi.org/10.1016/B978-0-12-811330-1.00008-9

Li J, Luo G, He L, Xu J, Lyu J (2018) Analytical approaches for determining chemical oxygen demand in water bodies: a review. Crit Rev Anal Chem 48(1):47–65

Liu X, Hu S, Sun R, Wu Y, Qiao Z, Wang S, Zhang Z, Cui C (2021) Dissolved oxygen disturbs nitrate transformation by modifying microbial community, co-occurrence networks, and functional genes during aerobic-anoxic transition. Sci Total Environ 790:148245. https://doi.org/10.1016/j.scitotenv.2021.148245

Long J Revd (1846) The Banks of the Bhagirathi. Calcutta Rev VI(XII):447

Majumdar D (1978a) District Gazetteer, Nadia, Government of West Bengal, p 7

Majumder D (1978b) West Bengal District Gazetteers Nadia. Government of West Bengal, p 7, 16, 25–34

Majaumder S (1985) Sri Chaitanya Janasthan Bitarka, Tar Truti O Samadhan. Nabadwip, Bengali, p 26

Mallik K (1910) Nadia Kahini, in Bengali, ed-Ray.M,1998, Pustak Bipani, Kol. p-250-300

Markovits C (ed) (2004) A history of modern India, 1480–1950. Anthem Press, p. 38. ISBN 978-1-84331-152-2

Mishra A (2006) The Life of Lord Chaitanya and his Sankirtana movement in India. Orissa Review, pp 33–36

Ministry of Jal Shakti (2023) Annex referred in the reply to Lok Sabha Unstarred Question No. 1268 answered on 09.02.2023 chrome-extension://efaidnbmnnnibpcajpcglclefindmkaj/https://sansad.in/getFile/loksabhaquestions/annex/1711/AU1268.pdf?source=pqals

Morgan JP, McIntire WG (1959) Quaternary geology of the Bengal Basin, East Pakistan and India. Bull Geolog Soc Am 70:319–342

Majumder SC (1941) Rivers of the Bengal Delta, ed-Biswas K.R.2001, Rivers of Bengal, Department of Higher Education, Govt. of West Bengal, p-17,18,54

Mukharji BB (1932) Final report on the survey and settlement operation in the district of Murshidabad 1924–1932. Bengal Secretariat Book Depot, Government of Bengal, pp 5–8, 74–76, 96, 125–127

Mukerjee RK (1938) The Changing face of Bengal, a study in riverine economy, C.U., reprinted in 2009, p 6–8

Mukherjee KN (1984) A Study for Sri Chaitanya's Birthplace, Indian Journal of Landscape Systems. Vol. 7, No. 11. https://gosai.com/writings/a-study-for-srichaitanyas-birthplace

Nakatani T (2003) A sacred place or tourist spot? J Japan Associat South Asian Stud 15:113–141

Nanson GC, Hickin EJ (1986) A statistical examination of bank erosion and channel migration in western Canada. Bull Geol Soc Am 97:497–504

Ness R (2004) Belt width delineation procedures, Toronto and Region Conservation Authority, Parish Geomorphic Ltd.

Nilsson C, Renöfält BM (2008) Linking flow regime and water quality in rivers: a challenge to adaptive catchment management. Ecol Soc 13(2). https://www.ecologyandsociety.org/vol13/iss2/art18/

Phillimore RH (1945) Historical records of the survey of India, vol I. 18th century. Survey of India. Dehra Dun p 18

Pringle, Kemm. Final report on the survey and settlement operations in the district of Nadia 1918–1926. Bengal Secretariate Book Depot. Calcutta Government of Bengal, pp 11–14

Ray KC (1876) Kṣitīśa banśābalī carita, natuna sanskr̥ta yantra, In Bengali (pp 85, 107–109)

Reaks HG (1919) Report on the physical and hydraulic characteristics of the rivers of the delta. In: Stevenson-Moor et al. Report on the hooghly river and its headwaters, vol I. The Bengal secretariat book depot, Calcutta, 1919, In Biswas (2001) Rivers of Bengal, vol II, Government of West Bengal, p 45–55, 61–75, 87,107

Rennell J (1780) A Bengal atlas containing maps of the theatre of war and commerce on that side of Hindoostan. London (edited by Kalyan Rudra (2016) and reprinted by Sahitya Samasad, Kolkata)

Rennell J (1781) Map of Bengal & Bahar. To The Honorable Warren Hastings Esqr. Governor General of the British Possessions in Asia; This Map of Bengal & Bahar, comprehending a tract of more extensive and populous than the British Islands, is respectfully inscribed in testimony of his distinguished abilities; And in gratitude for favours received: by his obliged h' servant Rennell.

Rudra K (2010) Banglar Nadikatha (in Bengali), Sahitya Sansad, Kolkata, p-33,39

Rudra K (2018) The Bhagirathi-Hugli River system. Rivers of the Ganga-Brahmaputra-Meghna Delta: A Fluvial Account of Bengal, 77–93

Sargaonkar A, Deshpande V (2003) Development of an overall index of pollution for surface water based on a general classification scheme in Indian context. Environ Monit Assess 89:43–67

Sarkar B, Islam A and Das BC (2020) Anthropo-Footprints on Churni River: A River of Stolen Water. In Das et al. (2020) eds. Anthropogeomorphology of Bhagirathi-Hooghly River System in India. Taylor & Francis Group. ISBN (eBook) 9781003032373. https://doi.org/10.1201/9781003032373

Sarkar B, Islam A, Das BC (2021) Role of declining discharge and water pollution on habitat suitability of fish community in the Mathabhanga-Churni River, India. J Cleaner Prod 326:129426

Sen H (2010) The drying up of River Ganga: an issue of common concern to both India and Bangladesh. Current Science 99(6):725–727

Singh P, Sharma S, Nisar S, Choudhary OP (2023) Structural stability and organic matter stabilization in soils: Differential impacts of soil salinity and sodicity. J Soil Sci Plant Nutr 23(2):1751–1773

Stevension-Moor CJ, Ryder CHD, Nandi MC, Law RC, Hayden HH, Campbel J, Murray AR, Adama-Williams C, Constable EA (1919) Report on the Hooghly river and its headwaters, in Biswas KR (2001), Rivers of Bengal, Vol-II, Govt. of West Bengal, pp 1–15

Sunitha V, Reddy YS, Suvarna B, Reddy BM (2021) Human health risk assessment (HHRA) of fluoride and nitrate using pollution index of groundwater (PIG) in and around hard rock terrain of Cuddapah, AP South India. Environ Chem Ecot Oxicol 4:113–123

Survey of India (1994)

Swarnkar S, Prakash S, Joshi SK, Sinha R (2021) Spatio-temporal rainfall trends in the Ganga River basin over the last century: understanding feedback and hydrological impacts. Hydrol Sci J 66(14):2074–2088. https://doi.org/10.1080/02626667.2021.1976783

Tassin JB (1840) Map of Tassin. collected from Rudra K (2010) Banglar Nadikatha (in Bengali), Sahitya Sansad, Kolkata, p-38,39

Tavernier (1666) 6th January noted that traveller Bernier was going overland from Rajmahal to Cossimbazar as the river route was impracticable

Transactions of the American Geophysical Union, vol 26, pp 415–429

Van-den-Brouck (1660) Source-Roy, Niharranjan (1972). Bangalir Itihas. in Bengali, reprint 2005, Dey's Publishing, p 13 Bankim Chatterjee Street, Kolkata, ISBN 81-7079-270-3. https://en.m.wikipedia.org/wiki/File:Van_de_Brook%27s_map.jpg

VTANR (2004) River Corridor Delineation Process, Stream Geomorphic Assessment Handbooks. Vermont Agency of Natural Resources. Cf. Islam et al. (2020). Assessing meander belt width of Bhagirathi-Jalangi river system in lower Ganga delta, India. European J Geogr 11(1)

Weihaupt JG (1977) Morphometric definitions and classifications of oxbow lakes, Yukon River Basin, Alaska. Water Resour Res 13(1):195–196

Willcocks W (1930) Ancient system of irrigation in Bengal and its application to modern problems, ed-Biswas K.R.2001, Rivers of Bengal, vol-I, p-11, 114, 116

Williams GP (1986) River meanders and channel size. J Hydrol 88(1–2):147–164. https://doi.org/10.1016/0022-1694(86)90202-7

Winternitz M (1981) A history of Indian literature, vol 3. Motilal Banarsidass Publication

The Churni River: Exploring the Past, Present, and Future

4

Abstract

This chapter articulates the historical significance, present condition, and prospective future of the Churni River. As the river encompasses the rich history and cultural legacy, the present chapter explores how the river has been changing throughout time by the natural processes and anthropogenic interventions and how it has been affecting the riverine communities depending on the river for their lives and livelihoods. The river has been experiencing several issues, such as decreasing discharge (environmental flow), chronic water pollution from point- and non-point sources like industrial waste water, urban runoff, agrichemicals from agricultural runoff, brick kiln industry, etc. Therefore, this chapter examines these urgent concerns, illuminating the causes of the environmental flow crisis as well as the consequences for the ecosystem and society. Moreover, this attempt will also offer a forward-looking view, outlining potential strategies for Churni River restoration and sustainable management. By fusing historical perspectives, contemporary issues, and promising future developments, the present attempt aims to outline the nature and extent of key issues which may serve as a helpful guide for academician, policymakers, and regional planners for implementing the necessary actions to ensure the ecological resiliency and long-term sustainability.

Keywords

Cultural legacy · Topophilia · Trans-boundary river · Water quality · Ecosystem services · River restoration · Mathabhanga-Churni River

4.1 Etymology

The Mathabhanga-Churni River, an important transboundary river in the Bengal delta, traverses through the Bangladesh and Nadia districts of India. The river branches off from the Padma River within the administrative boundary of Bangladesh, and after cross-cutting the international boundary several times between Bangladesh and India and traveling for 196 km, at Pabakhali near Majdia the river splits into branches: the Churni and Ichamati. The origin of the Churni, and the Ichamati from the river Mathabhanga is available in the historical literature. The Mathabhanga River carries its unique name attached to distinct Bengali etymologies.

Supplementary Information The online version contains supplementary material available at https://doi.org/10.1007/978-3-031-85971-7_4.

B. C. Das et al., *The Nadia Rivers*, Geography of the Physical Environment,
https://doi.org/10.1007/978-3-031-85971-7_4

The term 'Mathabhanga' (মাথাভাঙ্গা) combines two Bengali words: মাথা (Matha), meaning 'head' and ভাঙ্গা (Bhanga), meaning 'broken', therefore, Mathabhanga means 'beheaded'. Because of its characteristics offtake nature, i.e., the river regularly remains separated (beheaded) from the feeder river (the Padma) by the deposited sand bars (char) in-between, except during the monsoon season (Sarkar et al. 2021). Similarly, the name 'Churni', probably distorted form of 'Chorni' (চোরনী) has also been derived from the Bengali word চোর (Chōra) meaning 'male thief'. However, adding the suffix 'নী' (Nī), the river has got its feminity character, i.e., 'female thief', the plausible meaning of the river Churni (Das 2003; Sahitya Samsad 2000). Therefore, their unique etymology infused with myths and legends bears cultural or historical context of the region. In the many historical accounts and local literature, the river has been introduced as artificial canal (Ganguly 2021; Chattopadhyaya 2007; Thakur 2019). This prevalent myth argued that Churni River is an artificial canal, created under the direction of Maharaja Krishnachandra of Nadia. According to this belief, he purportedly redirected a segment of the Ichamati River's flow to create 'Churni'. The origin of this notion can be traced back to the historical work 'Kshitish Vansabali Charita' by Ray (1876), detailing Maharaja Krishnachandra's construction of a palace at Shibnibas on the Churni's bank, encircled by a moat and a canal connecting moat to the river. Remarkably, this historical account takes on a different hue in 'Nadia Kahini' by Mallik (1910). Besides, Chatterjee (2013), and Mondal and Bandyopadhyay (2022) supported the argument of artificial canal. However, these assertions of artificial canal lack scientific or logical underpinnings. In the recent scholastic works, for example, Sarkar et al. (2020) and Das (2023, 2024) propose the river's natural origin, injecting further complexity into the ongoing discourse on the Churni's origin, evolution, and its name. The debate is illustrated under the subhead 'Is Churni a natural river or an artificial canal?'.

4.2 The Basin and the Channel

4.2.1 The Course of the Mathabhanga-Churni River (MCR)

The MCR is an important watercourse in the Bengal delta (Hirst, 1915; Majumder 1942). The Bengal delta in southeast Asia is renowned for its rich plains with complex river systems and intricate eco-geomorphology (Fig. 4.1) (Kuehl et al. 2005; Allison and Kepple 2001; Guchhait et al. 2016). The MCR, as a part of this larger deltaic landscape, emerges as a complex hydrological entity that characterizes the complex dynamics of the Bengal delta (Biswas 2001; Sarkar et al. 2022). The river originates from the Padma River in Bangladesh. Particularly, Char Mahiskundi (24°03′34″ N and 88°43′48″ E), to the east of Char Madhugari village and northeast of CD Block Karimpur-I, is marked as the origin of the river from a small bypass channel of the Padma River. As a unique watercourse, the river runs for 196 km crossing and recrossing the international border between Bangladesh and India (Das 2024). The river, near Shibnibas village of Krishnaganj CD Block, Nadia (23°24′15″ N and 88°42′38″ E), get splitted into two channels—the Churni (right branch) and the Ichamati (left branch) (Sarkar et al. 2023). Stretching 56 km, the Churni River passes through Shibnibas, Hanskhali, Aranghata, and Ranaghat before joining the Hooghly River close to Shibpur at 23°7′59″ N, 88°30′9″ E (Sarkar and Islam 2019). The Churni River is divided into an upper and a lower reaches, which helps to better comprehend its hydrogeomorphological properties. The Anjana River empties into the Churni River near Byaspur village at 23°16′55″ N, 88°35′01″ E. Discharge of the Churni River increases downstream of the confluence due to the discharge contribution by the Anjana River, particularly during the monsoon season. Therefore, the upstream of the confluence (30 km) of the river is considered as upper stretch and downstream of the confluence

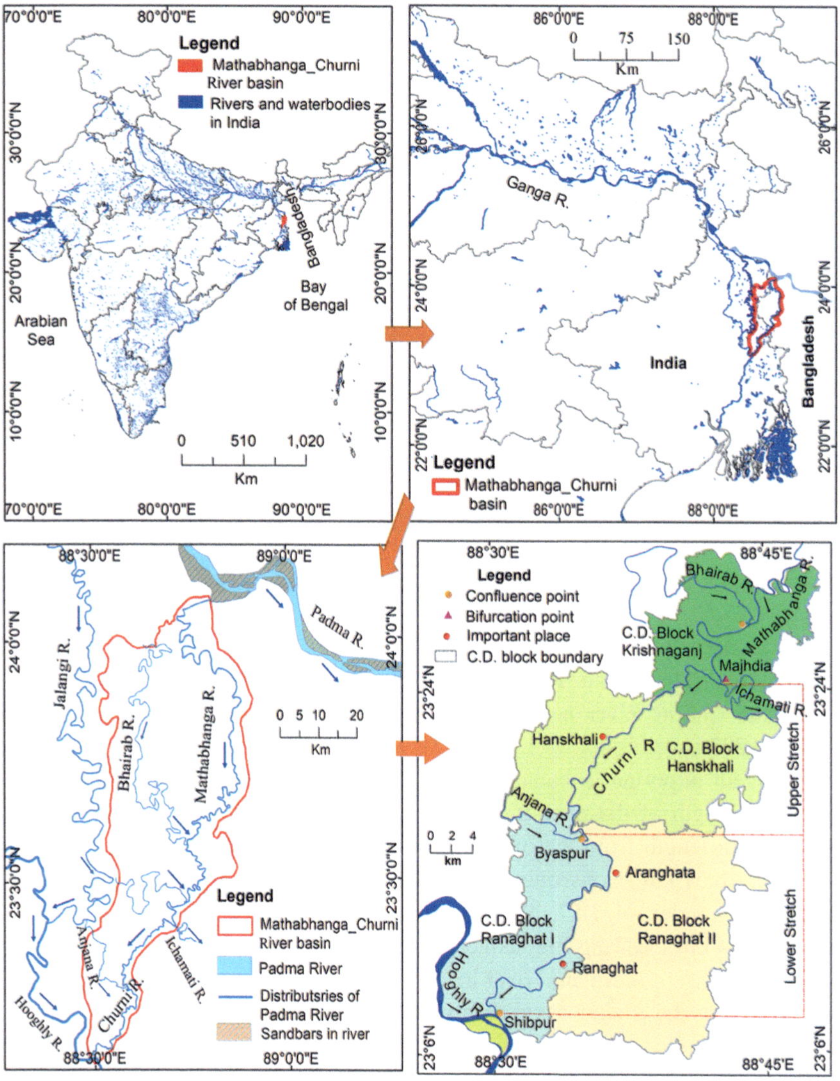

Fig. 4.1 Location of the study area

is considered (26 km) as the lower stretch of the river (Sarkar and Islam 2020).

4.2.2 Tributaries and Distributaries

The MCR system in the Bengal delta is a complex network of tributaries and distributaries that supports to the hydrology, ecology, and socio-economic life of its basin. The Ichamati and Anjana Rivers are two of its major watercourses (Das 2024).

The Ichamati River was an important distributary of the Mathabhanga River, branching off at Majhdia in the Nadia District. The river follows its path parallel to the international border between Bangladesh and India (Mondal et al. 2020). Presently the river gets disconnected from the Mathabhanga River because, except the monsoon months, the Ichamati's offtake zone remains dry due to the higher bed elevation of the Ichamati River compared to that of the Mathabhanga River (Das 2017). Therefore, the upper non-tidal part of the river, 155 km long, is degrading. The river Ichamati traverses Indian territory for 25 km and enters into Bangladesh at Mubarakpur and again returns to Indian territory at Sarahati, which is 55 km away from Duttaphulia. Bangaon town lies at a distance of 101.5 km from the source of the Ichamati River, and the river turns tidal a little downstream. The river is in India up to Goalpara (200.9 km), where it meets the Kalindi and Raimangal Rivers before emptying into the Bay of Bengal. Additionally, the entire stretch forms the Indo-Bangladesh Border.

The Anjana River flowed through the heart of the Nadia district for about 400 years (Das and Das 2020). Anjana functions as a tributary and distributary river in the drainage system of the largest Ganges–Brahmaputra delta in the world. After beginning from the Jalangi River in Krishnagar (23°25′7.5″ N and 88°29′17.5″ E), the river Anjana pasfses through the C.D. Blocks of Krishnagar-I, Hanskhali, and Ranaghat -I and joins with the Churni River near Byaspur (Sarkar and Islam 2020).

Therefore, it is an important tributary of the Churni River. The length of the river is 29 km, of which the straight length from the offtake to the confluence is 18.51 km. Consequently, the river has a sinuosity index of 1.57. The basin has an area of 89.54 km^2. During the sixteenth and seventeenth centuries, the Anjana River held such significance that the Nadia royal family chose to establish their city and royal residence on its banks.

4.2.3 Basin Characteristics

The geometry of a basin is one of the key factors affecting its hydrological behavior. As a result, measuring and carefully assessing the geometric features of basins are necessary for several hydrological modeling components. The four primary morphometric characteristics of a river basin are area, length, perimeter, and shape. The MCR traverses an area of around 2585 km^2 and has a perimeter of about 295.30 km (Fig. 4.2). Measuring along the main river, the basin is approximately 110 km long and 40 km wide at its widest point. Then, a complex yet harmonious relationship is observed in the basin characteristics in the Bengal delta where river basins of the Ganga Brahmaputra and Meghna dominate the entire deltaic region having their colossal basin area. MCR basin, though smaller in comparison, holds immense significance for the delicate equilibrium influencing the intricate environmental, hydrological, socio-economic, and cultural dynamics, of the Bengal delta. The MCR appears to be an elongated basin, characterized by the value of Horton's shape factor $F=0.21$ (Horton 1932). This elongation nature is also supported by the values of the lemniscate ratio ($K=3.67$), compactness index ($C=1.63$), basin elongation ($E=0.52$), and circularity ($c=0.37$) (Miller 1953; Schumm 1956) of the basin. This elongated basin character indicates reduced peak flow for extended periods of time, which has been clearly seen in the river (Sassolas-Serrayet et al. 2018).

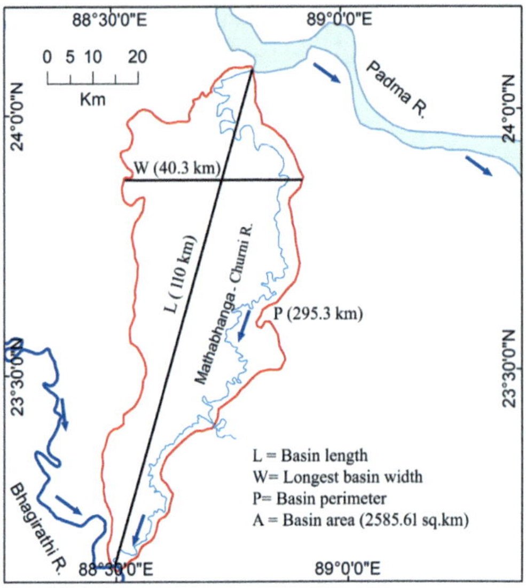

Fig. 4.2 Geometry of the MCR basin

4.2.4 Morphology and Sediment Characteristics

Because of its location within the Gangetic plain, the MCR Basin can be characterized physiographically as a monotonous deltaic plain with an imperceptible slope. However, upon closer inspection, the basin shows a gradual decrease in elevation from the northeast to the southwest. The highest elevation in the river basin is measured at ~ 16 m at Karimpur Block, in the northwestern part of MCR basin, and the lowest elevation is measured at ~ 6 m at Chakdah C.D. Block, in the southern part of the river basin. More specifically, the higher elevations in the river basin typically indicate levees, while the lowlands denote marshes (Fig. 4.3a and b). In the space between these two micro morphological features, riverine plains predominate. Furthermore, this prevailing natural setback has been subject to accelerated anthropogenic intervention through the building of artificial embankments.

As a crucial part of the vast Ganga–Brahmaputra Delta, the MCR basin represents the sediment strata alike to those of the GBM delta. The floodplain stratigraphy is mostly dominated by sand of different grain size. About 80% of the bed sediments in the Ganga–Brahmaputra Delta are classified as fine to very fine sand, with an average grain size of between 2.5 and 4 Φ (1/6 and 1/16 mm) (Datta and Subramanian 1997). The lowermost layers of this deltaic region are mostly composed of sands of different sizes, whereas the uppermost strata typically display silty to clayey silts (Garzanti et al. 2019; Guchhait et al. 2016; Jha and Bairagya 2011; Mukherjee et al. 2009; Weinman et al. 2008). The sediment stratigraphy in the research region may be understood by looking at borehole data. This borehole 1 (BH1) reveals that the first 35 m are mostly made up of different kinds of clay. From about 35–55 m, the sand becomes finer to medium, and below that is a layer of fine sand (Fig. 4.4). In BH2, fine sand covers the surface down to a depth of 55 m, making sand the predominant material throughout. Below 55 m, two separate layers of 5 m each, consisting of fine to medium sand and medium to coarse sand, are visible. Five thin clay

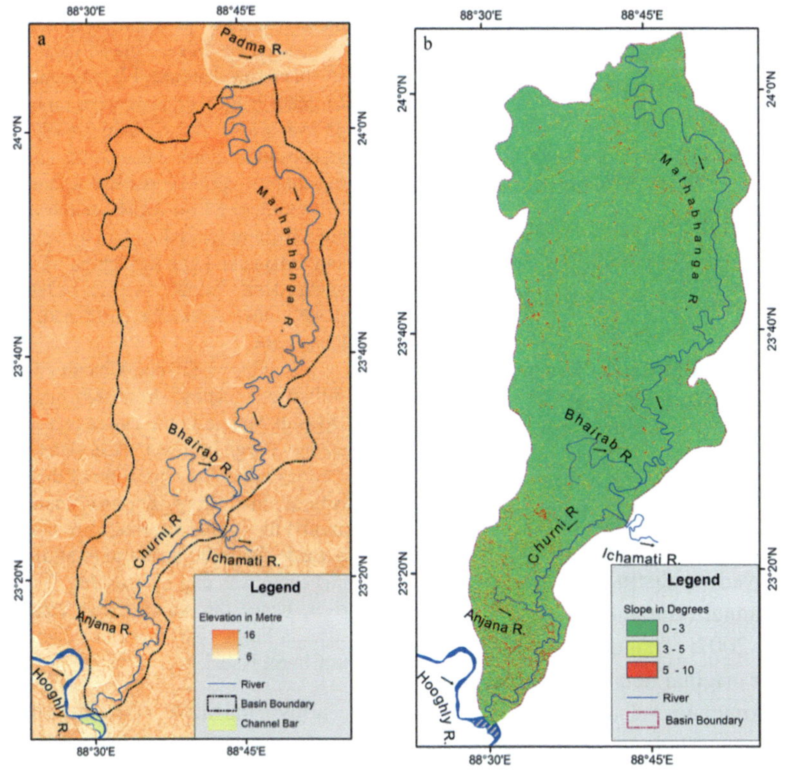

Fig. 4.3 Physiography and relief of the MCR basin (**a** elevation, **b** slope) (prepared from ASTER DEM, 2011)

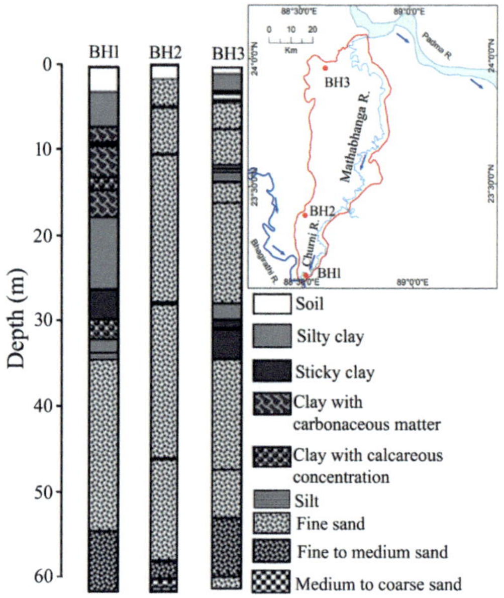

Fig. 4.4 Sediment stratigraphy in the floodplain of MCR river—BH1 (at Ranaghat), BH2 (at Krishnanagar), BH3 at (Karimpur) (based on Datta and Subramanian 1997)

layers are scattered among the fine sandy layers at different depths. BH3, which is located in the upper portion of the basin, has layers of sand and clay that alternate until a depth of 35 m, at which point a fining-upward sequence of sandy layers begins (Koenig 2011; Mukhopadhyay et al. 2006; Bhattacharya et al. 2001; Mukherjee and Fryar 2008; Ghosal et al. 2015).

4.3 The Changes Over Time

4.3.1 Is Churni a Natural River or an Artificial Canal?

In many documents including historical accounts of the Bengal and Nadia districts, local literature, and scientific research, the Churni River has received its distinctive character, i.e., an artificial channel (Ganguly 2021; Das 2022; Chattapadhaya 2007; Thakur 2019; Chatterjee 2013). For their original detailed Bengali works, please access Annexure - I to IV in the supplementary document. Their works placed a common story that Maharaja Krishnachandra of Nadia made a special attempt to divert a portion of the Ichamati River by creating this canal and Maharaja Krishnachandra named it 'Churni'. This incident, i.e., harnessing (stealing) water through the artificial channel gave birth to the name "Churni", meaning a female thief (Sahitya Samsad 2000; Das 2003). If their speculations are considered true, the construction of the Churni River represents an important anthropogenic signature of natural landscape alteration in this region. One of such story states -

"নদিয়ার রাজা কৃষ্ণচন্দ্র রায় সপ্তদশ শতকের শেষের দিকে বর্গি আক্রমণের হাত থেকে রক্ষা পাওয়ার জন্য তিনি তাঁর রাজধানী কৃষ্ণনগর থেকে মাজদিয়ার শিব নিবাসে স্থানান্তরিত করেন। এই রাজধানী স্থানান্তর করণের বিষয়ে ঐতিহাসিকদের বিভিন্ন মতবাদ পাওয়া যায়। কেউ কেউ মনে করেন, রাজা কৃষ্ণচন্দ্র তাঁর রাজধানীর সুরক্ষার জন্য রাজধানীর চারিদিকে পরিখা খনন করে দেন। এই পরিখাটি হাওলি নদীর মাথা থেকে তৈরি করা হয়। যেহেতু এই খালটি হাওলি নদীর মাথা চূর্ণ করে খনন করা হয়েছিল, তাই এই খালটির নাম দেন 'চূর্ণী'। .... ১৭৮০ সালের পর থেকেই নাকি এই খাল চূর্ণী নদী নামে পরিচিতি লাভ করে"। (গাঙ্গুলি 2021)।

Nadiẏāra rājā kr̥ṣṇacandra rāẏa saptadaśa śatakēra śēṣēra dikē bargi ākramaṇēra hāta thēkē rakṣā pā'ōẏāra jan'ya tini tām̐ra rājadhānī kr̥ṣṇanagara thēkē mājadiẏāra śiba nibāsē sthānāntarita karēna. Ē'i rājadhānī sthānāntara karaṇēra biṣaẏē aitihāsikadēra bibhinna matabāda pā'ōẏā yāẏa. Kē'u kē'u manē karēna, rājā kr̥ṣṇacandra tām̐ra rājadhānīra surakṣāra jan'ya rājadhānīra cāridikē parikhā khanana karē dēna. Ē'i parikhāṭi hā'ōli nadīra māthā thēkē tairi karā haẏa. Yēhētu ē'i khālaṭi hā'ōli nadīra māthā cūrṇa karē khanana karā haẏēchila, tā'i ē'i khālaṭira nāma dēna 'cūrṇī'. .... 1780 Sālēra para thēkē'i nāki ē'i khāla cūrṇī nadī nāmē pariciti lābha karē.

[Meaning: "Krishna Chandra Raya of Nadia shifted his capital from Krishnanagar to Siva Nivas at Majdia in the late seventeenth century to escape Bargi invasion. Historians have different theories about the relocation of this capital. Some believe that King Krishnachandra dug moats around the capital to protect his capital. This moat is made from the headwaters of Hawli River. As this canal was dug by crushing the head of Haoli river, hence the name of this canal is 'Churni'. .... From 1780 this canal came to be known as Churni Nadi".]

The above narrative contains ambiguities from several perspectives. First, historical records indicate that the events of Maharaja Krishna

Table 4.1 Chronological events of changes in the MCR

Year	Events	References
1820	Mr. Robinson removed 300 snags from the river which encumbered the channel and caused innumerable wrecks	Reaks (1919, p. 49)
1840	Mathabhanga was taking off from Jalangi at Bausmari	
1850	In November 1850, the primary flow of the Ganges diverted into a bend of the Mathabhanga above Jutpara, creating a separate offtake of Mathabhanga	Reaks (1919, p. 46)
1918	The entrance of Jalangi and Mathabhanga became united	Das (2013)
1821	A caisson and several old boats were deliberately submerged at the entrance of Kumar River, and a 1540-yard cut was made across a bend of the Mathabhanga River to divert water into it	Reaks (1919, p. 49)
1859	Referring to the five river cuts made in 1859 by Mr. J.W. Armstrong, Mr. Vertannes, the Superintending Engineer, stated in 1881 that four of these cuts have been deepened and are slightly narrower than the natural width of the river. These alterations have proven advantageous for navigation. However, he also expressed concern that these cuts may have significantly raised the flood level downstream, suggesting that the then damages to the Eastern Bengal State Railway are more attributable to this effect than to breaches in embankments above Murshidabad	Garrett (1910, p. 16)

Chandra and Shibnibas as described align more with the middle of the eighteenth century rather than the late seventeenth century (Samaddar 1925; Chatterjee 2008) (Table 4.1). However, the historical context should be reliable to justify the emerging and real meaning of the Churni River and the Shibnibas palace. Additionally, there is a lack of evidence supporting claims that Maharaja Krishna Chandra created the river Churni and the Hauli River was formerly known as Pangasi (Reaks 1919). Questions also arise regarding the authenticity of the narratives surrounding the naming of the rivers and moat around palace.

In this regard, three historical accounts 'Nadia Kahini by Mallik (1910), 'Kshitish Vansabali Charita by Ray Kartikeya Chandra (1876), and 'Sri Maharaj Krishnachandra Rayasya Charitang' by Mukhopadhyay Rajeev Lochan (1811) are the crucial primary sources and provide reliable and accurate understanding compared to other writings.

4.3.1.1 Nadia Kahini

Mallik (1910) chronicled an intriguing tale 'Nadia Kahini' in which Maharaja Krishna Chandra encountered a notorious bandit named Nasarat Khan gaining prominence in his kingdom. The Maharaja located Nasarat Khan's hideout in a dense forest on the east bank of the Churni River. Armed and ready, the Maharaja set camp to confront the bandits, ultimately suppressing them. Following their defeat, he spent the night there. The next morning, while seated by the river, a royal relative named Kripamoy Ray suggested that the Maharaja settle in that beautiful area to ensure his happiness. Coincidentally, fearing raids from the Bargis, Maharaja Krishna Chandra surrounded the area with a river resembling a bracelet and constructed a magnificent palace based on the advice of his Dewan Raghunandan. Additionally, he built his own residence along with two grand Shiva temples, installing two mighty Shiva Lingas and Rama Sita in the other temple (For original Bengali literature, see Annexure - V in supplementary document).

Analyzing the narrative from 'Nadia Kahini,' several points come to light: (i) Nasarat Khan's hideout was situated in a dense forest on the east bank of the Churni River, indicating preexistence of the river Churni; (ii) Maharaja Krishnachandra, after defeating the bandits, was depicted sitting by the riverbank in the morning, washing his face; (iii) the area (presumably where Nasarat Khan's hideout was located) was encircled by rivers in the shape of a bracelet (Kangkana). This narrative raises several questions: Which river was Maharaja Krishnachandra washing his face in? Was it the Churni River, the Ichamati River, or the oxbow lake surrounding Nasarat Khan's hideout in the dense forest? Did Maharaja Krishnachandra excavate a new trench around the capital or renovate the pre-existing deteriorated water enclosure or oxbow lake around the area, turning it into a moat? It's imperative to address these questions before concluding whether the Churni River is an artificial canal. However, 'Nadia Kahini' explicitly states that Nasarat Khan's abode was found in a deep forest on the east bank of the Churni River. This raises the question: Did the Churni River exist before the establishment of Shibnibas?

4.3.1.2 Kshitish Vansabali Charita

In his work Ray, Kartikeya Chandra (1876) discusses Maharaja Krishnachandra's involvement with Churni River, Sivanivasa, and Kankana but expresses uncertainty regarding the naming of the Churni River. Ray recounts how the Maharaja, after careful consideration, selected a location near the Ichamati River, distanced from Krishnanagar. This area, surrounded by water and lush forests, was known as Nasrat Khan Ber due to the residence of a fakir named Nasrat Khan. The king cleared the forest and established a city there. To facilitate the flow of water in moat surrounding the city, canals were dug—one stretching a thousand cubits eastward to link with the Ichamati River, and another spanning three korosha (equivalent to 6 miles) and joining the mouth of the Anjana River north of Hanskhali (For original Bengali literature, see Annexure - VI & VIII in supplementary document).

Maharaja Krishnachandra aptly named the moat Kankana and christened the city Shibnibas. Ray underscores that the area was encompassed by water bodies, likely horseshoe-shaped lakes, with no connection to feeder rivers, rendering them stagnant. To remedy this, Dewan Raghunandan orchestrated the restoration of a decayed river, extending a thousand cubits eastward and six miles westward, thus connecting the moat (Kankana) with the Ichamati River, fortifying the palace's defenses. Crucially, Ray clarifies that the moat's connectivity was achieved through canals cutting eastward and westward, linking with the Ichamati and Anjana Rivers, respectively.

Notably, there is no mention of Krishnachandra naming the canals 'Kankana' or 'Churni'; rather, it is affirmed that the moat was named 'Kankana.' Additionally, the canal from Hardham to Shibpurr, spanning half a krosha (approximately 1.61 km), was restored under Maharaja Krishnachandra's supervision. Ray's account raises the question of whether Krishnachandra excavated a new canal or restored an existing dilapidated river. He mentioned,

> 'হরধাম হইতে শিবপুরের সমীপস্থ ভাগীরথী অধিক দূরবর্তী ছিলনা; এ কারণ কৃষ্ণচন্দ্র গঙ্গাস্নানের সুগমের নিমিত্ত, হরধাম হইতে শিবপুর পর্যন্ত এক খাল কাটাইয়া দেন, তাহাতেই শিবপুরের সন্নিহিত ত্রিমোহিনীর সৃষ্টি হয়। শিবনিবাস হইতে শিবপুর পর্যন্ত যে নদী আছে, তাহার চূর্ণী নাম রাজা কৃষ্ণচন্দ্র দেন, কি এই নদীর যে অংশ পূর্বে ছিল তাহার এই নাম থাকে, তাহা নিশ্চয় জানিতে পারা যায় নাই' (Ray 1876)।
>
> *Haradhāma ha'itē śibapurēra samīpastha bhāgīrathī adhika dūrabartī chilanā; ē kāraṇa kr̥ṣṇacandra gaṅgāsnānēra sugamēra nimitta, haradhāma ha'itē śibapura paryanta ēka khāla kāṭā'iẏā dēna, tāhātē'i śibapurēra sannihita trimōhinīra sr̥ṣṭi haẏa. Śibanibāsa ha'itē śibapura paryanta yē nadī āchē, tāhāra cūrṇī nāma rājā kr̥ṣṇacandra dēna, ki ē'i nadīra yē anśa pūrbē chila tāhāra ē'i nāma thākē, tāhā niścaẏa jānitē pārā yāẏa nā'i.*
>
> ['Bhagirathi near Shivpur was not far from Hardham; this is because Krishna Chandra cut a canal from Hardham to Shivpur for the convenience of bathing in the Ganges, that is why Trimohini near Shivpur was created. Whether Raja Krishnachandra named Churni to the river that runs from Shivnivas to Shivpur, or whether the earlier part of this river has this name, is not possible to know for sure'].

Therefore, it is evident from the above-mentioned statement, that even during Ray's time, the narrative of Krishnachandra naming the Churni River persisted and Kartikeya Chandra Ray tried his best to know the truth, though conclusive evidence remained elusive. Thus, it is plausible that the Churni River existed in a deteriorated state before Maharaja Krishnachandra's relocation of his palace to Shibnibas, whereupon he might have undertaken its restoration.

4.3.1.3 Sri Maharaj Krishnachandra Rayasya Charitang

Sixty-five years prior to Kartikeya Chandra Ray's analysis, Mr. Rajeev Lochan Mukhopadhyay penned 'Kshitish Vansambali Charit' in 1811, offering a narrative about Maharaja Krishnachandra's hunting expedition. Mukhopadhyay vividly describes the Maharaja's discovery of a serene locale within a dense forest, characterized by a small island encircled by rivers, where abundant wildlife flourished. Acting on Krishnachandra's directives, his attendants erected a palace on the island, utilizing the surrounding river as a natural moat. Additionally, the Maharaja ordered the construction of a main road by damming the river to the south, fortifying it as a strategic defense position with cannons flanking its sides (For original detailed Bengali literature, see Annexure - VII in supplementary document).

Mukhopadhyay's account prompts scrutiny of several key elements: (i) the island's isolation amidst rivers, (ii) the repurposing of the surrounding river as a defensive moat, and (iii) the utilization of the riverbed for constructing a roadway.

These details suggest that the Shivanivas palace site was originally encompassed by a river, repurposed as a protective moat. However, it remains uncertain whether the adjacent water body was fed by a flowing river or constituted a stagnant oxbow lake. The absence of canal excavation from 'Kankana to the Ichamati and Anjana Rivers in Mukhopadhyay's narrative is noteworthy. Mukhopadhyay wrote the history 69 years after the event took place, whereas 'Kshitish Vansambali Charit' and 'Nadia Kahani' were composed 134 and 177 years later the event. Therefore, Rajiv Lochan Mukhopadhyay's may be considered more authentic than later two.

Even today, we find several meander scars and oxbow lakes around the area indicating ancient river sources. Hence, it's quite plausible to consider that Maharaja Krishnachandra might have constructed a moat around Shibnibas after revitalizing the oxbow lake. He likely revived the river from its deteriorated state, only for it to degrade again 25 years later, as noted during Rennel's survey.

Fast forward to 2024, much like Bhairab and Jalangi, the Mathabhanga now stands detached from the river Padma. Looking ahead, unless a miraculous intervention occurs, the River Churni could inevitably meet a fate akin to Jalangi, perhaps in the foreseeable future.

4.3.1.4 Anjana Theory

The Anjana Theory posits that the stretch of the Churni River extending from Hanskhali to its confluence with the Hooghly River at Shibpurr was originally part of the Anjana River (Fig. 4. S1). According to this theory, Maharaja Krishnachandra renamed this section of the river 'Churni' after extending the Churni canal from the Ichamati River to Hanskhali, thereby connecting it with the moat surrounding Shibnibas palace. Regarding the Anjana River, Ray (1876) stated that- 'এই স্রোতস্বতী জলঙ্গী নদীর একটি শাখা। ইহা কৃষ্ণনগরের পশ্চিম দিয়া দক্ষিণাভিমুখে গমন পূর্বক যাত্রাপুর গ্রামের অনতিদূরে দ্বিধারা হয়। এক ধারা, জয়পুর, জালালখালি, ধর্মদা, বাদকুল্লা প্রভৃতি গ্রামের নিকট দিয়া আরংঘাটা সন্নিহিত মামজোয়ান গ্রামের নিকট যাইয়া দক্ষিণ বাহিনী হয়; অপর ধারা যাত্রাপুর ও বেৎনা প্রভৃতি কতিপয় গ্রামের নিকট দিয়া, হাঁসখালি গ্রামের সমীপস্থ হয়, এবং তদনন্তর, দক্ষিণ মুখে মামজোয়ানের নিকট যাইয়া পূর্ব ধারার সহিত মিলিত হইয়া যায়'। .... 'ওই নদী শেষে হরধামের উত্তর দিয়া গিয়া চাকদহের নিকট গঙ্গার সহিত মিলিত হয়'। Key points for analysis include the origin and textual references of the theory, as well as its factual plausibility (For original Bengali literature, see Annexure - V in supplementary document).

(i) Villagers downstream from Hanskhali are not familiar with the names 'Anjana' or 'Kankana' linked to their part of the Churni River. (ii) The Churni River's depth (4.5 m) contrasts significantly with the Anjana River's shallower depth (1.96 m), casting doubt on the existence of the latter. (iii) The Anjana's flow against the south and southwest slope of its basin raises questions about its viability, especially considering its smaller size. The presence of Haler Khal, the northern branch of the

Anjana, adds complexity to this anomaly. (iv) The presence of an oxbow lake prior to the capital's relocation suggests the pre-existence of a river in the area, possibly originating from the river Mathabhanga near Shibnibas. Dewan Raghunandan's efforts for the restoration of the river further support this assumption. (v) The Anjana River may have been comparable in scale to the Churni during that era, explaining why Kartikeya Chandra Ray referred to the downstream Churni as the Anjana.

4.3.1.5 Principles of River Science

(i) In our research, we conducted a comparative analysis of the meandering patterns of the Bhagirathi, Jalangi, and Mathabhanga-Churni Rivers utilizing the sinuosity index developed by Brice in 1964. The Churni River exhibits a lower sinuosity index (1.30) in contrast to the Bhagirathi River (2.96) and the Jalangi River (2.67). Bandyopadhyaya (2007) has attributed this phenomenon to the Churni's alignment with a fault line, which diminishes its sinuosity. Additionally, surveys have revealed that the soil composition within the Churni basin possesses a high stickiness factor, impeding the river from adopting a more pronounced meandering pattern (Table 4.2).

(ii) In the reach spanning from Pabakhali to Chowgancha, along both banks of the Churni River, the presence of at least five horseshoe-shaped lakes and meander scars is conspicuous. Moreover, in the vicinity of Mayurhat-Tarak Nagar, Mamjoan, and Parua to the east, numerous meander scars are discernible. These geographical traits serve as compelling evidence of the natural genesis and evolution of the Churni River, without any detected anomalies or irregularities. It is imperative to delve into the correlation between the oxbow lakes, meander scars, and the Churni River. To ascertain this connection, we employed the doubt index (Das et al. 2020), which yielded results indicating a 90% likelihood that these features indeed originated from the Churni River.

(iii) In the examination of river characteristics, it is common to observe channel asymmetry, where the river bed displays non-symmetrical features. Employing Knighton's (1981) method, we evaluated the asymmetry of the Churni River, Bhagirathi River, and Jalangi River. The findings reveal that the Churni River exhibits comparatively lower levels of asymmetry (ranging from ±0.01 to ±0.48) in contrast to the Bhagirathi River (ranging from ±0.21 to ±0.62) and the Jalangi River (ranging from ±0.12 to ±0.33). Although the Churni River does not demonstrate symmetrical attributes akin to those of an artificial canal, it showcases a reduced degree of asymmetry, likely influenced by tidal flows in its lower stretch, which tend to mitigate asymmetry.

(iv) Natural deltaic channels, unlike static artificial canals, are characterized by dynamicity. The planform geomorphology of the Churni River demonstrates this dynamism, as shown in Table 4.3. Parameters such as wavelength, radius of curvature, arc angle, and amplitude of

Table 4.2 Comparison of morphometric parameters of Bhagirathi, Jalangi, and Mathabhanga-Churni

Subject	Bhagirathi	Jalangi	Churni
Sinuosity index	2.96	2.67	1.30
Presence of oxbow lakes	High	Moderate	Less
Channel asymmetry (Knighton 1981)	±0.21 to ±0.62	±0.12 to ±0.33	±0.01 to ±0.48
Channel shifting	High	Moderate	Less
Width/depth ratio	~100	~41	~19

Table 4.3 Planform morphometric variability (SD and CV) indicating toward natural evolution of the river (after Sarkar 2021)

Descriptive statistics	Wavelength (m)		Radius of curvature (m)		Arc angle (degree)		Amplitude (m)	
	Year 1984	Year 2018	Year 1984	Year 2018	Year 1984	Year 2018	Year 1984	Year 2018
Average	1896.33	1877.44	471.58	463.91	127.55	126.76	681.12	687.54
SD	931.19	963.02	464.82	484.69	39.18	37.87	514.15	536.14
CV	49.10	51.29	98.57	104.48	30.72	29.87	75.49	77.98
IQR	735.84	838.06	441.03	419.92	37.42	34.63	269.25	281.75
Kurtosis	1.89	5.71	9.01	13.13	0.22	− 0.02	11.26	10.75
Skewness	1.28	1.84	2.83	3.29	0.23	− 0.05	2.64	2.61

meanders exhibit significant variability. Unlike canals, which tend to be straight, the Churni River features numerous meanders, highlighting its natural origin and evolution.

(v) While deltaic rivers are prone to channel shifting, the Churni River shows less migration compared to the Bhagirathi and Jalangi Rivers. However, near 'Dakshin-Majdia' village southwest of Ranaghat, the Churni River has shifted up to 671 m southeastward. This notable movement provides additional evidence of the river's natural origin and evolutionary processes.

The river science analysis strongly supports the natural origin of the Churni River, refuting its classification as an artificial canal. According to Rudra (2010), the Churni River existed disconnected with the Mathabhanga River during Rennell's survey from 1766 to 1780, indicating its natural formation. The description of Shivanibaas, surrounded by water, suggests the presence of a river before it became the capital in 1742, with the palace moat likely representing remnants of its decayed channel. Despite restoration efforts by Krishnachandra Raja in 1742, the Churni River deteriorated again by 1767, detaching from the Mathabhanga River. The Stevension-Moore Committee's report of 1919 refers to the 'Modern Churni,' suggesting the existence of an 'Ancient Churni' prior to the late eighteenth century. Captain Colebrooke's account in the report further supports this notion, highlighting changes in the navigability of the Mathabhanga River over time (Colebrooke 1803; Stevension-Moor et al. 1919; Rudra 2010). These findings imply that Maharaja Krishnachandra possibly restored the pre-existing ancient natural Churni River, as indicated by historical records and survey data.

4.3.1.6 A Cost-benefit Analysis

Maharaja Krishna Chandra's decision to excavate a new canal for the Churni River raises questions about its necessity and feasibility. The meandering path observed in the course of Churni suggests its natural origin, not an artificial canal which is generally straight for a considerable length. Analysis of the labor and time required for such an excavation reveals significant challenges. For instance, to excavate the estimated volume of soil employing 300 workers daily for the canal, approximately 6,315,943 laborers would be needed over a period of 57 years and 7 months. Alternatively, employing 3000 workers daily could shorten the timeline to around 5 years and 10 months (Das 2023, 2024). However, such a massive investment of time, labor, and resources seems questionable, especially considering the lack of purpose or relevance in absurd stories about the project. Moreover, it seems unlikely that a discerning Dewan like Raghunandan would have made such a decision. A more cost-effective and efficient approach would likely involve restoring the pre-existing decayed course of the river, rather than undertaking such a monumental excavation.

Based on an analysis of historical records, the geomorphology of the river channel and paleo-channels, as well as logical reasoning, it becomes evident that the Churni River is a naturally formed waterway. However, by the mid-eighteenth century, it had degraded significantly, prompting Maharaja Krishnachandra to undertake its restoration to safeguard his newly established capital at Shibnibas. Whether Krishnachandra himself bestowed the name "Churni" upon the river remains uncertain.

4.3.2 Before Independence

Since the first quarter of the nineteenth century, Nadia Rivers (Bhagirathi, Jalangi and Mathabhanga-Churni) were of prime concern of British Raj and different artificial efforts were adopted to keep them navigable (Table 4.4). Abnormally winding offtakes of these rivers and huge sand deposit closing offtakes made Bhagirathi and Jalangi less suitable in comparison to Mathabhanga for navigation and channel training.

Several cuts across the winding channel of the river to facilitate navigation through the straightened course were made during the first half of the nineteenth century. The navigation route of the MCR was shortened by 10.46 km (12.3–1.84 km) by a single cut of 1.84 km just above Pangasi offtake during this period.

4.3.3 After Independence

The MCR basin has experienced significant changes after the independence of India, mostly as a result of road-stream crossings growing industry, and modifications to agricultural practices (Rudra 2010, 2014, 2018). The growth of riverside agriculture, which resulted in more on-bed farming, has brought notable changes in channel morphology (Sarkar and Islam 2020). The installation of bridges over river channels has altered the flow of water and process of sedimentation, resulting in modifications to the hydrology and morphology of rivers. Additionally, the river has been suffering greatly as a result of the rapid growth of the brickfield industry near the river (Sarkar et al. 2021). The rapid development of these industries within the riverbanks due to easy access to clay and water has also added the extra complexity within the riverbed and river bank. Additionally, although not directly threatening the Mathabhanga River, the Farakka Barrage on the Ganges River has had an extensive indirect impact on the hydrology and morphology of the MCR. The Farakka Barrage regulates the discharge pattern of the downstream Padma River. In the dry months the Padma River gets insufficient discharge and the water level in the river remains lower compared to the bed level of the Mathabhanga River. Therefore, the Mathabhanga River has experienced more sedimentation at its offtake point resulting in seasonal separation between the two rivers. Moreover, this phenomena also reduces the discharge in the MCR, increases sedimentation process, and modifies the depth, width, and channel form of the river. The details have been outlined in the following sections.

Road-stream crossings: distribution, types, and geometry

A variety of stream crossings, including culverts and bridges (with rail and foot roads), have been built in the MCR basin to improve transportation accessibility and connectivity (Fig. 4.5). These crossings range in terms of their dimensions, structures, and orientation. Many crossings are often found along the MCR, which goes through a densely populated area. The hydromorphological responses to stream crossings are very complex (Roy and Sahu 2017; Roy and Sahu 2018; Roy 2013; Surian and Rinaldi 2003; Wang et al. 2014). Therefore, the entire river has been gauged from this perspective.

Three types of bridges have been seen over the Churni River: road (5), rail (2), foot-bridge made of concrete (2), and bamboo (9) bridges (Figs. 4.6 and 4.7a–c) (Table 4.5). Although the foot bridge has less influence over the morphology of the river channel, it is nevertheless recognized as an anthropogenic mark on

Table 4.4 Chronological arrangement of anthropogenic signatures on *MCR* channel during the nineteenth century

Year	Superintending engineer of Nadia River division	River training scheme had taken	Result	Sources
1819	Mr. C. K. Robinson was appointed Superintendent and Collector Mathabhanga River and Nadia River Division of Public Works Department was originated	An effort to close the offtake of River Kumar by 'bandals' (a branch of the river Mathabhanga was diverting 5/6th of the water to the southeast) was adopted	Unsuccessful attempt	Reaks (1919)
1820 June	Mr. May	Attention was given to clear snags	–	Reaks (1919)
1821	Mr. May	Cleared many snags including 300 timbers sunken, many wrecked boats and cut down leaning trees on banks likely fall into the river	Cleared	Garrett (1910), Reaks (1919)
1821	Mr. May	To close the offtake of the river Kumar which was still taking away 3/4th of the Mathabhanga supply	A caisson and some old boats were sunk across the offtake of the river Kumar Kumar offtake was closed	Reaks (1919)
1821	Mr. May	Straightening of the channel of the river Mathabhanga to attract water into the main channel	A cut of 1540 yards (1408.176 m) was made across a bend of the river Mathabhanga to attract the water Mathabhanga was navigable throughout with 3feet draft	Reaks (1919)
1821	Mr. May	Closing of the offtake river Pangasi, the second branch of the river Mathabhanga takes off at 23°44′09″ N and 88°52′02″ E	Unsuccessful attempt	
1823	Mr. May	Dredging of Mathabhanga offtake	A dredging machine drawn by bullocks was employed but owing to deteriorated channel and severely closed offtake by sand deposit, the attempt was unsuccessful	Reaks (1919)
1824	Mr. May	Due to better condition of Jalangi and Bhagirathi, attention was given to those two rivers also and Mr. May was given additional charges	–	Reaks (1919)
1825–30	Mr. May	Two attempts were taken to divert the water flowing away through the Pangasi	Attempts were unsuccessful	Reaks (1919)

(continued)

Table 4.4 (continued)

Year	Superintending engineer of Nadia River division	River training scheme had taken	Result	Sources
1828–29	Mr. May	Maintaining minimum depth for navigation	A steam dredge was employed below Kumar entrance without much result due to unsuitable machinery and heavy draft (6 feet) of the dredging boat	Reaks (1919)
1831–34	Mr. May	To have a satisfactory solution of Pangasi problem, a proposal was made to divert the current into a loop taking off the right bank of the Mathabhanga above the Pangasi offtake	–	Reaks (1919)
1852	Captain Smyth	Construction of a channel for 10 miles below the Pangasi by artificial banks with the use of kodalis	A passage was maintained for small boats	Reaks (1919)
1913	Mr. F.A.A. Cowley, Superintending Engineer, Central Circle and Deputy Secretary, Irrigation Department	Opined that only Mathabhanga out of all the Nadia Rivers is worthy to be kept artificially navigable because its offtake is just above the Sara bridge and not abnormally winding as is common for two other Nadia Rivers	–	Reaks (1919)

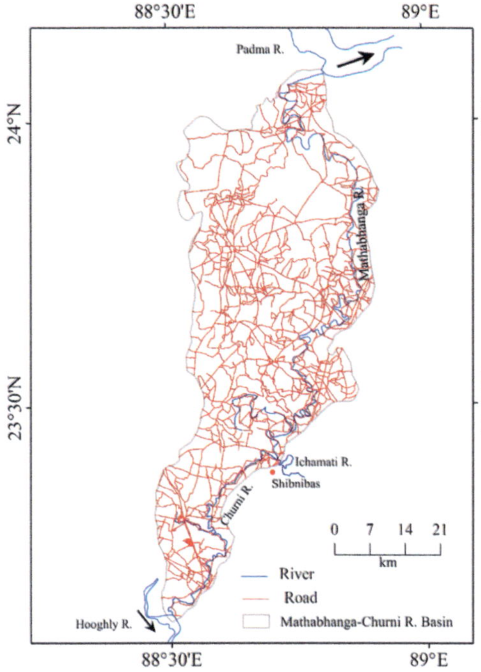

Fig. 4.5 Road network in the MCR basin

the river channel. The Shibnibas road bridge, which has ten piers in each group, is the ideal illustration of grouped piers. The piers are five meters in circumference and have a cylindrical form. Two of the four piers in the group are bank-attracted, while the other two are located extremely close to the center of the river channel. Since the newly constructed railway bridge at Kalinarayanpur destroyed the previous bridge, the river has undergone extensive modification during the last five years (2014–2019). Currently, three bridges—two rail and one foot—exist relatively close to one another and are essential to the river channel contraction. Every rail bridge is supported by four piers: two are circular, with a diameter of 32 m, and they are situated at the riverbed where, during dry seasons, the flow depth is maintained at least 1.15 m; the other two are rectangular, with a diameter of 32 m, and they are situated at the river on the river's bank.

Road-stream crossings and channel morphology

It is evident how the bridge has affected the Churni River's channel morphology. With respect to the Shibnibas Bridge, there is a general tendency in the changes in average depth and cross-sectional area as the distance from the bridge piers increases. Cross-sectional area values for both the upstream and downstream portions of the bridge rise with increasing distance from the structure. The cross-sectional area measurements for the downstream portion are 107, 141, and 160 m^2 at every 30 m interval, whereas the upstream portion's values are 118, 165, and 208 m^2.

A further observation has revealed that the cross-sectional area values in the upstream segment are higher than in the downstream region. This is because, in general, the downstream of

Fig. 4.6 Bamboo foot bridge connecting Chandan Nagar and Shibnibas

Fig. 4.7 Geometry of road-stream crossing **a** bridge at Kalinarayanpur, near Ranaghat, **b** bridge at Shibnibas, **c** development of scouring at Kalinarayanpur bridge

the bridge is more morphologically susceptible to changes brought about by the construction of bridges than is the upstream portion. Furthermore, the cross-sectional area and average depth appear in the same way. The average depth (downstream portion) rises with increasing distance (d=2.1 m at 30 m, d=2.8 m at 60 m, and d=2.95 m at 90 m downstream of the bridge). The upstream portion shows the similar trend (d=2.3 m at 30 m, d=3.3 m at 60 m, and d=2.95 m at 90 m downstream of the bridge). However, the morphological changes in the vicinity of the Kalinarayanpur bridge (near Ranaghat) appear differently from those of the Shibnibas bridge. The value of the cross-sectional area (downstream of the bridge) decreases with increasing distance (A=569 square meters at the bridge's immediate vicinity, A=306 square meters at 30 m below the bridge, and A=273 square meters at 60 m below the bridge). This decreasing trend is caused by the fact that at the bridge site, the formation of local and contraction scouring is a major factor in raising the average depth of the channel, which in turn causes the cross-sectional area value to be higher than that of the distant cross-sections. Similar findings have also been obtained for the bridge's upstream sections (A=286 square meters right up the bridge, A=193 square meters 30 m downstream). Additionally, the average depth at the downstream bridge site is 6.84 m, yet from a 50 m distance, it seems to be 4.32 m. The scouring factor is the reason for the high average depth value observed at the bridge location. Furthermore, it has been observed that the average depth in the upstream section increases naturally with increasing distance (d=3.36 m at 30 m, d=3.96 m at 60 m, and d=4.28 m at 150 m distance).

It is evident from the asymmetry's nature that asymmetry is greatest right up and below the bridge and diminishes with increasing distance. Regarding the Shibnibas bridge, every asymmetry value appears to be positive. A*, A1, and A2 are determined to be, respectively, 0.27, 0.67, and 0.37 immediately under the bridge, and they appear to be, respectively, 0.23, 0.61, and 0.29, 90 m below the bridge.

Table 4.5 Distribution of road, rail, and bamboo bridges over the MCR

Bridge name	Descriptions	Location	Length (m)	Width (m)
Krishnaganj MathaBhanga River	Majhdia road	23°24′58.92″ N 88°42′19.08″ E	59	7
Churni Bridge	Bajitpur-Shibnibas Road	23°24′57.08″ N 88°41′21.91″ E	132	7.5
Hanskhali Bridge	Hanskhali Bridge	23°21′19.86″ N 88°36′25.72″ E	83	6.8
Bridge over river Churni (Jugol Kishor, Aranghata)	River Churni (Jugol Kishor)	23°14′43.98″ N 88°36′24.50″ E	138.5	7.5
Churni Bridge, Ranaghat	NH12	23°10′39.98″ N 88°33′30.04″ E	130	14
Cable Suspension Foot Bridge over River Churni, Ranagha	Suspension Foot Bridge over River Churni Road	23°10′45.43″ N 88°33′42.76″ E	102.5	2.4
Kalinarayan rail bridge	Connecting Kalinarayanpur to Ranaghat	23°12′21.11″ N 88°34′31.78″ E		20
Kalinarayan foot bridge	Connecting Kalinarayanpur to Ranaghat	23°12′21.11″ N 88°34′31.78″ E	120	1–2
Raghunathpur	Bamboo foot bridge connecting Raghunathpur and Ghoraghata	23°13′59.91″ N 88°34′55.32″ E	54 m	1–2
Byspur	Bamboo foot bridge connecting Byspur and Hajrapur	23°16′53.29″ N 88°35′4.45″ E	58	1–2
Mamjoan Ghat	Bamboo foot bridge connecting Mamjoan and Takshali	23°17′56.01″ N 88°34′41.50″ E	61	1–2
Bhairavchandrapur	Bamboo foot bridge connecting Bhairavchandrapur and Benali	23°23′5.43″ N 88°38′46.86″ E	41	1–2
Chandannagar Ghat	Bamboo foot bridge connecting Chandan Nagar and Shibnibas	23°24′41.96″ N 88°40′47.80″ E	53	1–2
Shibnibas	Bamboo foot bridge connecting Majdia and Shibnibas	23°24′50.74″ N 88°41′38.39″ E	64	1–2
Swarnakhali Ghat	Village road connecting Swarnakhali village itself	23°27′20.15″ N 88°41′36.48″ E	40	1–2
Taldaha	Bamboo foot bridge connecting Taldaha and Mahesh Chandrapur	23°27′11.61″ N 88°44′19.34″ E	40	1–2
Gobibdapur	Village road connecting Gobibdapur village itself	23°28′14.68″ N 88°44′45.06″ E	41	1–2

Similarly, A1, A2, and B appear as 0.28, 0.70, and 0.36 immediately upstream of the bridge A*, and as 0.03, 0.35, and 0.17, respectively, 90 m upstream of the bridge. Furthermore, it is evident that the upstream portion has a higher reduction in asymmetry with increasing distance than the downstream section. This is because, in comparison to the upstream portion, the channel below the bridge has been more severely affected across a longer distance. The bridge in Kalinarayanpur (Ranaghat), on the other hand, has both positive and negative asymmetries. An analogous finding to the Shibnibas bridge is that the upstream portion experiences a greater reduction in asymmetry with increasing distance than the downstream portion. Near the bridge, the values of (A*), (A1), and (A2) have been estimated to be 0.23, 0.15, and 0.09, respectively, but they appear to be – 0.02, 0.18, and 0.08, respectively, 90 m above the bridge. Furthermore, A*, A1, and A2 have been estimated to be, respectively, 0.27, 0.48, and 0.21 just below the bridge, but they appear to be, respectively, – 0.12, 0.06, and 0.02, 90 m below the bridge (Fig. 4.8a, b).

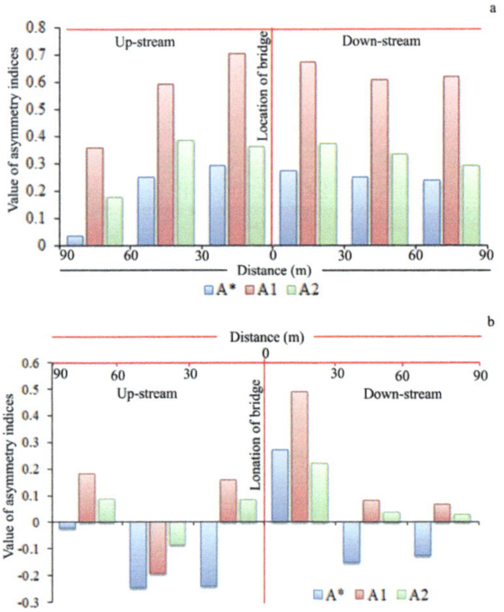

Fig. 4.8 Nature of channel asymmetry at bridge sites **a** at Shibnibas, **b** at Kalinarayanpur

4.4 Channel Geomorphology

4.4.1 Plan View Morphology

4.4.1.1 Sinuosity Index

The meandering of a river or stream in relation to a straight line is measured by channel sinuosity. The computation involves dividing the length of the channel by the straight-line distance (Leopold and Wolman 1957; Schumm 2005). Understanding the historical development of river channels depends on the channel sinuosity index. The channel may be dynamic if there are changes in sinuosity over time (Mueller 1968; Morisawa 1985). The sinuosity index that Brice (1964) offered was used in this investigation. We utilized Das's (2017) calculations to obtain data on the channel's sinuosity prior to 1984. Furthermore, photos from Google Earth were used to calculate the data after 1984. From the 1850s onward, maps with reliable scales and bearings are accessible for the computation of sinuosity. As a result, the difference in SI was computed from the most recent time to the earliest, spanning 1849–1844. Nonetheless, the sinuosity hasn't altered much since 1925 (standard deviations for the Mathabhanga and Churni Rivers are 0.020 and 0.035, respectively). Table 4.6 shows that the river has become less active and is in a condition of moribund degradation. Its sinuosity has remained unaltered or static, and it has not had a meander cut-off since 1979.

4.4.1.2 Meander Cut-off: Distribution and Pattern

The meander cut-off serves as a crucial gauge for the dynamics of rivers. The magnitude of channel dynamics increases with meander cut-off emergence rate. In the past, the MCR was hyperdynamic, leaving behind a number of abandoned channels that are characteristics of fluvial origin. There are ten clearly marked cut-off points along the river; seven of them happened between 1849 and 1855 and between 1916 and 1828. The significant average discharge—101.95 $m^3 s^{-1}$ in 1915—made the river channel more active. The river grew more

Table 4.6 Sinuosity index of the MCR

Years of surveying	Number of oxbow lakes formed in the entire course of the Mathabhanga-Churni River since the preceding survey	Sinuosity index	
		Part of the Mathabhanga River	Part of the Churni River
1849–55	3	2.18	1.50
1916–28	4	2.21	1.64
1979	0	2.24	1.71
1989	1	2.19	1.71
2001–07	0	2.21	1.71
2020–21	0	2.20	1.69

Computed from Das (2017) and Google Earth Image (18 June 2020)

sinuous between 2.18 in 1849–55 and 2.21 in 1916–28, despite these cut-off structures, suggesting the creation of more meander loops (Das 2017). The abandoned loops along the Mathabhanga River are longer than the other cut-offs at Baot (M1: 9.9 km) and Asmankhali (M4: 9.8 km). On the other hand, the cut-offs that exist in India (M7: Harishnagar cut-off and M8: Bijoypur cut-off) are the shortest. The greatest cut-off, Munshiganj (M5), has an arc that is 3.7 km shorter than the other. The largest abandoned meander loop in the Mathabhanga River is the dried-up Munshiganj cut-off in Bangladesh, measuring 132.4 m at its widest point. The smallest one is known to exist in the Khalishakundi, measuring 35.5 m at its widest point. Here has only been one meander cut-off in the course since 1950 and that was at Khalishakundi. This observation suggests that from the 1950s, the MCR became nearly steady and less active (Table 4.7).

4.4.1.3 Meander Geometry

An examination of the anatomy of meander geometry can provide a better understanding of the spatiotemporal variation of the meander planform (Leopold and Maddock 1953; Leopold and Wolman 1960; Das 2014; Morisawa 1985). Any alluvial channel has inherent instability in the meander. Throughout a river's whole course, meander behavior can range from regularly fluctuating to steady. When compared to historical data, all of the meander's basic variables (λ, A_m, r_c, and θ_a) have bigger magnitudes now than they had in the past (Sarkar et al. 2020). Furthermore, the ANOVA statistics ($p<0.05$) and absolute values of the meander geometry variables on 118 meander loops demonstrate that the meander geometry has been constant since 1984. For instance, the meander shape index for 1984 showed that, for the year, the proportion of acute, regular, and open meanders was 43.22%, 22.88%, and 33.98%, respectively

Table 4.7 Meander cut-off of the MCR system

ID	Name of the oxbow lake/nearest locality	Relative position	Formation period	Width at the apex (m)	Nearest channel width(m)	Length of the abandoned loop (km)
1	Boat	Right bank	1849–55; 1921–28	100.2	34.8	9.9
2	Khalishakundi	Right bank	1979–89	35.5	31.1	6.6
3	Morbhanga	Left bank	1849–55; 1921–28	124.6	41.4	3.7
4	Asmankhali	Right bank	1849–55; 1921–28	100.3	42.6	9.8
5	Munshiganj	Left bank	1849–55; 1916–18	132.4	34.7	3.7
6	Darshana	Left bank	Before 1849	124.7	55.7	8.3
7	Harisnagar	Left bank	Before 1849	97.5	40.2	3.2
8	Bijoypur	Left bank	Before 1849	118	42.6	3.0
9	Bishnupur	Right bank	Before 1849	248	40.6	9.10
10	Khal Boalia	Right bank	1916–18	90.7	42.0	3.54

Compiled from Das (2017)

(Fig. 4.9). In contrast, for the year 2018, it was 41.52%, 27.11%, and 31.35%, respectively. In addition, the meander form index values for 1984 and 2018 show that there were 45.49% acute, 19.83% regular, and 34.78% open-type loops, respectively, compared to 42.88% acute, 25.36% regular, and 31.76% open-type loops in 1984. Consequently, the distribution of loops according to SmI and FmI values indicates that meander loop nature has been constant since 1984, which is further supported by ANOVA statistics (Fig. 4.10).

4.4.2 Cross Profiles

4.4.2.1 Channel Width

Since 1849, the channel width has been steadily decreasing. Prior to 1849, the channel width was around 200 m, with the meander cut-off No. 9 having a maximum known width of 248 m (meander cut-offs No. 4, 5, 6, 9, 10) (Fig. 4.9a–g). The average channel widths of the Churni and Mathabhanga Rivers are still considerably less than they were in the past, at 63.5 m and 70 m, respectively (Fig. 4.11a–d).

As a result, the current channel width decrease as compared to historical channel widths shows the gradual degradation of the channel. The lowest channel widths have been observed, especially in the vicinity of the bridge locations (52 m at Shibnibas, 1 km downstream of the Churni River; 56 m at Hanskhali bridge, 15 km downstream; 60 m at Aranghata bridge, 35 km downstream; 43 m at Kalinarayanpur bridge, 45 km downstream; 49 m at Ranaghat bridge, 50 km downstream). The longitudinal channel width research indicates a few unique trends based on fourth-degree polynomial curves, with the trend being almost the same for both the pre-monsoon and monsoon seasons. The trend is upward for 15 km downstream, then drops for 40 km downstream before rising again to the mouth of the Churni River. The cut-off sections of the river illustrate how much wider the river was in the past than it is now. Surveying 200 channel width data points—100 from 10 abandoned channels and the remaining 100 from the nearby current channel—has confirmed this assertion. An analysis of the width data, followed by an ANOVA, revealed considerable width variation between the cut-off channels and their nearby current course (Table 4.8).

4.4.2.2 Channel Depth Analysis

Historical sources (Reaks 1919) concerning the depth at the river's entrance and Hanskhali reveal a complex pattern. 19th-century government documents show a steadily falling tendency between 1840 and 1853 for both the highest and lowest river depths. Conversely, Fig. 4.11b shows a pattern of constant expansion for both the greatest and lowest depths between 1902 and 1918. At the entrance, measurements have shown that the lowest depth was 0 m for a number of years, while the greatest depth was 8.38 m in 1846. An growing tendency in depths is seen in the first and second decades of the twentieth century, from 1903 to 1917, presumably as a result of the favorable conditions during offtake. The Mathabhanga River was navigable year-round when Captain Colebrooke studied it in 1795 (Reaks 1919). Mathabhanga's lowest depth at its offtake in 1819–1820 was 0.8–1 m. This depth allowed for large trading boats that typically weighed 250–300 maunds (1 maund = 37.3242 kg) to pass (Biswas 2001; Sarkar et al. 2021). The appointment of Mr. C. K. Robinson as the administrator of the Nadia Rivers division in 1819–18 is important (Sarkar et al. 2020). The river was navigable through the end of March 1820–1821. From 1821 to 1822, the river was open for navigation year-round (Garrett 1910). Between 1913 and 1918, there was a downward tendency in the river's depth near Hanskhali. The lowest depth of 0.91 m was reported in 1917, considerably for the boats of 112 quintals, while the highest depth of 6.4 m was recorded in 1913. Based on the available records, it appears that although the river saw some decline throughout the nineteenth century, its offtake and the remaining portion of the channel remained navigable for vessels with a minimum depth of around one meter (Biswas 2001). The river's offtake was, nonetheless,

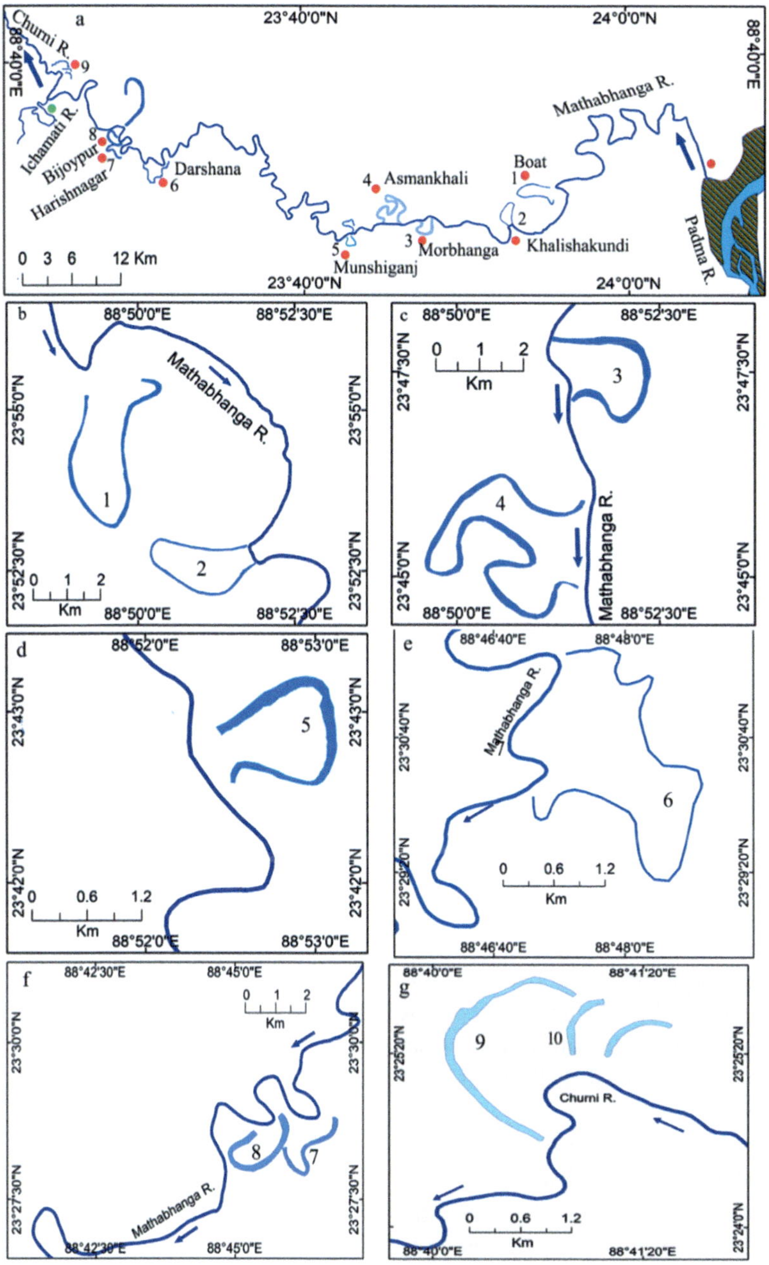

Fig. 4.9 History of meander cut-offs. **a** Distribution of meander cut-offs along the course of the MCR; **b** cut-offs of Boat and Khalishakundi; **c** cut-offs of Morbhanga and Asmankhali; **d** cut-offs of Munshiganj; **e** cut-offs of Darshana; **f** cut-off of Harisnagar and Bijoypur and Bishnupur; **g** cut-off of Bishnupur and Khal Boalia (*Note* The ID number of the oxbow lakes are mentioned in Table 4.7

closed off for many months in the late 20th and early twenty-first centuries. Figure 4.12c mentions similar observations on the decline in the channel depth. This trend suggests that the cross-sectional area and channel depth have significantly decreased. More specifically, the Padma River heavily silts up the entry to the MCR (Sarkar and Islam 2020).

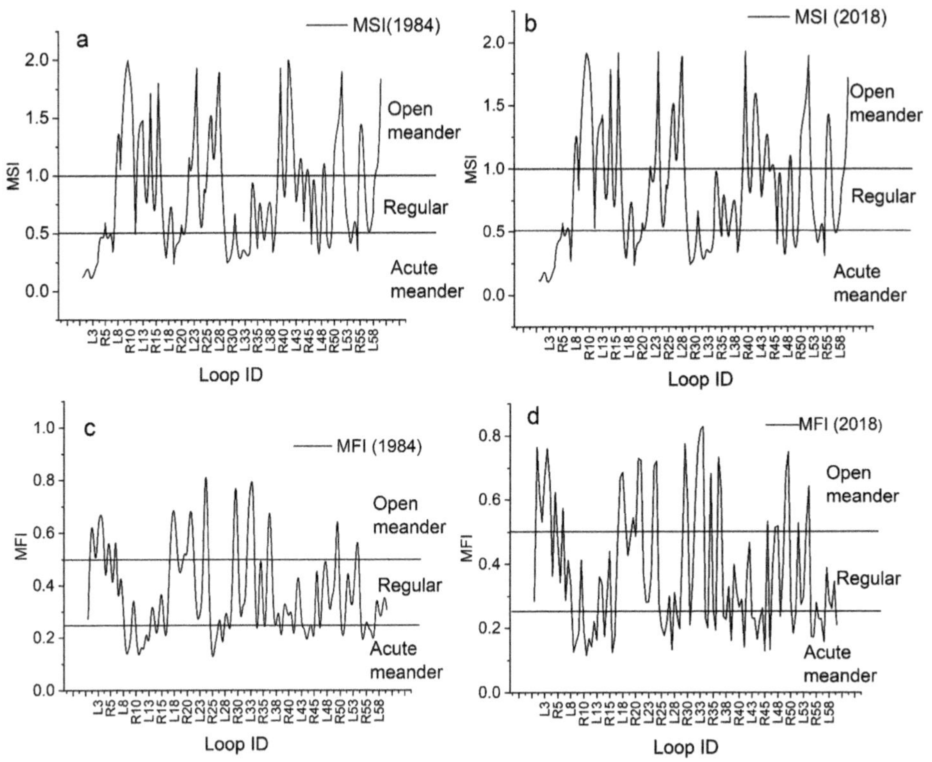

Fig. 4.10 Dynamics of the nature of meander of the MCR. **a** Meander shape index of 1984 and **b** Meander shape index of 2018. **c** Meander form index of 1984 and **d** Meander form index of 2018 (Prepared from Google Earth Image, 31 December 1984; 15 September 2018)

As a result, the MCR is year-round unconnected, with the exception of the two monsoon months of August and September. The MCR is not able to carry large quantities of silt from the Padma River over extended distances. The Mathabhanga River's upper course is where the most silt has gathered. As a result, the Churni River has considerably lost depth as well.

An analysis of the longitudinal fluctuations in the present river's depth (from the source to the mouth) indicates that the bankfull average depth (monsoon) is declining as the distance increases, and the pre-monsoon season has also demonstrated a similar trend. The Churni River has a rising trend from its source down to 15 km, a declining trend down to 40 km, and then an increasing trend up to the river mouth, according to the bankfull average depth (monsoon) polynomial curve. The pre-monsoon season average depth of up to 40 km exhibits a pronounced inverse pattern (Fig. 4.12a–d). However, during the pre-monsoon and monsoon seasons, the maximum depth (d_{max}) has been declining.

4.4.2.3 Cross-sectional Area

River width and depth decrease over time, resulting in a declining trend in the cross-sectional area of the river channel (Fig. 4.13a and b). At the Hanskhali gauge station on the Churni River, for instance, the cross-sectional area was measured by British investigators in 1890 and was 631.74 m^2 (Reaks 1919) (Fig. 4.13a and b). 1916 saw a further decrease in the cross-sectional area at the same place to 464.52 m^2 (Biswas 2001). The notable aspect of this time

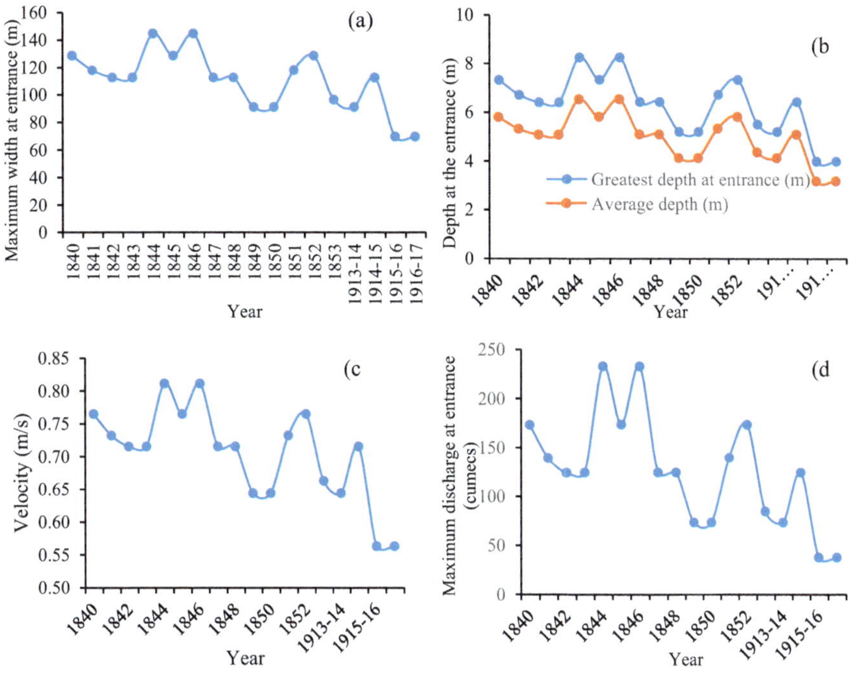

Fig. 4.11 Past channel morphology and hydraulics at the entrance of the river during 1840–1916-17. **a** Maximum channel width, **b** channel depth, **c** velocity, and **d** maximum discharge at the entrance

Table 4.8 ANOVA signifying significant difference in width between the past and present channels

Name of the adjacent abandoned channels	*F* value	*P* value	*F* crit*
Boat	103.7374	6.72911E−09	4.4138
Khalishakundi	36.2852	1.07359E−05	4.4138
Morbhanga	52.0757	1.03002E−06	4.4138
Asmankhali	23.0635	0.000142631	4.4138
Munshiganj	36.0107	1.12493E−05	4.4138
Darshana	26.6187	6.58455E−05	4.4138
Harisnagar	211.5499	2.15643E−11	4.4138
Bijoypur	37.8088	8.31986E−06	4.4138
Bishnupur	85.9611	2.82311E−08	4.4138
Khal Boalia	30.3218	3.14416E−05	4.4138

* For 162 degrees of freedom at 0.05 significance level

is the pace of decline (6.19 m/year). According to the current study, the cross-sectional area at Hanskhali in relation to bankfull width is now 340 m^2. Values from 47 cross-sectional areas throughout the whole Churni River display the similar picture as Hanskhali. With the greatest and lowest measured at 429 m^2 and 150 m^2, respectively, the average cross-sectional area is 285 m^2 (Fig. 4.13a and b).

4.4.2.4 Width/depth Ratio

Investigation on the Churni River indicates that the W/D ratio, or width/depth ratio, is greater in the pre-monsoon than it is in the monsoon. The regional distribution of the W/D ratio further demonstrates that the upper stretch produced both the lowest (10.48) and highest (42.51) values. The upper section of the river maintains a higher average W/D ratio (23.85) in comparison to the lower segment (18.51) (Fig. 4.12d). For the first 15 km of the river's downstream course, the spatial (longitudinal) pattern is rising; for the following 35 km downstream, the trend is downward; and in the downstream part of the river,

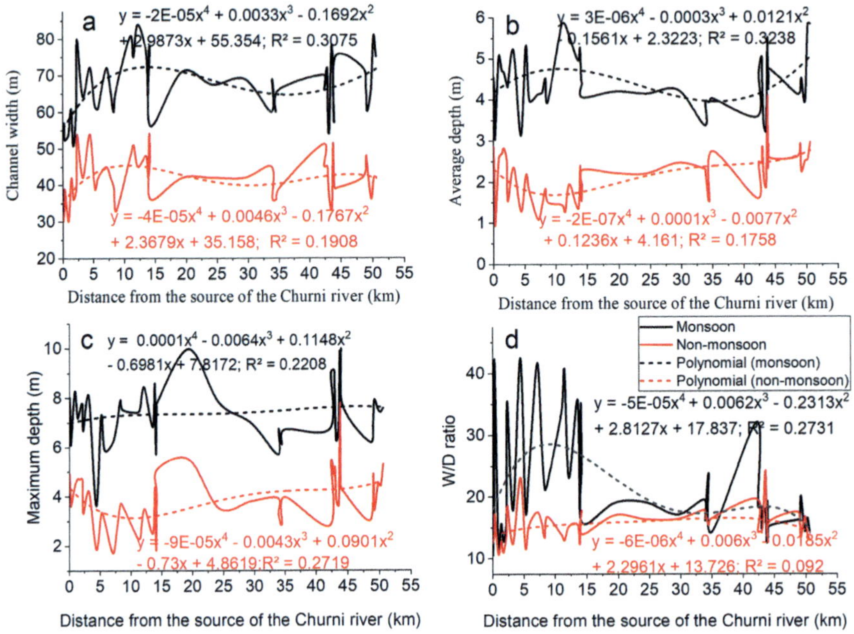

Fig. 4.12 Seasonal variation of cross-sectional forms of the Churni River with polynomial fit curves. **a** Channel width. **b** Average depth. **c** Maximum depth (d_{max}). **d** Width-depth ratio of the Churni River during pre-monsoon and monsoon (*Source* Field survey, 2018–2020) (*Note* Pre-monsoon width is measured with respect to the wetted perimeter while monsoon width is measured as bankfull stage)

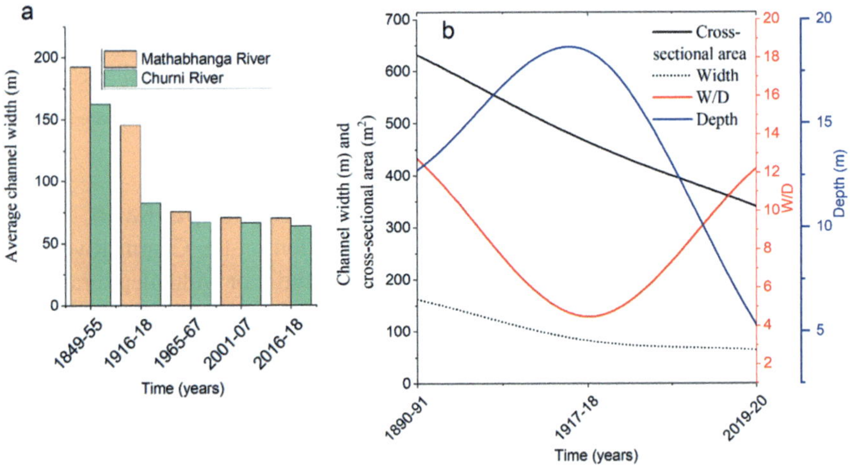

Fig. 4.13 Gradual decrease in channel dimensions, **a** average channel width, **b** cross-sectional area, W/D, and depth (based on Biswas 2001)

the trend is upward once more. Despite having a high *W/D* ratio, the Churni River has not experienced lateral instability from erosion since the turn of the twentieth century because of insufficient water flow in the channel.

4.4.2.5 Channel Asymmetry

A significant sign of a dynamic channel is channel asymmetry. Compared to a stable channel, a dynamic channel exhibits more asymmetry (Knighton 1998, 1981; Das 2015).

A meandering reach of an active channel that has more asymmetry is more dynamic than a symmetric length of the channel (Das and Islam 2015). Therefore, using the same 47-gauge stations and the same time period, the channel behavior of the MCR has been examined using the indices of channel asymmetry as suggested by Knighton (1981). The empirical findings on the channel asymmetry reveal that the river is less asymmetric in nature; however, the seasonality is observed in it. For example, in the monsoon months the bankfull channel asymmetry of the river inclined to be left asymmetric for 57% of the cross-section (A^* ranging from − 0.002 to − 0.024) while the rest cross-sections (43%) represent right asymmetric (A^* = 0.006–0.24). Besides, the non-monsoon season, when asymmetry is measured with respect to wetted channels, 49% of cross-sections appears to be left asymmetric (A^* ranging from − 0.005 to − 0.045), while the rest are figured out as right asymmetric for 51% (A^* = 0.008–0.37) (Fig. 4.14a–d). The range of A^* therefore signify that the river is very close to the ideal symmetric channel. The findings also indicate that the river is less dynamic for long time. Moreover, the values of A_1 and A_2 also affirm the symmetrical character of channel (Fig. 4.14a–d).

4.5 The Flow

The hydraulic processes prevailing in the MCR depict the same character that the other dying rivers like Jalangi, Bhairab, Saraswati are exhibiting. The first two basic variables, namely velocity (v) and discharge (Q), have been measured seasonally (monsoon, i.e., August when the river gets bankfull discharge due to higher recipient of rainfall, and post-monsoon, i.e., January when river discharge is minimal) in order to understand open channel hydraulics [Froude number or Fr and Reynolds number or Re and their effect on stream power (Ω) and shear stress (τ_0)] (Bizzi and Lerner 2015; Brookes and Shields 1996), which are responsible for erosion and sediment transportation in the river (Brookes 1987). Therefore, the important

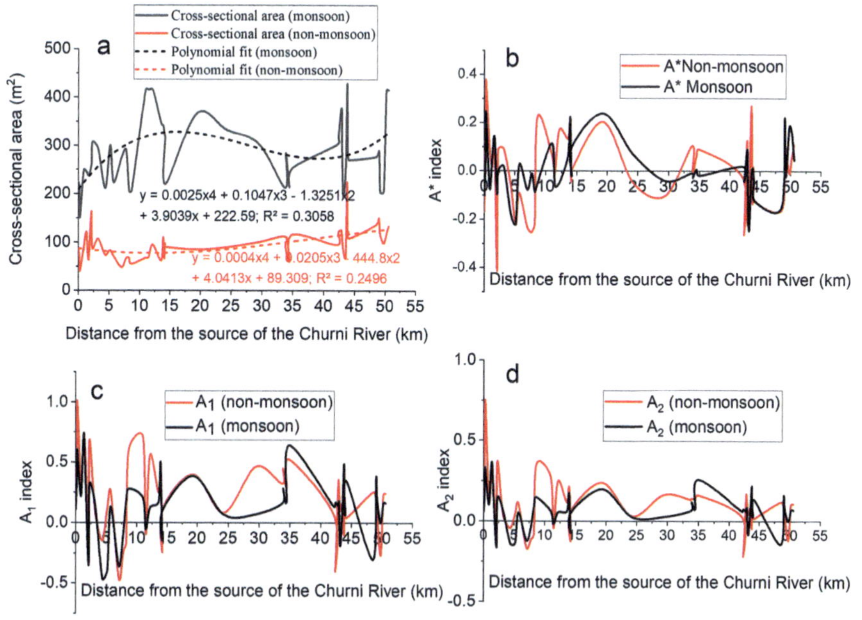

Fig. 4.14 Cross-sectional area with polynomial fit curve and channel asymmetry of the Churni River, **a** cross-sectional area, **b** A^* index, **c** A_1 index, **d** A_2 index of the Churni River during pre-monsoon and monsoon

hydraulic character of the river and their significance in modifying the channel morphology has been discussed in the following sections.

4.5.1 Velocity

In 1840, the estimated velocity at the study river's entrance was more than 0.75 m per second; by the 1920s, it had decreased to less than 0.50 m per second. At Hanskhali gauge station, velocity was still greater than 0.30 m s^{-1} in the 1920s, as seen in Fig. 4.15a. The velocity of the Churni River changes significantly during the pre-monsoon season, from 0.09 to 0.32 m $s-^{1}$ with an average of 0.15 m s^{-1}, and during the monsoon season, from 0.14 to 0.51 m $s-^{1}$ with an average of 0.25 m s^{-1}, according to the current study. The stream power and bed and boundary frictional forces of the Churni River are now insufficient to alter the river's course laterally or vertically at this abnormally low velocity. Once again, two notable peaks can be seen directly downstream of stream crossings when examining the longitudinal velocity curve. Because the channel has narrowed as a result of bridge construction, the velocity is higher at the locations directly downstream of the bridges. Moreover, the fourth-order polynomial curves for both seasons exhibit a similar pattern, increasing up to 10 km downstream, decreasing up to 25 km, increasing up to 45 km, and then decreasing up to the Churni River mouth (Fig. 4.15a).

4.5.2 Discharge

The falling trend of discharge is the principal issue of the MCR which is experiencing for a century (Sarkar and Islam 2022, 2024). The offtake of the MCR is suffering from high sedimentation in the nineteenth century resulting in continuous depth lowering. The average discharge has therefore sharply dropped since then (Fig. 4.15c). In 1915, the average discharge at

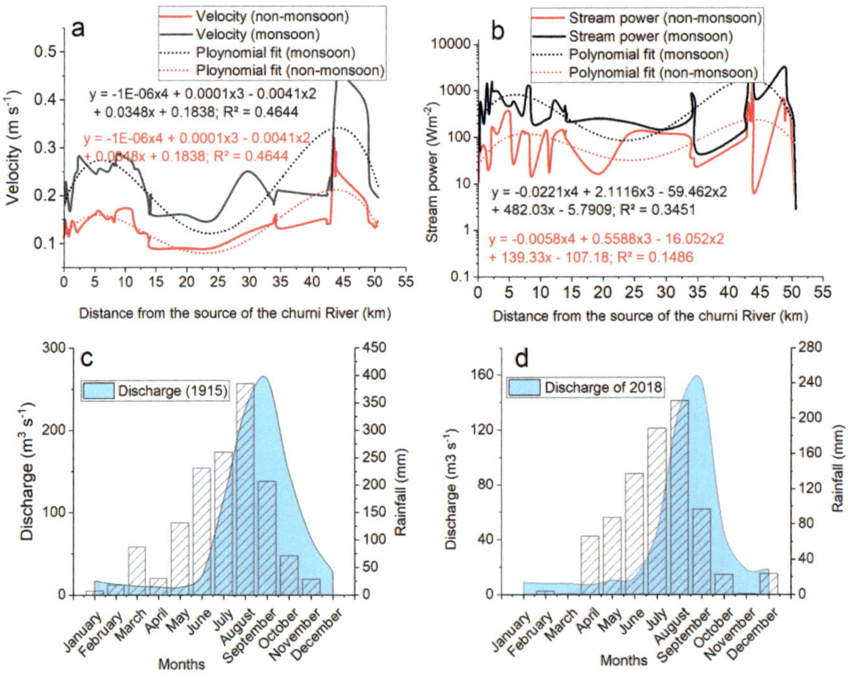

Fig. 4.15 Variation in velocity in the Churni River during pre-monsoon and monsoon, **b** variation in stream power in the Churni River during pre-monsoon and monsoon, **c** discharge hydrograph of the Churni River at Hanskhali in the year 1915 (*Source* Biswas 2001), and **d** in the year 2018 (*Source* Field survey, 2018)

the Hanskhali gauge station was 101.95 $m^3\ s^{-1}$; however, in 2015, it dropped to 54.65 $m^3\ s^{-1}$ (Sarkar et al. 2024). An average discharge measurement conducted in 2018 was around 67.96 $m^3\ s^{-1}$. According to the analysis of maximum discharge, the Churni River would have seen its greatest flow of around 339 $m^3\ s^{-1}$ at the beginning of the twentieth century (Biswas 2001). At Hanskhali, the river does, however, receive its maximum discharge of around 141 $m^3\ s^{-1}$ as of 2018. Furthermore, the discharge hydrograph from the Hanskhali gauge station in 1915 showed a wider peak, meaning the river would receive its maximum discharge during all monsoon months. In contrast, the discharge hydrograph from the same station in 2018 indicates a very narrow peak, meaning that the river only receives its maximum discharge during one monsoon month (Fig. 4.15c and d).

4.5.3 Pattern of Fluvial Hydraulics

In the pre-monsoon season, the Churni River flow pattern varied throughout time, with a maximum recorded value of 1209.18 and an average Reynolds number of 425.77. This implies that the flow is more laminar (Re<2000) than transitional (2000<Re 4000) (Fig. 4.16a). Therefore, one may claim that the river channel is stable during the pre-monsoon season since there is no frictional drag force from turbulent flow. The highest recorded Reynolds number was 2860, with an average of 1247, during the monsoon season, when flow and velocity are at their highest<Re 2500. An argument that the flow has turned more transitional than turbulent may be made based on this observation. The river's downstream portion has the highest value, particularly in a brief section near the Kalinarayanpur bridge where there is a high velocity because to channel constriction by the building of road and rail bridges across the Churni River. Furthermore, the calculated Froude number values for the whole Churni River's course are less than 1 during both the pre-monsoon and monsoon seasons, suggesting that subcritical flow is present throughout the entire river's course. Consequently, the current river's flow pattern indicates that the river bed and bank are less susceptible to lateral and vertical erosion.

In the pre-monsoon period, specific stream power (SSP) value along the Churni River ranges from 0.08 to 225 $Wm{-}^2$, with an average of 9.19 Wm^{-2} (Fig. 4.16b). The SSP values for the 95% cross-section are below the threshold, meaning that during the pre-monsoon season, the river's fluid cannot erode the river laterally or vertically. However, during the monsoon season, the value of SSP fluctuates from 0.03 to 743 $Wm{-}^2$. Even during the monsoon season, when the velocity and discharge are relatively high, the Churni River's fluid power is not favorable for the river's vertical and lateral erosion, even though the average is 46.78 Wm^{-2}. The SSP values for 82% of the cross-section are below the threshold. Furthermore, channel restriction at the Kalinarayanpur bridge location has resulted in abnormally high SSP.

With a few exceptions, the Churni River has very little shear force throughout the pre-monsoon and monsoon seasons. The pre-monsoon and monsoon seasons show considerable shear stress, particularly in the area just downstream of the bridges. As a result, the river's reduced shear stress also suggests a decreased likelihood of being a dynamic channel. On the other hand, the fact that the river has a large number of cutoffs with highly variable characteristics indicates that the previous channel was unstable and dynamic.

4.6 The River and Man

4.6.1 Man-River Interaction Before Independence

Historical evidence illustrates a deep and dynamic relationship between human communities and the Churni River in the pre-independence era. As a distributary of the Mathabhanga River and part of the wider Ganga-Padma fluvial system, the Churni provided the ecological and hydrological foundation upon which human

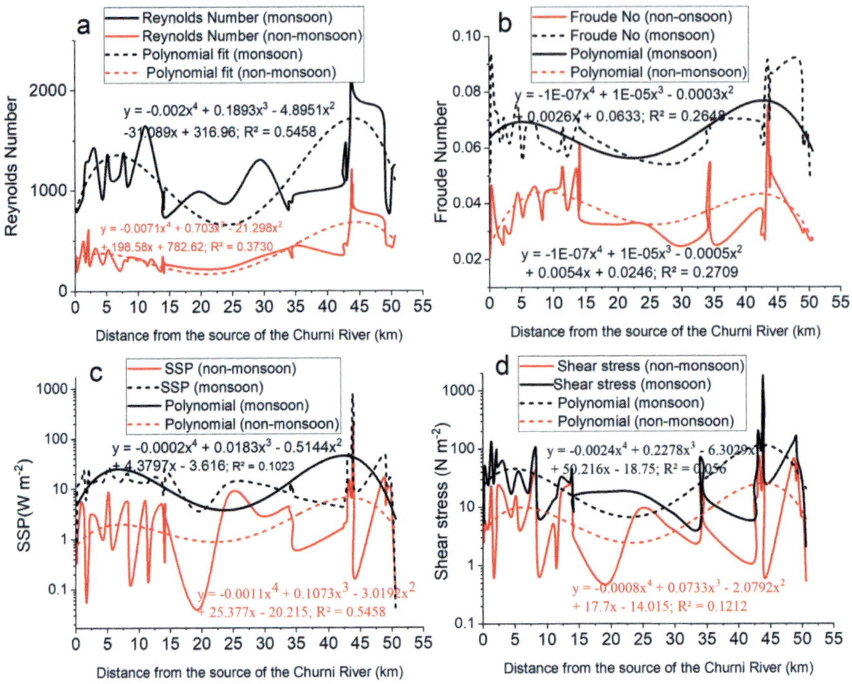

Fig. 4.16 **a** Variation in Reynolds number in the Churni River during pre-monsoon and monsoon, **b** variation in Froude number in the Churni River during Pre-monsoon and Monsoon, **c** variation in specific stream power (SSP) in the Churni River during pre-monsoon and monsoon, **d** variation in shear stress in the Churni River during pre-monsoon and monsoon (*Source* Field survey, 2018–2020)

settlements in the Nadia district of West Bengal developed and thrived. This river was not merely a geographic feature but an active agent shaping the rhythms of daily life, patterns of settlement, agricultural cycles, and socio-cultural practices. Settlements such as Shibpur, Birnagar, Aranghata, Hanskhali, Shibnibas emerged and expanded along the Churni's banks due to the river's dependable supply of freshwater, nutrient-rich alluvial soils, and seasonal flow regimes that supported paddy cultivation and inland fisheries. The British-era cartographic documents such as general map of Nadia river system (Map 2, Part 2) and River system of Bengal and old courses of certain river (map 1, sheet 2scale 1:4400) of Bengal District Gazetteers depict these settlements as densely inhabited clusters. Beyond utilitarian roles, the Churni River was also central to the spiritual and cultural life of local communities. Temple records and folk traditions document the use of the river for ritual bathing, immersion of idols, cremation rites, and religious festivals. The man-river interaction was developed not only as a material resource but as a sacred space, embedded in the collective consciousness and identity of the people.

4.6.2 Man-River Interaction After Independence

In 1975, the construction of the Farakka Barrage on India's Ganga River (Fig. 4.17a) brought about significant alterations to the hydrological patterns downstream, particularly affecting the Padma River. Mirza (1997) noted substantial changes in downstream hydrology following the erection of the Farakka Barrage. Mirza (1997) highlighted a concerning decrease in discharge during dry months and an increase during the monsoon. Rahman and Rahaman (2018), focusing on discharge patterns at the Hardinge Bridge gauge station, presented compelling data on the Padma River's hydrological dynamics

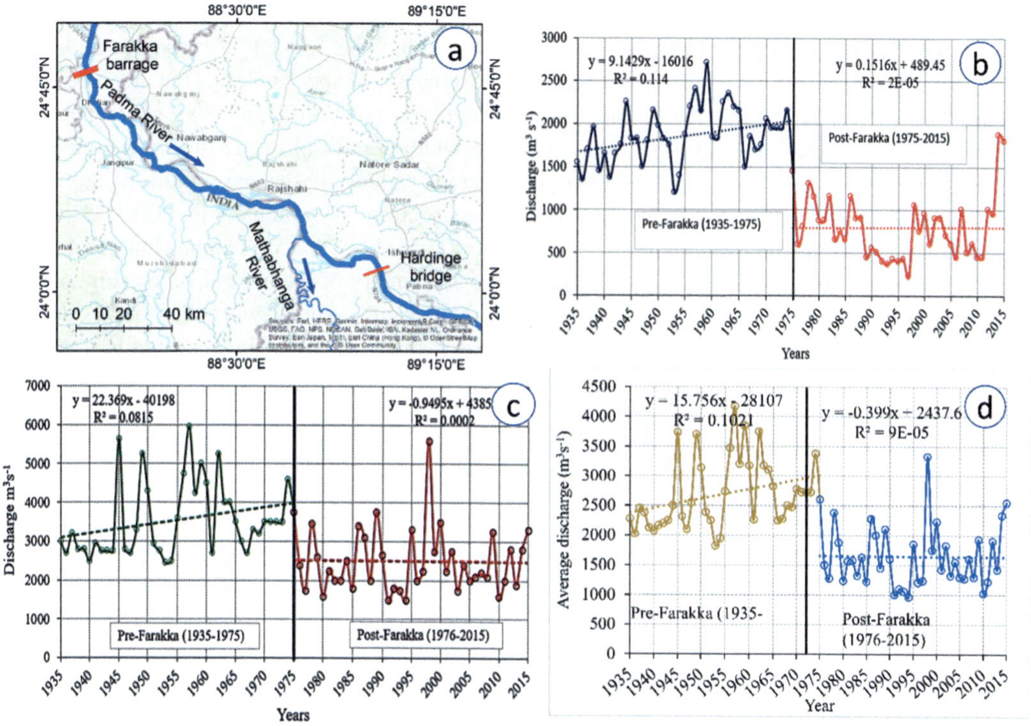

Fig. 4.17 Farakka Barrage's influence on the Padma River and its distributary, the MCR, examined through dry season discharge patterns (January to May) at Hardinge Bridge: **a** Hardinge Bridge location, **b** minimum discharge, **c** maximum discharge, **d** average discharge (Rahman and Rahaman 2018)

post-Farakka Barrage. They observed a notable decrease in minimum, average, and maximum discharge from the pre-Farakka period (1935–1975) to the post-Farakka period (1976–2015) (Fig. 4.17b–d). Particularly noteworthy was the drop in average discharge from 2696 $m^3\ s^{-1}$ to 1617 $m^3\ s^{-1}$ post-Farakka Barrage. Similar trends were evident for minimum and maximum discharge (Fig. 4.17b–d). This significant decrease underscores how the Farakka Barrage's construction has profoundly altered downstream hydrological conditions, especially evident in the Padma River's discharge pattern.

The MCR, a distributary of the Padma River, faces challenges with outflow, particularly during dry seasons, when it occasionally detaches from the Padma River. The river's hydrology is further complicated by reduced flow during dry seasons. Understanding the connection between the Padma and MCR is crucial for comprehending the latter's discharge dynamics, which significantly influence channel morphology. Additionally, Sarkar et al. (2020) observed that with the installation of the Farakka Barrage, the Padma River's water level during dry months consistently remains significantly lower than the MCR's bed level, resulting in continuous low flow. This exacerbates issues in tributary rivers, such as decreased depth and pollution from various sources.

4.6.3 Man-River Interaction: Recent Scenario

The anthropogenic interventions in the form of road-stream crossings, agricultural practice, and brick kiln industries, and discharge of industrial wastewater have induced significant changes in the hydrogeomorphic behavior in the Churni River Basin.

4.6.3.1 Anthropogenic Transformation of River Bed to Agriculture

About 5.8% (3.8% left bank, 2.0% right bank) of the Churni River is utilized for on-bed agriculture, mostly in the pre-and post-monsoon seasons. However, due to valley inundation during the monsoon, this share is much lower (Fig. 4.18a–d). The fertile valley of the river is used to grow different kinds of vegetable throughout the winter. Mustard, onion, brinjal, pumpkin, carrot, etc., are prominently cultivated. During the field verification, it was noted that there are tendency to cultivate many crops on a single farm. The primary irrigation method is river lift. The river lift irrigation method irrigates agricultural fields on their beds using both tiny water pump sets and traditional manual methods. The nature of the bank differs from the natural bank when agricultural tracts are prepared, resulting in slide alteration. The agriculturally dominated bank is less steep compared to the antipodal bank. Additionally, these modifications may also affect width, average depth, cross-sectional area, asymmetry, and other characteristics of the channel (Fig. 4.19a and b). Hence, six cross-sections have been selected to examine the impact of agricultural land on the channel shape. Because the average width and depth are 73.66 and 3.99 m, respectively, the width-depth ratio is 18.46. Furthermore, the average values of the agriculturally dominated cross-section asymmetry ($A^*=0.13$, $A_1=0.34$, $A_2=0.14$) show that it is somewhat greater than the asymmetry of the natural valley (Fig. 4.20).

Numerous scholarly studies have demonstrated that the use of high pesticide, fertilizer, and insecticide inputs in modern agriculture is a crucial element in increasing the nutritional concentration in nearby rivers. Therefore, several studies (Adusumilli et al. 2010; Chimwanza et al. 2006; Dunca 2018; Moss 2007; Mutisya and Tole 2010; Toner 1986) considered agriculture as the dominant non-point cause of pollution in a river basin. Runoff from on-bed agriculture considerably regulates the water status during the pre-monsoon and post-monsoon periods, when the discharge in Churni River is minimum due to the separation from the Ganga River. Furthermore, another factors influencing pollution during the monsoon season are

Fig. 4.18 In-bed and on-bank agriculture **a** Land preparation for the upcoming crop, **b** vertical farming method, **c** varied crop planting scheme, **d** monocropping practice

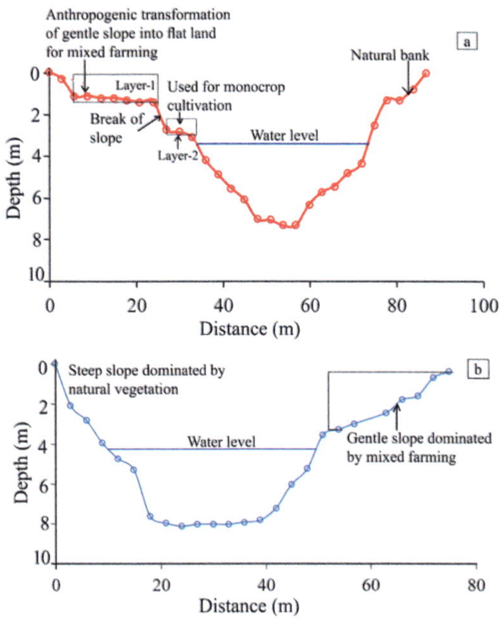

Fig. 4.19 Cross-sections over agriculture-dominated valley **a** stepped agricultural practice modifying the natural slope near Shibnibas, **b** slight modification of valley slope and practice of agriculture

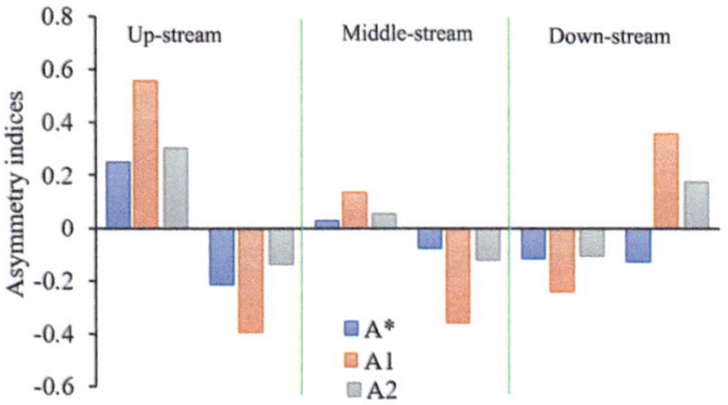

Fig. 4.20 Channel asymmetry in agricultural dominated area

pollutants from off-bed agriculture, which occurs during periods of high rainfall intensity and quantity. The major non-point source of pollution in the river basin is agricultural activity, which is defined by the overuse of pesticides and fertilizers. This study provides a detailed assessment of the role that on-bed agriculture practices have played in raising the concentration of these nutrients in the Churni River. Phosphate concentration in uncontaminated water is normally between 0.01 and 0.03 mg/L, but between 0.03 and 0.08 mg/L, plant growth in the water is greatly accelerated, and when cross 0.1 mg/L, high eutrophication is unavoidable due to the accelerated growth of aquatic macrophytes.

Moreover, a phosphate content exceeding 0.1 mg/L in water signifies that the primary sources of phosphate contamination include sewage from both rural and urban areas, agricultural fertilizer application, and industrial wastewater. Majhdia (the upper stretch of the river) exhibits peak concentrations of 0.109 mg/L during the post-monsoon season, 0.096 mg/L during the monsoon season, and 0.084 mg/L during the pre-monsoon season. Similarly, Ranaghat (the lower stretch of the river) registers the highest phosphate concentration during the post-monsoon period at 0.119 mg/L, trailed by the pre-monsoon (0.112 mg/L) and monsoon (0.110 mg/L) seasons (Fig. 4.21a–c). Consequently, phosphate levels in both upstream and downstream segments of the river surpass the optimal threshold of 0.03 ng/L. Consequently, certain river sites exhibit accelerated growth of rooted aquatic macrophytes.

Furthermore, as soon as these macrophytes perish, lowering dissolving oxygen leads to an unfavorable environment for river ecosystem. Sarkar and Islam (2020) observed that concentrations of dissolved oxygen in the pre-monsoon, monsoon, and post-monsoon (2.22 ± 2.80, 3.49 ± 1.79, and 1.61 ± 1.35 at Majhdia, and 3.80 ± 1.73, 3.41 ± 1.86, and 3.53 ± 0.90 at Ranaghat, respectively) are extremely low causing an alarming decline in fish species diversity and productivity. Moreover, N-containing fertilizers like urea and ammonia are also polluting the river water. The standard nitrate loading in rivers is 0.2 mg/L; however, in Majhdia, the post-monsoon has the greatest nitrate concentration measuring 1.21 mg/L, followed by the monsoon (0.59 mg/L) and pre-monsoon (0.56 mg/L). Similarly, at Ranaghat, post-monsoon is recorded as 1.14 mg/L, followed by pre-monsoon measurements (0.94 mg/L) and monsoon (0.85 mg/L). As a result, the nitrate content in both stations and every season is higher than desired, which damages the aquatic ecosystem of the river.

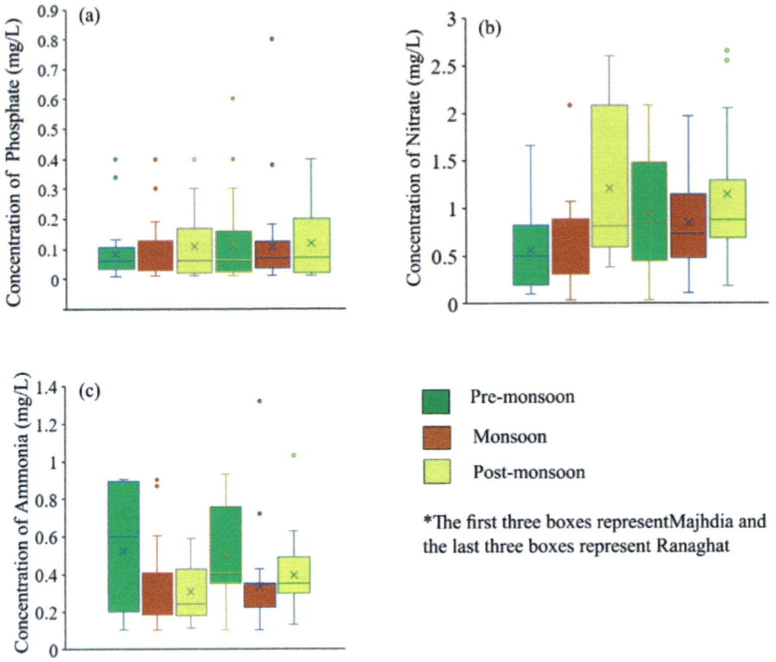

Fig. 4.21 Concentration of **a** phosphate, **b** nitrate, and **c** ammonia in the Churni River (Data source: WBPCB 2020)

Besides, the ammonia content in the Churni River has been observed beyond the acceptable threshold of 0.1 mg/L. Pre-monsoon (5.33 mg/L) has the greatest average concentration at Majhdia, followed by monsoon (032 mg/L) and post-monsoon (0.31 mg/L). Likewise, pre-monsoon shows 0.50 mg/L at Ranaghat, followed by post-monsoon (0.40 mg/L) and monsoon (0.34 mg/L). Fish species in the Churni River are losing their resistance to illness as a result of this unfavorable ammonia concentration. Therefore, it is unavoidable to suffer from many diseases caused by ammonia poisoning, such as gill damage, osmoregulatory imbalance, and the gradual death of mucous-producing membranes (Sarkar and Islam 2020).

4.6.3.2 Brick kiln on the River Bank: Clearance of Riparian Vegetation and Bank Excavation

At both sides of the MCR basin, thirty-four significant brick kiln industries are observed (Fig. 4.22a–e). Among them, sixteen are located on the banks of the Churni River (from the bifurcation point at shibnibas to the confluence point with Hooghly River at Payradanga), while rest of them are along the Mathabhanga River (from the source to the bifurcation point of Ichamati and Churni Rivers at Shibnibas). River bank is the important source of sand for the majority of industries for bricks production. Therefore, the shape of the channel and hydraulic behavior have been significantly changed by the illicit and unscientific practice of bank excavation. The brick kiln of Betbaria (Khulna division, Bangladesh) has detected the remarkable phenomenon of channel change in the upper section of the Mathabhanga River basin. A pond around 3000 m^2 is created by the ongoing operation of bank excavation, which has designated an area of 10,000 m^2 as the bank excavation area. This pond remains linked to the river's flow even during the dry season. In the vicinity of Goalgram, there is another instance of bed excavation by a brick kiln (Daulatpur police station, Bangladesh). Situated around 200 m from the river, the industry has resulted in the excavation of 1000 m^2, causing the valley's width to increase from 30 to 55 m. The middle course of the river, Hanskhali,

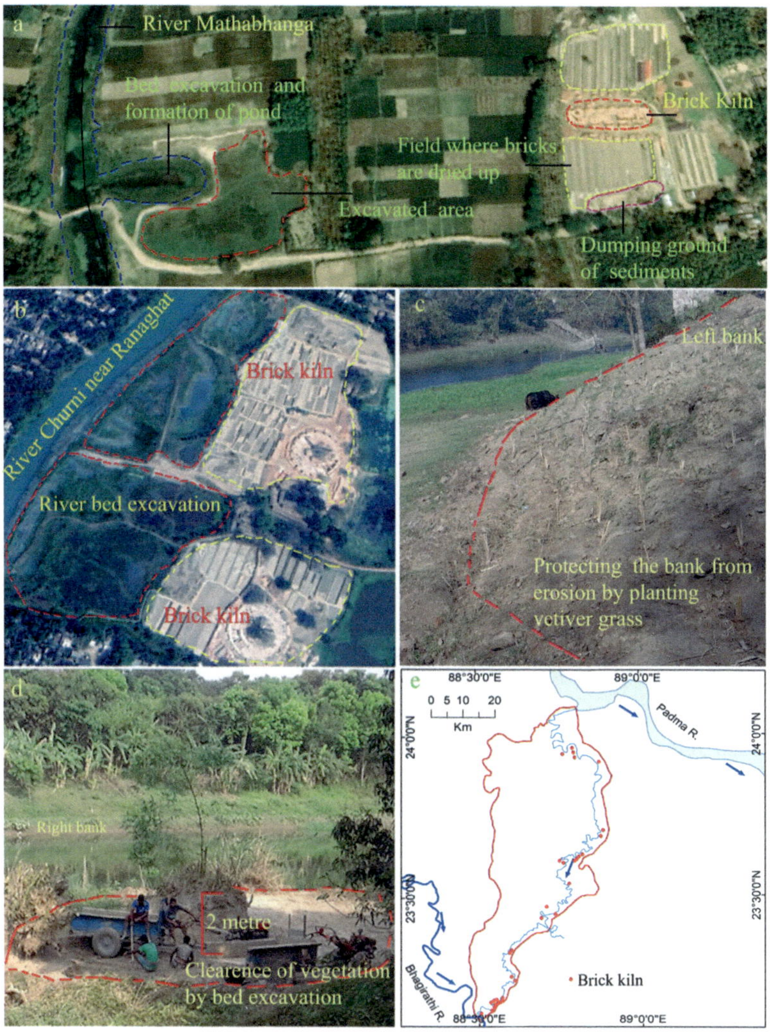

Fig. 4.22 Collection of sediments from bank for brick production **a** formation of pond by bed excavation, **b** massive destruction of river bank by sediment collection for brick production near Ranaghat, **c** protection of bank by planting vetiver grass near Hanskhali brick kiln (middle stream), **d** clearance of vegetation by bank excavation, **e** location of brick kiln in the MCR basin

is home to a brick kiln that has significantly altered the valley's shape in the Churni River basin. The valley's natural width in this region is from 40 to 50 m, but it reaches 99 m near the brick kiln. Additionally, the average depth has grown from about 3.3 m to approximately 4 m, and the total cross-sectional area has expanded from around 300 to approximately 430 m^2. Furthermore, because the excavation is being done on the left bank of the river, the left side's cross-sectional area from the valley's center line is larger than the right side's. As a result, the character of the channel has grown increasingly asymmetric. The average values of A^*, $A1$, and $A2$ for Hanskhali are 0.230, 0.204, and 0.102, respectively, according to the asymmetry research. Ultimately, the downstream stretch of the river near Ranaghat municipality has undergone significant modifications compared to the upstream and middle courses. This is due to the presence of eleven brick kilns in close proximity to one another, as well as the excavation of

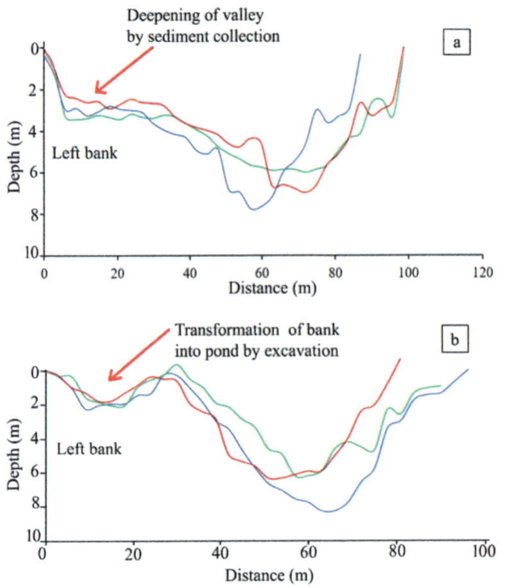

Fig. 4.23 Modification of cross-section near brick kiln industry **a** near Hanskhali, **b** near Ranaghat

bank ponds by all industries, which has had an astounding cumulative effect on the channel. A^*, $A1$, and $A2$ in Ranaghat had average asymmetry values of 0.238, 0.431, and 0.262, respectively (Fig. 4.23a and b) (Table 4.9).

4.6.3.3 Anthropogenic Signature Over the Basin: Land Use and Land Cover Change

The LULC change over the past three decades has been noticeable. Vegetation accounted for 51.02% of the total share in 1991 and decreased to 21.80% in 2018. In addition, the waterbodies have decreased from 7.56% in 1991 to 1.8% in 2018. Therefore, this significant decrease is the outcome of the rapid expansion of agricultural and settlement areas. In 2018, the dominant LULC is agriculture, growing from 23.22% in 1991 to 53.37%. Comparably, from 18.18% in 1991 to 22.95% in 2018, the settlement area has likewise grown. Accordingly, the shift in LULC type shows that over the course of three decades, areas of waterbodies and vegetation have decreased alarmingly, while settlement and agricultural areas have expanded more quickly. The quantity of basin runoff and sediment discharge into the river has significantly risen as a result of these modifications. In addition, the runoff from agricultural land has raised the concentrations of phosphate, nitrate, ammonia, and other chemicals that cause eutrophication in rivers, raising BOD and COD levels and lowering DO in the water. Furthermore, the extent and intensity of pollutants have also grown as a result of runoff from both urban and rural regions (Sarkar and Islam 2020).

4.6.3.4 Discharge of Industrial Wastewater

Industrial water pollution significantly impacts the degradation of fish habitats in the MCR. Fish species exhibit severe adverse reactions to the effluents released by Carew and company, notably the sugar and chemical industry. While the lower stretch of the river is situated farther from this industry, the upper stretch is in closer proximity, resulting in a more pronounced impact of industrial water contamination. The oxygen demand of the river experiences the primary and foremost effect of industrial effluent. Certain bacteria and microorganisms responsible for breaking down organic materials lead to a decline in dissolved oxygen levels (by 0.2 mg/L) and an increase in BOD levels (by 68 mg/L). Additionally, it was observed that during industrial pollution, primary productivity plummeted from 18.4 mg C/M3/hour to zero, while the concentration of free CO_2 surged from 12.8 mg/L to 31 mg/L (Ghosh 1991). Consequently, the deterioration of fish habitats, abnormal fish behavior, and the sudden extinction of fish species are all directly linked to industrial water pollution.

4.6.3.5 Discharge of Urban Wastewater and DO Sag Analysis

The continuous disposal of municipal sewage into the Churni River poses another significant point-source of pollution, threatening aquatic ecology, particularly the quality of fish habitat (Chakrabarty and Das 2006; Sanyal et al. 2015). Only the downstream portion of the Churni River (RE3) faces the risk of untreated effluent from the Ranaghat Municipality

Table 4.9 Comparison of channel morphology

Channel type		Mean and SD	CV	Range	IQR	Skewness
Natural	A^*	0.08 ± 0.09	116.71	0.29	0.13	0.15
	$A1$	0.26 ± 0.13	53.12	0.43	0.19	0.35
	$A2$	0.11 ± 0.05	50.47	0.15	0.09	− 0.12
	w	64.44 ± 7.65	11.87	24	9	0.46
	D_{average}	4.61 ± 0.67	14.59	1.81	1.03	0.48
	CS area	303.81 ± 66.27	21.81	187.94	74.05	0.17
Agricultural	A^*	0.13 ± 0.08	61.28	0.22	0.11	0.21
	$A1$	0.34 ± 0.14	42.04	0.42	0.12	− 0.01
	$A2$	0.14 ± 0.08	57.44	0.25	0.06	0.27
	w	73.66 ± 10.26	13.94	26	15	− 0.11
	D_{average}	3.99 ± 0.67	16.86	1.84	0.48	1.73
	CS area	292.23 ± 33.68	11.52	86.85	12.24	− 2.29
Road-stream crossing	A^*	0.19 ± 0.09	46.82	0.26	0.10	− 1.16
	$A1$	0.39 ± 0.24	62.50	0.63	0.43	− 0.10
	$A2$	0.20 ± 0.13	67.95	0.35	0.25	0.06
	w	59.66 ± 8.84	14.82	27	7.50	1.19
	D_{average}	3.57 ± 1.37	38.61	5.45	1.52	0.92
	W/D	18.77 ± 7.09	37.81	25.72	7.67	1.51
	CS area	231.45 ± 127.14	55.04	462	137.61	1.80
Brick field	A^*	0.22 ± 0.11	49.96	0.320	0.10	− 0.41
	$A1$	0.34 ± 0.18	53.43	0.420	0.29	0.47
	$A2$	0.18 ± 0.14	78.86	0.312	0.23	0.66
	w	87 ± 66.38	17.31	39	15.75	− 1.38
	D_{average}	3.33 ± 0.67	20.31	1.63	1.02	− 0.12
	CS area	338.75 ± 15.05	19.59	177.37	74.05	− 0.82

seeping into the river throughout its entire length (Fig. 4.24a–c). The projected annual production of municipal solid waste (MSW) stands at 9490 tons (TPA). Additionally, the anticipated total sewage generation and water usage for this municipality are 10.14 million liters per day (MLD) and 15.49 MLD, respectively. Currently, there is no sewage treatment plant (STP) in operation for wastewater recycling. Hence, the gap of 10.14 represents the number of towns without STPs and sewage treatment. According to the report, the Ranaghat Municipality has nine significant point sources of water contamination, all located closely to each other. Among these drains, Sreenathpur (2.799 MLD) and Basko Khal (3.638 MLD) contribute significantly to sewage water with high BOD (48 mg/L for Sreenathpur and 48 mg/L for Basko Khal) and fecal coliforms (900,000 MPN/100 mL for Sreenathpur and 230,000 MPN/100 mL for Sreenathpur) (Table 4.10). Consequently, water pollution and habitat destruction have intensified due to these outfalls.

The explanation of the DO shortage resulting from the combination of urban wastewater and industrial effluents is further assisted by the examination of the DO sag curve. In addition to failing to attain saturation conditions over the whole river, the curve shows a progressive fall in DO downstream to the point source of industrial effluent mixing (at Darshana). Gobindapur has the greatest deficit (critical DO), suggesting that the rate of deoxygenation is greater than the rate

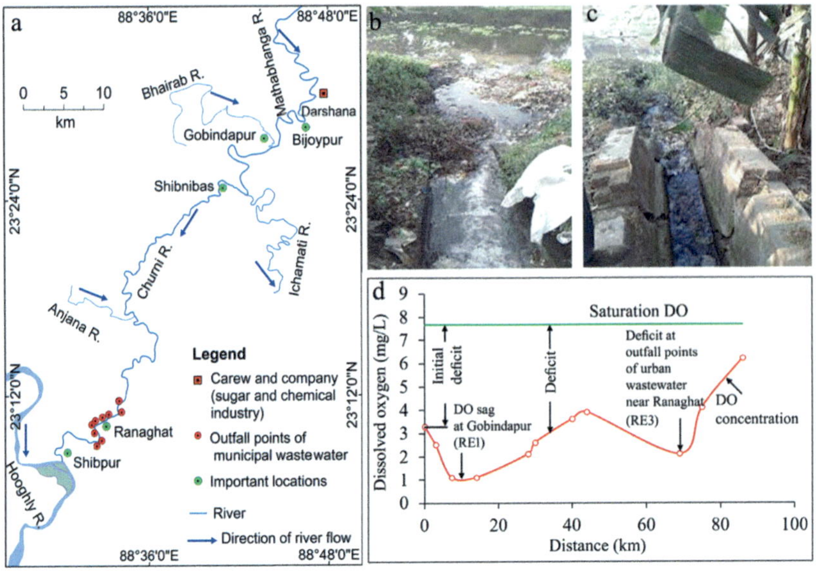

Fig. 4.24 Discharge of untreated urban wastewater from the Ranaghat municipality into the Churni River is a significant concern. **a** The location of drains or outfalls, **b** drain at Radharani Seba Sadan, **c** drain at Churipara. Additionally, the impact of this untreated wastewater discharge can be observed through the oxygen sag curve on the Churni River

of reoxygenation (Fig. 4.24d). As a result, RE1 denotes inadequate fish habitat and poor water quality. From RE1, the graph shows an upward trend over 45 km, which indicates that the river naturally purifies itself through chemical, biological, and hydrological processes (DeLoach 1972). The Ranaghat Municipality's untreated municipal sewage is mixed in with the water, causing the curve to drop once again and form a trough near Ranaghat. The high concentration of BOD (92 mg/L) at the urban sewage outfalls is indicative of the observed reduction in the river's diluting capacity and the generation of organic pollution due to the growing discharge of urban wastewater (Table 4.10).

4.6.4 Changes in the River Ecosystem

4.6.4.1 Habitat Variables and Their Spatio-temporality

Spatio-temporality of variables considered under the habitat suitability assessment has been analyzed considering upper and lower stretch of the Churni River (Table 4.11) and pre-monsoon, monsoon, and post monsoon season as discussed in the following sections.

For freshwater fish species, turbidity has a profound effect on all phases of their life cycles. According to Alabaster and Lloyd (1982), clean water with turbidity levels up to 40 Nephelometric Turbidity Units (NTUs) is ideal for fish feeding. Sigler et al. (1984) observed that coho salmon thrived most quickly in environments with turbidities between 40 and 50 NTU. But when developing habitat suitability index, many experts have determined that the ideal range of turbidity for freshwater fish survival is less than 47 NTU (Bain and Jia 2012). The average turbidity concentrations are well below 40 NTU at all the study reaches and for all seasons (13.02 NTU in pre-monsoon, 25.39 NTU in monsoon, and 27.39 NTU in post-monsoon at upper stretch); and 9.57 NTU in pre-monsoon, 13.85 NTU in monsoon, and 23.49 NTU in post-monsoon at lower stretch. Thus, it can be argued that the concentration of turbidity in the MCR does not have a significant impact on the life activities of fish species.

Table 4.10 Hydrochemical attributes of municipal wastewater at the outfall points

Landmark of drain/ outfall	Locational coordinate		Dimension (mm)		Water depth (mm)	Velocity (M/Sec)	Current flow	Projected flow (2035)	BOD (mg/L)	Fecal coliforms (MPN/100 mL)
	North	East	Width	Depth			MLD	MLD		
Chaitanya Ghat	23°10′58.94″	88°33′58.08″	2500	1750	247	0.068	0.518	0.682	59.5	300,000
Radharani Seba Sadan	23°10′48.49″	88°33′50.48″	1000	1650	15	0.4	0.333	0.497	47.82	1,300,000
Nadia Samaj Pally	23°10′48.45″	88°33′50.49″	600	600	15	0.429	1.296	1.46	54.24	210,000
Chhoto Najar	23°10′44.41″	88°33′43.77″	1200	1350	25	0.5	0.091	0.255	71.3	190,000
Churipara	23°10′39.04″	88°33′33.10″	350	600	15	0.2	0.822	0.986	70	160,000
Thanapara-I	23°10′38.83″	88°33′17.64″	900	1500	141	0.075	0.187	0.351	92	220,000
Thanapara-II	23°10′36.21″	88°33′15.64″	450	1200	12	0.4	0.456	0.62	58	140,000
Sreenathpur	23°10′25.03″	88°33′12.07″	600	1350	22	0.4	2.799	2.963	48	900,000
Basko Khal	23°11′2.63″	88°33′54.96″	1800	1500	36	0.5	3.638	3.802	68	230,000

Source CPCB (2020)

The pH level significantly influences the physiological activities of freshwater fish species. Alabaster and Lloyd (1982) observed that the optimal pH range for the growth of freshwater fish species is between 6.8 and 8.6 while below 6.0 deteriorates the environment and below 5.0 becomes dangerous for fish life. In both the study reaches, pH concentrations fall within the desired range: 7.18–7.91 in the pre-monsoon, 7.22–7.95 in the monsoon, and 7.46–8.42 in the post-monsoon at the lower stretch; and 7.7–8.12 in the pre-monsoon, 7.20–8.02 in the monsoon, and 7.38–8.18 in the post-monsoon at the upper stretch. Therefore, the fish species in the MCR do not face threats due to pH levels hindering their daily activities.

The significant concentration of dissolved oxygen (DO) required for normal fish physiological functioning is 5 mg/L. Fish typically alter their behavioral patterns in response to a significant reduction in DO levels. The vertical and horizontal habitat shift. The increased frequency of air breathing, and digestion and absorption issues are all observed during the acute fall of DO. According to Bhatnagar et al. (2004), fish growth is sub-lethal at DO concentrations between 1 and 3 mg/L, lethal between 0.3 and 0.8 mg/L, and fatal beyond 14 mg/L. The concentration of DO in the MCR is insufficient to sustain fish life, posing a significant concern for the fish species. Particularly, the average DO values for all seasons and reaches—1.19 mg/L in the pre-monsoon, 2.24 mg/L in the monsoon, and 2.35 mg/L in the post-monsoon at the upper stretch; and 2.53 mg/L in the pre-monsoon, 2.01 mg/L in the monsoon, and 3.56 mg/L in the post-monsoon at the lower stretch—are significantly below the desirable limit of 5 mg/L.

Armantrout (1998) highlighted the significance of pool for fish stating that a pool is an aquatic environment deeper and broader than a riffle, with a gradient of less than 1%. These pool regions are preferred by most species due to their sluggish current. For freshwater fish species, a pool velocity of less than 8 cm/s is considered excellent, with an optimal range of 0–24 cm/s, as suggested by the majority of

Table 4.11 Location and description of study reach

Reach name	Geographical location		Reach characteristics
	Start point	End point	
Shibnibas upper stretch	23°24′37.78″ N 88°41′58.18″ E	23°24′52.30″ N 88°41′29.68″ E	The reach can be characterized by average width of 66.56 m, average depth of 4.38 m, a maximum depth (d_{max}) reaching 6.54 m, and the average cross-sectional area (266.25 m^2). The average velocity is recorded to be 0.27 m/s. The sinuosity of the reach is measured as 1.39. Orchards decorate the slopy right bank. A single road-stream crossing is present. Occasionally, the river receives effluent from the sugar mill complex in Darshana via the Mathabhanga River
Ranaghat lower stretch	23°10′42.86″ N 88°33′35.05″ E	23°10′22.10″ N 88°32′50.06″ E	This reach is characterized by the average width of 69.96 m, average depth of 4.74 m, maximum depth (d_{max}) of 7.27 s, and average cross-sectional area 301.13 m^2. Besides, the average velocity remains at 0.18 m/s. The sinuosity of the reach is measured as 1.28, and it extends into a tidal reach. There are three road-stream crossings and one rail-crossing present. Additionally, soil cutting from the banks occurs for the brick kiln industry

habitat suitability models. The seasonal average values in all study samples are concentrated within the desired range, although a significant portion of samples have just exceeded the threshold limit for the optimal range. The highest pool velocity, recorded during the monsoon season at the lower stretch, was 18 cm/s. Therefore, pools in the MCR offer an excellent habitat for fish species with optimal velocities.

Another crucial factor in assessing environmental suitability is the velocity at riffles. Most freshwater fish species do not prefer breeding in areas with minimal or no flow. Habitat suitability models indicate that the ideal riffle velocity falls between 30 and 60 cm/s. There is significant seasonal variation in the riffle velocity of the MCR. During the pre-monsoon season, the average riffle velocity does not reach the lower bound of the desired range. Similar conditions are observed in the post-monsoon period, albeit to a lesser extent. The highest average riffle velocity during the pre-monsoon season was recorded at the lower stretch (29.17 cm/s), followed by the upper stretch (20.17 cm/s). A comparable situation was also observed during the post-monsoon season (27.67 cm/s at the upper stretch). Additionally, no sample has been found to exceed the upper limit of the desired range in either the pre-monsoon or post-monsoon seasons.

The majority of fish species use riffles, which are shallow sections of rivers with a gradient of 1% to 4%, as breeding grounds. Fuiman (1978) found that the optimal depth for spawning is between 20 and 25 cm. Many models of the habitat suitability index have determined that the ideal riffle depth is between 15 and 31 cm. The empirical investigation shows that the average riffle depth is 35 cm, 37 cm, 44 cm, and 43 cm at upper stretch and 31 cm, 34 cm, and 31 cm at lower stretch in the pre-monsoon, monsoon, and post-monsoon, respectively. As a result, the analysis of riffle depth shows that it is marginally over the ideal range in all reaches and throughout the year.

The term "pool cover" describes the logs, brushes, and other items that are in the pool area and provide the fish with places to eat and relax. Therefore, while determining the appropriateness of a habitat, the percentage pool cover is crucial. 35–65% is the ideal range for pool cover. The Churni River study reaches exhibit pool cover that falls within the ideal range during the pre-monsoon at RE1, the pre-monsoon and monsoon at RE2, and the monsoon at RE3. For fish species, the existence of a river pool is essential since it is the only place where they can carry out their unique life functions. Numerous academic studies have determined that the ideal pool (%) range is between 44 and 70. The MCR's upper length has a pool (%) that is within the ideal range, whereas the lower section's pool (%) is always below the ideal range.

4.6.4.2 Effects of Habitat Degradation on Fish Species

Fish habitat deterioration has a negative impact on all aspects of fish life in the MCR, including feeding, reproduction, and migration. In addition, there has been a noticeable drop in fish species as a result of habitat destruction. As a result, the following sections provide a rigorous analysis of every element pertaining to the impact of habitat deterioration on fish community structure.

Fish Species Availability and Distribution According to the IUCN Red List

There are 44 fish species in the Churni River overall, classified into 7 orders and 17 families in 1983. There are currently just fifteen fish species at RE1 (six as carnivores, four as omnivores, one as a benthic feeder, and four as planktivores). At RE2, there were 16 individuals (5 planktivores, 1 benthic feeder, 4 omnivores, and 6 carnivores) and 23 at RE3 (six planktivores, three benthic feeders, six omnivores, and eight carnivores) (Table 4.12). As a result, fish variety is sharply declining in all of the river's lengths, and the upper stretch is seeing a higher degree of species removal than the lower stretch. In addition, scores for habitat orientation and trophic level are computed to gauge the river's ecological deterioration. According to the trophic level score, the most common fish species in 1983 were benthic feeders (11.36%), carnivores (29.54%), omnivores (27.27%), and planktivores (31.81%). Currently, the highest trophic level score for the upper stretch stands at 37.5%, followed by planktivores (31.25%), omnivores (25%), and benthic feeders (13.14%). In the lower stretch, the highest score is 34.78%, followed by planktivores (26.08%), omnivores (26.08%), and benthic feeders (13.04%). In 1983, there were 35 species classified as least concern (LC), 3 as vulnerable (VU), 2 as near threatened (NT), 2 as critically endangered (EN), and 1 as data deficient according to the IUCN Red List. As of 2020, there are only 11 fish species in the LC category in the upper stretch and 17 in the lower stretch. Similarly, species classified as critically endangered in 1984 have disappeared from all study areas by 2020.

4.6.4.3 Fish Species Behavioral Responses to Severe Water Pollution

Behavioral alterations in fish species subjected to acute water contamination from industrial effluents are a general fact observed in many scientific research works (Das and Chakraborty 2007; Sarkar et al. 2021). The deficiency in dissolved oxygen (DO) and high concentration of biochemical oxygen demand (BOD) severely affect fish digestion and food absorption, leading to suppressed hunger. Despite eating during these periods, fish are compelled to eject immediately (Ghosh 2015). Their reduced rate of food consumption due to acute respiratory issues and loss of appetite. Acute drops in DO concentration, particularly during industrial pollution events (0.4 mg/L), further exacerbate respiratory discomfort. Fish species such as *Puntius sophore* and *Cirrhinus mrigala* are forced to surface during these periods to compensate for the DO deficit in the water in the river. Ghosh (1991) observed that severe industrial water pollution results in a thin covering of suspended particles on fish gills, significantly impeding gaseous exchange through the osmotic mechanism. Additionally, elevated blood CO_2 levels exacerbate respiratory conditions. To obtain atmospheric oxygen, certain fish species, including *Puntius sophore* and *Cirrhinus mrigala*, have

Table 4.12 Nature of trophic level, habitat orientation of the available fish species at the study reaches

Systematic position	Trophic level nature	Trophic level score			HON	Habitat orientation score			Maximum size captured (kg)		
		RE1	RE2	RE3		RE1	RE2	RE3	RE1	RE2	RE3
Order: Cypriniformes, family: Cyprinidae, subfamily: Cyprininae											
1. *Catla catla* (Hamilton-Buchanan)	PL	31.25	31.25	28.57	P	50	50	47.61	0.40	0.40	0.50
2. *Labeo rohita* (Hamilton-Buchanan)	PL	31.25	31.25	28.57	G	37.5	37.5	38.09	0.50	0.50	0.70
3. *Labeo bata* (Hamilton)	PL	Elim	Elim	Elim	G	Elim	Elim	Elim	Elim	Elim	Elim
4. *Labeo calbasu* (Hamilton)	BE	Elim	Elim	Elim	B	Elim	Elim	Elim	Elim	Elim	Elim
5. *Cirrhinus mrigala* (Hamilton-Buchanan)	BE	Elim	Elim	14.28	B	Elim	Elim	19.04	Elim	Elim	Elim
6. *Puntius sarana sarana* (Hamilton-Buchanan)	PL	31.25	31.25	28.57	P	50	50	47.61	0.02	0.02	0.02
7. *Puntius sophore* (Hamilton-Buchanan)	PL	31.25	31.25	28.57	P	50	50	47.61	0.02	0.02	0.02
8. *Puntius ticto* (Hamilton-Buchanan)	PL	31.25	31.25	28.57	P	50	50	47.61	0.01	0.01	0.01
Order: Cypriniformes, family: Cyprinidae, subfamily: Cultrinae											
9. *Chela laubuca* (Hamilton-Buchanan)	OM	Elim	Elim	Elim	P	Elim	Elim	Elim	0.03	0.03	0.03
Order: Cypriniformes, family: Cyprinidae, subfamily: Rasborinae											
10. *Amblypharyngodon mola* (Hamilton-Buchanan)	PL	Elim	Elim	Elim	P	Elim	Elim	Elim	Elim	Elim	Elim
11. *Esomus danricus* (Hamilton-Buchanan)	PL	Elim	Elim	Elim	P	Elim	Elim	Elim	Elim	Elim	Elim
12. *Mystus aor* (Hamilton)	OM	Elim	Elim	28.57	G	Elim	Elim	38.09	Elim	Elim	0.05
13. *Mystus seenghala* (Skyes)	OM	Elim	Elim	Elim	G	Elim	Elim	Elim	Elim	Elim	Elim
14. *Mystus vittatus* (Bloch)	OM	25	25	28.57	G	47.61	47.61	38.09	0.03	0.03	0.03
15. *Mystus bleekeri* (Day)	OM	Elim	Elim	Elim	G	Elim	Elim	Elim	Elim	Elim	Elim
16. *Rita rita* (Hamilton)	OM	Elim	Elim	Elim	B	Elim	Elim	Elim	Elim	Elim	Elim
Order: Cypriniformes, family: Siluridae											
17. *Ompok pabo* (Hamilton)	CA	Elim	Elim	Elim	G	Elim	Elim	Elim	Elim	Elim	Elim
18. *Wallago attu* (Schneider)	CA	Elim	Elim	38.09	G	Elim	Elim	38.09	0.40	0.40	0.40
Order: Cypriniformes, family: Schilbeidae											
19. *Ailia coila* (Hamilton)	PL	Elim	Elim	Elim	P	Elim	Elim	Elim	Elim	Elim	Elim
20. *Eutropifchthys goongwaree* (Skyes)	CA	Elim	Elim	Elim	P	Elim	Elim	Elim	Elim	Elim	Elim
21. *Eutropifchthys vacha* (Hamilton)	CA	Elim	Elim	Elim	P	Elim	Elim	Elim	Elim	Elim	Elim

(continued)

Table 4.12 (continued)

Systematic position	Trophic level nature	Trophic level score			HON	Habitat orientation score			Maximum size captured (kg)		
		RE1	RE2	RE3		RE1	RE2	RE3	RE1	RE2	RE3
Order: Cypriniformes, family: Clariidae											
22. *Clarias batrachus* (Linnaeus)	OM	25	25	28.57	P	50	50	47.61	0.4	0.4	0.4
Order: Cypriniformes, family: Heteropneustidae											
23. *Heteropneustes fossilis* (Bloch)	OM	25	25	28.57	P	50	50	47.61	0.4	0.4	0.4
Order: Perciformes, family: Gobiidae											
24. *Glossogobius guiris* (Hamilton-Buchanan)	OM	25	25	28.57	B	12.5	12.5	19.04	0.1	0.1	0.1
25. Gobias striatuis (Day)	OM	Elim	Elim	Elim	B	Elim	Elim	Elim	Elim	Elim	Elim
Order: Perciformes, family: Anabantidae											
26. *Anabas testudineus* (Bloch)	OM	Elim	Elim	Elim	G	Elim	Elim	Elim	Elim	Elim	Elim
Order: Perciformes, family: Belontiidae											
27. *Trichogaster fasciatus* (Bloch)	OM	Elim	Elim	28.57	P	Elim	Elim	Elim	Elim	Elim	0.02
28. *Nandus nandus* (Bloch)	CA	Elim	Elim	Elim	G	Elim	Elim	Elim	Elim	Elim	Elim
Order: Perciformes, family: Channidae											
29. *Channa striatus* (Bloch)	CA	37.5	37.5	38.09	G	47.61	47.16	38.09	0.50	0.50	0.50
30. *Channa punctata* (Bloch)	CA	37.5	37.5	38.09	G	47.61	47.16	38.09	0.30	0.30	0.30
31. *Channa gachua* (Bloch)	CA	Elim	Elim	Elim	G	Elim	Elim	Elim	Elim	Elim	Elim
32. *Channa marulius* (Bloch)	CA	37.5	37.5	38.09	G	47.61	47.16	38.09	0.30	0.30	0.30
33. *Channa orientalis* (Bloch)	CA	37.5	37.5	38.09	G	47.61	47.16	38.09	0.20	0.20	0.20
Order: Perciformes, family: Mastacembelidae											
34. *Mastacembelus armatus* (Lacepède)	BE	Elim	Elim	Elim	B	Elim	Elim	Elim	Elim	Elim	Elim
35. *Macrognathus pancalus* (Hamilton-Buchana)	BE	Elim	Elim	14.28	B	Elim	Elim	19.04	0.02	0.02	0.02
Order: Perciformes, family: Synbranchidae											
36. *Amphipnous cuchia* (Hamilton-Buchanan)	BE	0.62	0.62	14.28	B	12.5	12.5	19.04	0.30	0.30	0.30
Order: Osteoglossiformes, family: Notopteridae											
37. *Notopterus notopterus* (Pallas)	CA	37.5	37.5	38.09	P	50	50	47.61	0.20	0.20	0.20
38. *Notopterus chitala* (Pallas)	CA	Elim	Elim	38.09	P	Elim	Elim	47.61	Elim	Elim	0.20

(continued)

Table 4.12 (continued)

Systematic position	Trophic level nature	Trophic level score			HON	Habitat orientation score			Maximum size captured (kg)		
		RE1	RE2	RE3		RE1	RE2	RE3	RE1	RE2	RE3
Order: Osteoglossiformes, family: Ambassidae											
39. *Chanda nama* (Hamilton-Buchanan)	CA	37.5	37.5	38.09	P	50	50	47.61	0.01	0.01	0.01
Order: Clupiformes, family: Clupeidae											
40. *Tenualosa illsha* (Hamilton)	PL	Elim	Elim	Elim	G	Elim	Elim	Elim	Elim	Elim	Elim
41. *Gadusia chapra* (Hamilton)	PL	Elim	Elim	Elim	P	Elim	Elim	Elim	Elim	Elim	Elim
Order: Clupiformes, family: Engraulidae											
42. *Setipinna phasa* (Hamilton)	PL	Elim	Elim	Elim	P	Elim	Elim	Elim	Elim	Elim	Elim
Order: Mugiliformes, family: Mugilidae											
43. *Rhinomugil Corsula* (Hamilton)	PL	Elim	Elim	Elim	P	Elim	Elim	Elim	Elim	Elim	Elim
Order: Beloniformes, family: Belonidae											
44. *Xenentodon cancila* (Hamilton)	PL	Elim	Elim	28.57	P	Elim	Elim	Elim	Elim	Elim	0.02

HON Habitat orientation nature, *Elim* Eliminated, *PL* Planktivore, *BE* Benthic feeder, *OM* Omnivore, *CA* Carnivore, *P* Pelagic, *G* Generalist, *B* Benthic
Based on Das and Chakrabarty (2007a, b)

been observed leaping out of the water and projecting their snouts over the water's surface. This behavior is another aspect addressed in the current study. Furthermore, significant respiratory issues can cause a decline in swimming ability, leading to death. It is worth noting that fish in the lower stretch migrate to the Hooghly River, where the Churni River converges, to avoid high levels of industrial pollution. Once the pollution subsides, the fish return to their natural habitat in the lower stretch. Consequently, industrial water pollution also impacts fish species migration in the lower stretch.

4.6.4.4 Loss of Growth and Reproduction

Fish cannot grow in starvation due to lack of enough food. Fish such as *Puntius sophore* may have impaired growth rate and appetite loss as a result of industrial effluents' impact on their enzymatic activity. Thus, in the MCR, the impact of water pollution on the decline in fish productivity has become a serious issue. The present study has observed that the productivity of certain fish species—such as *Labeo rohita, Puntius sarana sarana, Puntius ticto, Puntius sophore, Mystus vittatus, and Channa striatus*—is declining at a rate of more than 10% annually, while other species such as *Heteropneustes fossilis, Clarias batrachus, Channa punctate, Glossogobius guiris, Channa marulius, Amphipnous cuchia, and Channa orientalis* have been experiencing a fall rate of less than 5% annually (Fig. 4.25a and b). The field study also reveals that a small number of species, including *Notopterus notopterus, Channa striatus, and Glossogobius guiris*, have greater fall rates throughout the year.

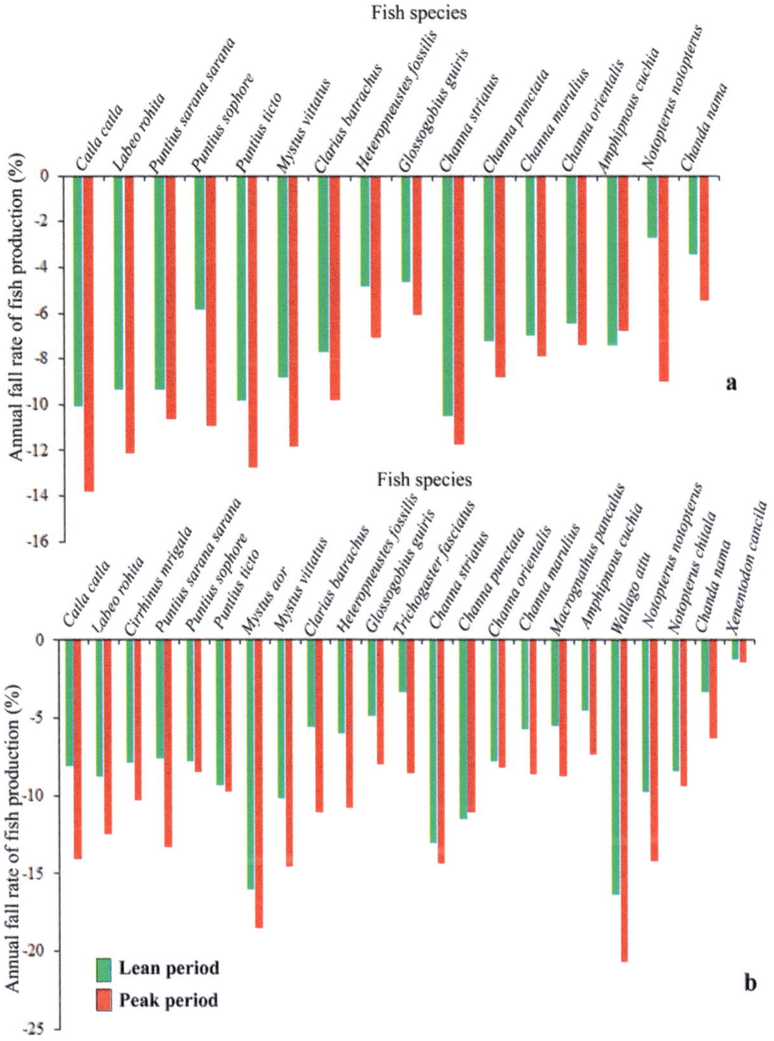

Fig. 4.25 Species-wise annual fall rate of production

4.6.5 Impact of River Decay on Economy and Livelihood

4.6.5.1 Sampling Design for the Selection of the Study Villages and Households

The river plays an important role in fishing and agricultural pursuits. The major economic activities of the villages adjacent to the Mathabhanga-Churni River are fishing and agriculture. There are 49 villages adjacent to the Churni River; out of which 31 are fishing dominated and 18 are agriculture dominated villages. Both the fishing and agricultural economy of the river adjacent villages are affected by the pollution of the Mathabhanga-Churni River; however, the severity of the fishing economy is more tell-tale. Therefore, we intend to deal with the economic status of the four selected fishing dominated villages (2 from the upper stretch and another 2 from the lower stretch) following a purposive sampling technique (Fig. 4.26). For determining the sample size from the target population, the total households of each village (631 of Shibnibas, 1281 of Bablaban, 755 from Raghunathpur, 1790 of Ghoraghata) have

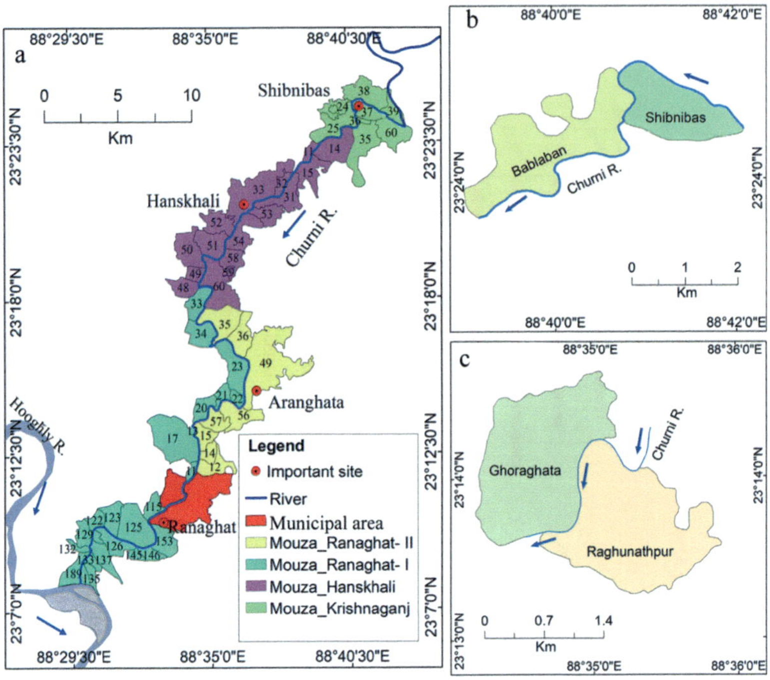

Fig. 4.26 Selection of the study villages for investigating the impact of river decay on the economy and society; **a** location of the villages adjacent to the Churni River; **b** the location of the upper stretch villages, Shibnibas (No. 37) and Bablaban (No. 25); **c** the location of the lower stretch villages, Raghunathpur (No. 20) and Ghoraghata (No. 57)

been obtained from the Census Handbook of Nadia District, 2011. Finally, 10% of the households from each village (63 for Shibnibas, 128 for Bablaban, 75 for Raghunathpur, and 179 for Ghoraghata) have been selected using a simple random sampling technique.

4.6.5.2 Changes in River-dependent Livelihoods

The monthly per capita income of the selected villages varies in space and time. The first-generation average monthly per capita income was highest in Ghoraghata (INR 2090), then in Raghunathpur (1570), Shibnibas (INR 1467), and Bablaban (INR 1347). The results for the second and third generations (INR 1873 for Ghoraghata, INR 1444 for Raghunathpur, INR 1366 for Shibnibas, and INR 1239 for Bablaban) also show a similar trend. The second generation shows up INR 1842 for Ghoraghata, INR 1449 for Raghunathpur, INR 1415 for Shibnibas, and INR 1315 for Bablaban. As a result, it has been shown that the average per capita income for each the village is on the decline. In particular, the declines from the first to the second generation and the second to the third generation are, for Shibnibas, – 0.51% and – 0.49%; for Bablaban, – 0.31% and – 0.75%; for Raghunathpur, – 1.20% and – 0.5%; and for Ghoraghata, – 2.48% and 0.31%. Because of this, the distribution of per capita income suggests that, from a historical perspective, the first generation's economic situation was better than that of the second and third generations during the period when the Churni River was less contaminated.

Comparably, from a geographical perspective, it has been discovered that the lower stretch's (Ghoraghata and Raghunathpur) average per capita monthly income is larger than that of the upper stretch's villages. According to headcount ratio, about 60% of all households in both villages of the upper stretch—Shibnibas and Bablaban—live below the poverty line, according to an empirical study on the state of poverty. The lower stretch villages have a smaller percentage of households living below the poverty line (about 45% in Raghunathpur and 27.5% in Ghoraghata) than the upper stretch villages. Regarding the poverty gap ratio, Bablaban has the largest gap at 8.52%, followed by Shibnibas at 6.38%, Raghunathpur at 3.73%, and Ghoraghata at 2.84%. Comparably, Bablaban has about 1.67% of families living in extreme poverty according to the poverty severity index, with Shibnibas (1.03%), Raghunathpur (0.48%), and Ghoraghata (0.47%) following.

4.6.5.3 Changes in Human Psychology

It is evident from the investigation that the people of the study villages are suffering from many socio-psychological problems due to the pollution of the MCR. The main pillars of the society, i.e., literacy and health of the fishing community have been worst-hit. In every study village, the literacy rate and health of the fishing community are significantly lower than that of their respective village. Besides, the social instability in the form of domestic violence, alcohol consumption, family disagreement, and physical unrest has been found to a large extent. In addition, the stress, anxiety, and depression (Figs. 4.26, 4.27 and 4.28) brought on by the environment have psychologically affected the fishermen. In general, the lower stretch villages are in a better position than the higher stretch villages, indicating the spatiality of the detrimental impact of environmental stress among the study communities. On the other side, there have been positive outcomes where the villagers are adopting several low- and high-level pro-environmental behaviors to enhance the river's quality and have altered their psychological characteristics. In addition, the villagers have united to voice their concerns and resist the pollution of the MCR.

The village-wise pattern of perceptions regarding the parameters is presented in Fig. 4.27.

Natural environments provide an array of values, both material and emotional, and as a result, people become increasingly attached to them over time (Fig. 4.29). To gauge the degree of involvement to a location, several researchers have developed various psychological measures (Sullivan and Young 2020; Gibbs 2002). Place identification (PI), place dependency (PD), place affect (PA), and place social bonding (PSB) are the four main sub-constructs that make up place attachment (Perkins 2010; Williams & Vaske 2003; Ramkissoon et al. 2013a). PI is defined as the self-dimension of place attachment, encompassing an individual's behavior, feelings, emotions, self-identity, and symbolic connection to a place (Raymond et al. 2010). PD is a component of the self-dimension that pertains to an individual's functional attachment, or goal-directed connection, to a location (Tonge et al. 2015). According to Ramkissoon and Mavondo (2014), PA implies a strong sense of belonging to a certain area. Finally, PSB shows that individual-place interactions greatly impact an individual's connecting with their community, neighbors, family, and even faraway people (Ramkissoon et al. 2013b). Place satisfaction is a more advanced form of place attachment that is influenced by personal opinions of a location's quality and the services received. When a location repeatedly encounters external threats, diverse behavioral reactions result. Prolonged threats to the environment can negatively impact topophilia in several ways, including breakdown of place-based identity, loss of mental and emotional attachment, and strong emotions of powerlessness over the dangers lowering a place's quality. Additionally, a severe consequence that is indicative of people's rejection of a location as a result of acute environmental degradation is tropophobia. The notion that people's attachment to a place increases when it is threatened or polluted and they take different protective measures by performing different PEB for that place has made the study of topophilia even more intriguing (Estrada et al. 2017; Buta et al. 2014; Fielding et al. 2008; Halpenny 2010; Vesely et al. 2020; Jugert et al. 2016). Because of this, human behavior in a given location is extremely complicated, and modern place researchers make an effort to investigate the many facets of topophilic behavior while

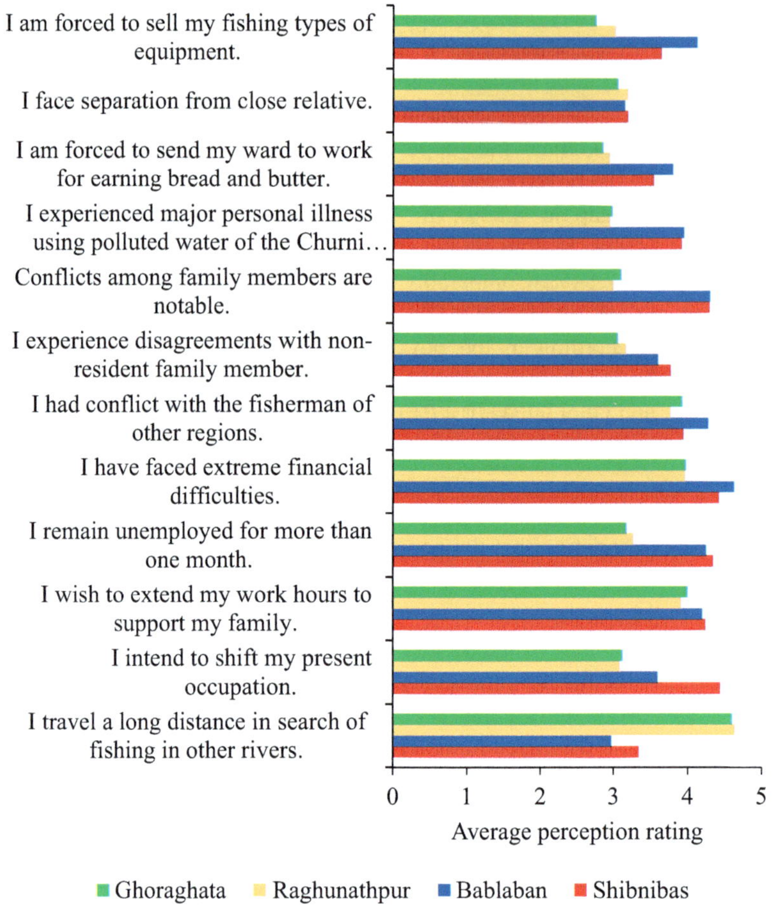

Fig. 4.27 Pattern of the parameters of the stressful life events (*Source* Field survey, 2018–2020)

taking the particular location of the world into consideration.

For villagers, the Churni River in the Bengal delta provides a substantial source of income. Additionally, people bind themselves to the river through a variety of sociocultural pursuits, including sports, boating, fishing, picnicking, and religious celebrations and rituals (Sarkar et al. 2021). As a result, the inhabitants have created a strong emotional bond with the river. In addition, the Churni River is experiencing issues with ongoing deterioration due to water pollution, a decrease in depth, and a decline in discharge (Sarkar and Islam 2020). As a result, these serious issues interfere with the social and economic life of the locals to the point that they are eager to find solutions (Das and Chakrabarty 2007a, b). Consequently, because of their location connection, the problem's long-term persistence has led to PEB. The study area's residents carry out many PEB to revitalize the river. The river's pollution and the locals' knowledge of the environment have a major influence on the kind of topophilic behavior. The community at large has grown highly conscious as a result of the increasing vulnerability of the local fishermen's livelihoods owing to water contamination (Sarkar and Islam 2020). Annually, the river suffers extensive industrial contamination episodes, according to the local fisherman (lasting approximately one week following the injection of industrial effluent). At certain seasons, the river's water is unfit for human consumption in any way (Sarkar and Islam 2022). A 5-item scale

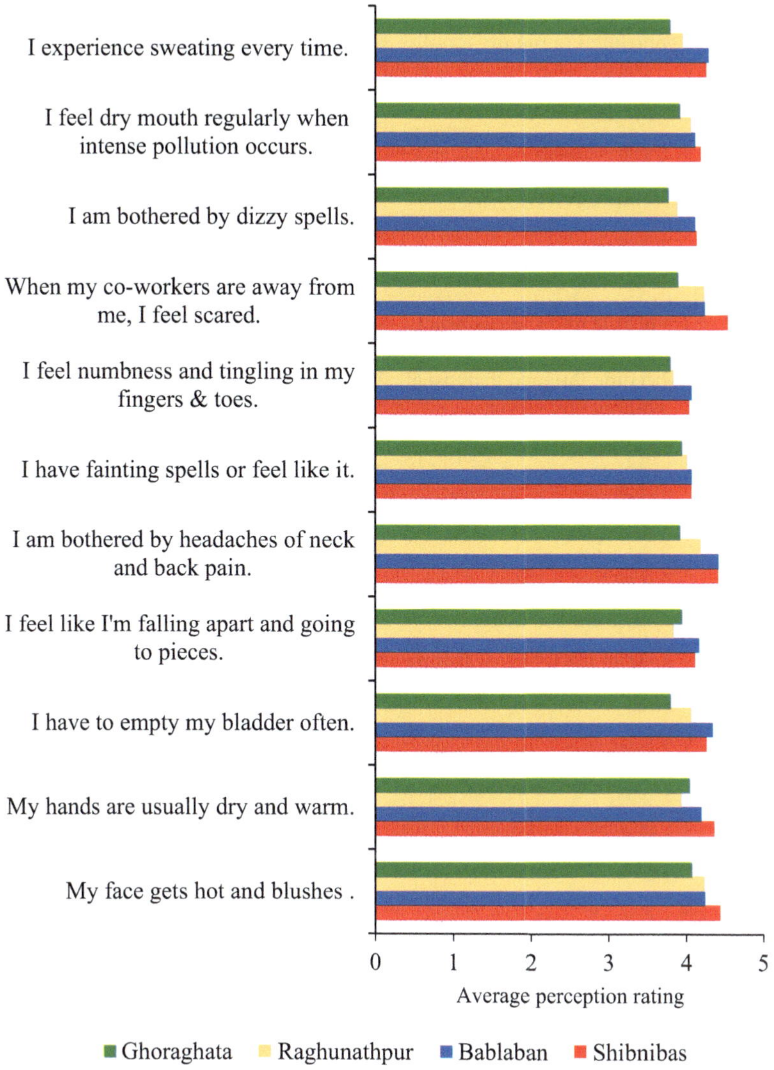

Fig. 4.28 Pattern of the parameters of anxiety (*Source* Field survey, 2018–2020)

has thus been used to gauge the local fishermen's awareness.

EA2 and EA3 show that individuals are looking for regular updates on fish species that are at risk of extinction and declining fish productivity, respectively, while EA1 shows people's interest in river water pollution. EA4 and EA5 are similarly related to individuals using social media to routinely get up-to-date information on the Churni River. The average values of EA1, EA2, EA3, EA4, and EA5 are 4.9, 4.8, 4.2, 3.9, and 3.6, respectively, based on the empirical data. Furthermore, the substantial correlation between E4 and E5 (R2 = 0.7) demonstrates the importance of social networks in informing the public about pollution. The correlations between E1 and E2 (R2 = 0.48), E3 and E4 (R2 = 0.42), and E3 and E4 are the other significant ones. The results of the field study also show that local fisherman often file complaints about water contamination with various authorities, such as the government, local government, and environmental organizations. Their actions and behaviors that support the environment demonstrate their

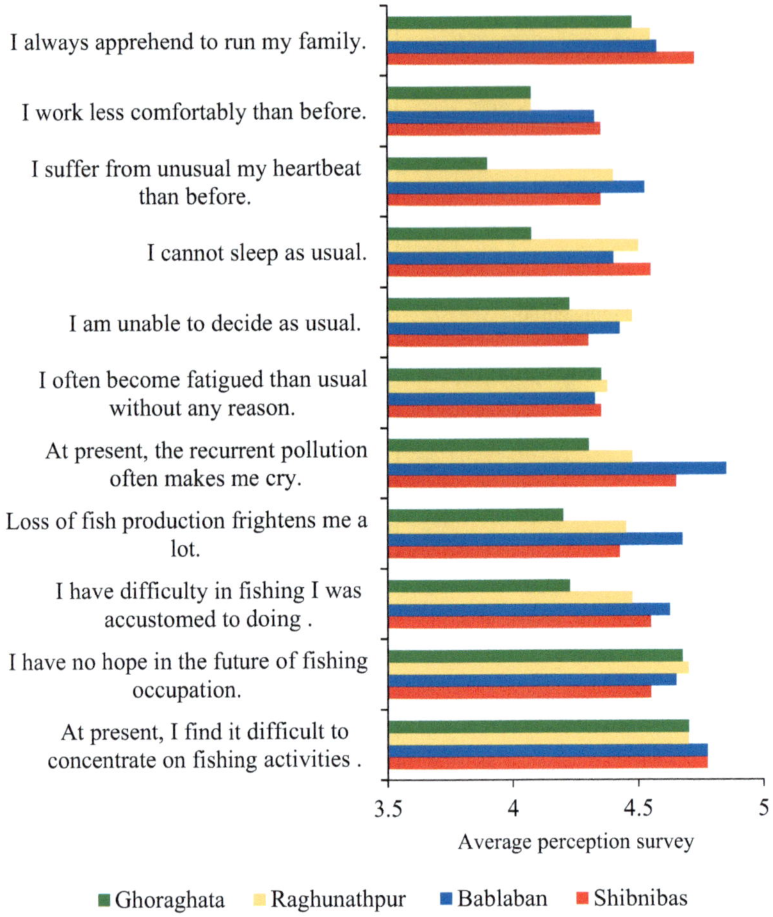

Fig. 4.29 Pattern of the parameters of depression (*Source* Field survey, 2018–2020)

knowledge of the declining riverine ecosystems. The locals implore the various parties involved to look into the problem and come up with a solution. In addition, the residents often plan events in collaboration with NGOs and government representatives to increase public awareness of the river's water contamination.

Out of the four dimensions of place attachment, place social bonding has the highest mean score (3.84) in the general pattern of topophilia. Place affect (3.78), place identity (3.76), and place reliance (3.70) are the next top scores. The data indicates that the average location satisfaction score was 3.86. Furthermore, it was discovered that the mean scores for low and high PEB were, respectively, 4.58 and 4.69. The average score across all topophilic behavior categories indicates that people have a strong emotional connection to the river and, as a result, a strong desire to protect it from external dangers by engaging in low- and high-PEB behavior. Furthermore, the Likert scale classes show that all place attachment dimensions range from moderate to high classes, with the lowest class being 1–1.5, the lowest being 1.51–2.50, the moderate class being 2.51–3.50, the high class being 3.51–4.50, and the highest class being 4.51–5. Whereas place dependence (35.4%) was the most common response for respondents with a moderate level of place attachment, place affect, place identity, and place social bonding were the most common responses for those with a high level of place attachment (84.47%), place social bonding (35.4%), and place affect

(35.4%). Furthermore, the low and HighPEB Likert scale classes range from high to very high classes. 52.8% and 47.2% of the LowPEB respondents, respectively, fall into the high and very high classes, whereas 26.7% and 73.3% of the HighPEB respondents, respectively, fall into the high and very high classes.

Social and demographic factors also influence the dynamics of topophilic behavior. Place identification (84.85%) indicates a higher degree of place attachment in females, but place dependency (82.81%) and place social bonding (87.5%) indicate a higher level of place attachment in males. It is not always true that males are more environmentally conscious than women in terms of worries, attitudes, and behaviors, according to a number of scholarly publications on gender variations in environmental views and behavior (Luchs and Mooradian 2012; Scannell and Gifford 2013). However, several studies (Gifford and Nilsson 2014; Islam and Guchhait 2018) have confirmed the notable gender disparity in environmental beliefs and behavior. According to the results of the current study, there is a distinct gender difference in the intention to conduct HighPEB, with 21.21% of the total female respondents ($N = 33$) and 72.65% of the total male respondents ($N = 207$) completing HighPEB. The male respondents' daily fishing activities in the MCR, where they earn their bread and butter, may account for this discrepancy. As a result, the problems of declining discharge and recurring river water contamination brought on by the discharge of industrial wastewater are of greater concern to them. In addition to taking care of home responsibilities, women are also involved in creating the fishing gear that the male family members utilize. They are therefore not in the forefront of the perilous circumstance involving the river due to pollution. Men and women's psychological makeup differs as a result of this. Furthermore, the image of topophilic behavior varies among age groups. The respondents who belonged to the 40–60 age group showed a high degree of place attachment through place identification and place social bonding; in contrast, the 40–60 age group's high place dependency is indicative of a high degree of place attachment. In addition, 75% of responders in the 40–60 age range are using HighPEB. Through place identification, place affect, and place social bonding, the respondents with only a primary education demonstrate a higher level of place attachment. Furthermore, a sizable portion of them is discovered to have HighPEB purpose. Finally, respondents with monthly per capita incomes below INR 2000 (1 USD = 71.9222 INR on December 16, 2018) demonstrate the highest involvement in HighPEB, whereas respondents with monthly per capita incomes above INR 2000 (1 USD = 71.9222 INR) demonstrate the highest level of place attachment through place identity, place affect, and place social bonding.

For the restoration of the riverine ecosystem and the enhancement of socio-psychological well-being, topophilic behavior has proven important when considering people's views and attitudes regarding changes in the riverine landscape (Davenport and Anderson 2005; Buijs 2009). The empirical database has proven the community's and individual's commitment to the Churni River and the creation of their active PEB for river restoration. The primary source of worry for them is not natural river degradation, but rather human contamination, particularly from Carew and Company Ltd., a wine, chemical, and sugar producing company based in Darshana, Bangladesh. The highly contaminated MCR is regularly the recipient of massive quantities of untreated wastewater from the industries. Sarkar and Islam (2020) reported that there was a notable decrease in DO (0.2 mg/L) and a high BOD (68 mg/L) during industrial water pollution. In the upper Churni River, there is also evidence of the catastrophic extinction of fish species due to industrial pollution. As a result, the local fishermen are struggling to make ends meet and have a significant grievance against the industrial water contamination incidents (Sarkar and Islam 2022).

According to Tuan (1974), there are two ways to identify a place: (1) its spatial characteristics, which include its physical qualities and geographical identification; and (2) its human connection, which includes its mental and emotional attachment. In his topophilia, Tuan (1974) emphasized the second sense of location (Tuan 1974; Stedman 2002). There are other aspects of human connection to the Churni River—such as livelihood, culture, and recreation—which lend credence to the topophilia idea. The findings of Hawthorne et al. (2022) further corroborate the importance of the local community's emotional bond with the Indian River Lagoon. The Churni River is the primary source of income for the local fisherman, which makes their love of the river even stronger. Khakzad and Griffith (2016) provided more evidence about the contribution of sociocultural advantages to the local fishermen's sense of place-making. In addition, emotional attachment to the river has been a family problem for a number of generations. Their articulation throughout the field survey interview reflects this. "We started living along the bank of the Churni River following the heritage of our predecessor that river is not only a site for a livelihood alternative but also an emotion associated with it for many cultural proposals," for instance, said 65-year-old male Ramesh Halder. Rivers are connected to cultural elements including festivals centered around rivers and rowing competitions that are headed by women every August 15, which is also known as Indian Independence Day. Still, societal rebellion has been noted, because the river is an integral element of their cultural activities. Because of the river's declining state, many cultural events are seldom planned. Therefore, the anti-degradation sentiment of the locals toward the Churni River is represented in impromptu environmental demonstrations that validate topophilia ideas. Hogg and Terry (2000) provided an explanation of exogenous dangers to a place that is similar to our study. They said that these risks might cause a range of psychological reactions, such as an increase in place identification and a community's defense mechanism against the common threat.

According to the field survey, there are primarily two types of protests: (1) large-scale politico-environmental movements organized by environmental groups in partnership with the local people, involving approximately 500 people, and (2) protests organized by the local people themselves, involving about 200 people. While the activities of environmental organizations are noticeable to be few and far between, those of the local population are regular in their travels. Consequently, the movement "of the people by the people and for the people" is the only form of initial expression. For instance, on April 10, 2018, 350 local fishermen organized a demonstration against Churni River pollution in front of Krishnaganj's Block Development Officer (BDO) office. Furthermore, a significant portion of the local population spontaneously participates in the second method of expression, despite being organized by a small number of organizations such as the "Save Mathabhanga and Churni River Committee," "Chakdaha Science and Cultural Organization," and "Dakshinbanga Matsyajibi Forum" (DMF). For instance, on October 10, 2018, some 500 fishermen gathered in Majhdia to demonstrate against Churni River pollution, using the placard "Save River, Save Fish, Save Fisher People." Therefore, the voices of the locals are the consequence of topophilia, or love for the Churni River; this movement is not a so-called "sponsored movement" that is managed from the outside.

Similar to topophilia, the locals' perspectives on the Churni River reinforce Relph's theory of placeness, as well as the notions of place insideness and depth of place (Relph 1976). Relph used the term "insideness" to describe how individuals felt when they were (i) comfortable rather than anxious, (ii) surrounded rather than exposed, and (iii) secure rather than vulnerable (Relph 1976; Canter 1977; Seamon and Sowers 2008). According to the current study, despite the fact that ongoing industrial water pollution puts the locals' livelihoods in jeopardy, they do not feel cut off from the river or inside out (the opposite of feeling inside out). Instead, they have a deep sense of place that helps them acquire the environmental awareness, attachment, and commitment needed to solve the current issue. Consequently, their depth of place is defined by their profound sense of place identity.

4.6.5.4 Environmental Awareness and Citizen Activism

A pollution incident having significant damage potential to the economy, human health, and the environment bears a critical impact on the psychology of an individual and a community (Van Rooij 2010; Sarkar et al. 2023). When pollution incidents happen frequently, people become immediately aware of the situation and attempt to solve it at all costs. Sometimes, they decide to choose the path of legal movement (Kamau 2005; Van Rooij 2010). In this regard, two concepts are very popular in the domain of environmental psychology—(1) environmental awareness and (2) citizen activism (Vaske and Kobrin 2001). Moreover, how the former influences the latter has become an interesting topic in environmental psychology (Gifford and Nilsson 2014; Klöckner 2013; Whitmarsh and O'Neill 2010). It is pertinent to mention here that all types of pollution do not induce grievance among the citizens. Pollution incidents, which have a higher risk potential for the environment, health, economy and society, give birth to strong grievances in the mind of the citizens' psyche (Felstiner et al. 1980). The swiftness of grievance development is determined by the factors that how quickly citizens perceive the negative consequence of a particular pollution incident and how fast they develop awareness among themselves (Alford et al. 2002). Moreover, a full-fledged grievance emerges when a particular pollution incident occurs at regular intervals (Sarkar and Islam 2022). As a consequence of full-fledged grievance, to overpower the pollution incidence, people are bound to select the path of activism or to participate in the environmental movement on a local or large scale (Sarkar et al. 2022). The actions generally taken by the people are of different types. The legal actions are litigation against the pollution generating agencies and the administrative litigation against the authority for their negligence while taking care of environmental management (Prasad 2004). Besides, political actions are required to file petitions to the enforcement governmental authorities regarding the risk of pollution, and if it doesn't work petitions need to be filed to the higher level of governmental authority.

Therefore, declining discharge coupled with severe water pollution brings about ecological stress reflected in the disappearance of fish species and declining production of the existing fish species. For example, Das and Chakrabarty (2007a, b) observed 28 fish species (out of 44) extinct from the Churni River since 1983. Sarkar and Islam (2020) also found that 16 fish species at the upper stretch and 23 species at the lower stretch were present having a high annual fall rate of production of more than 10% (e.g., *Puntius sarana sarana, Clarias batrachus, Heteropneustes fossilis, Glossogobius guiris, Channa punctate*). The disappearance of fish species and declining production of the existing fish species bring about the problem of a livelihood crisis in the fishing community. Sarkar and Islam (2022) explained the severity of poverty in the fishing community as a consequence of severe water pollution of the Churni River. Many researchers (e.g., Sarkar and Islam 2019) have addressed the mixing of industrial effluents from Carew and Company (Bangladesh) as one of the key factors responsible for severe water pollution. Therefore, being united against industrial pollution, the local people are performing various pro-environmental behavior (Sarkar et al. 2022).

Therefore, environmental awareness has a significant role in the development of citizen activism or environmental movement. However, environmental awareness does not alone determine victims' involvement in the environmental movement. For example, one having a strong knowledge about water pollution incidence and its potential damageability may not help him/her to get involved in a movement unless he/she has self-efficacy, self-responsibility, self-identity, and community value (Estrada et al. 2017; Dono et al. 2010).

It is generally known that Carew and Company Ltd., a business that makes sugar, wine, chemicals, and other products in Darshana, Bangladesh, discharges a sizable volume of wastewater onto the Bangladeshi bank of the Mathabhanga River, badly damaging the

Churni River. Thus, the understanding of the local riparian people on the detrimental effects of pollution has swiftly increased. Therefore, the awareness of the riparian people was measured using a 5-item scale. EA1 measures 'people's interest in river water pollution', and E2 and E3 reflect 'people's desire to receive ongoing updates on vulnerable fish species and diminishing fish productivity', respectively. The E4 and E5 are concerned with the 'people's use of social media to regularly obtain updated information about the Churni River'. The strong association between E4 and E5 among all the awareness-related variables (r = 0.7) depicts that social networks are necessary to aware the public about pollution. The other noteworthy relationships are between E1 and E2 ($r = 0.48$), E3 and E4 ($r = 0.42$). The average values of E1, E2, E3, E4, and E5 are 4.9, 4.8, 4.2, 3.9, and 3.6, respectively. The average values reflect how concerned the locals are about water pollution and vulnerable fish species. Additionally, complaints against water contamination made to local authorities, the government, and environmental organizations have served as a gauge of public grievance (PG1). The 5 items have been considered in this measurement. PG1 pertains to complaints made in writing or over the phone to local officials, PG2 to the Panchayat Pradhan (head of the political body acting as the village cabinet in India), PG3 to the Block Development Officer, PG3 to the Sub Divisional Officer, and PG4 to non-governmental organizations acting as agents for environmental health. The public grievance with the highest average value is recognized as PG1 (4.63), followed by PG2 (4.60), PG3 (4.2), PG5 (4.06), and PG4 (3.90). The findings show that the locals strongly urge the local authorities to tackle the issue. Additionally, the locals make an effort to include the administrative officials in the community by presenting them with demands and keeping an eye on the river. The locals make an effort to include NGOs in their efforts to investigate the issue and find a solution. Numerous socio-demographic parameters are analyzed alongside the complaints to better comprehend how the pattern of complaints differs with distinct socio-demographic factors. The correlation matrix reveals that income has a negative correlation with all types of public grievances (− 0.03 with PG1, − 0.38 with PG2, − 0.37 with PG3, − 0.47 with PG4, and − 0.37 with PG5), indicating that a decline in income (especially from fishing) prompts people to speak out against pollution. Similar results have been found for occupational diversity (− 0.05 with PG1, − 0.27 with PG2, − 0.28 with PG3, − 0.33 with PG4, and − 0.33 with PG5), suggesting that not many permanent fishermen have been turned into occupational fishermen and declining occupational diversity is also prompting the public to voice their complaints to local officials, administrators, and NGOs. Additionally, all of the PG items show extremely weak relationships with age, education, and dependence ratio.

4.7 Conservation

4.7.1 During the British Occupation

The rivers in the Bengal delta were subjected to continuous monitoring during the British period because the rivers were crucial to the colonial economy. The primary objective was to keep the rivers navigable throughout the year. Therefore, in this period British engineers had emphasized on few management measures, i.e., (i) river training (for preventing river bank erosion and flooding through structuring riverbanks), (ii) dredging (the process of removing bed sediment and debris from the river channels), and (iii) depth monitoring (continuous investigation on river water depth for ensuring navigable for large cargo vessels). Moreover, much evidence is also the construction of embankments, canals, and flood control structures for preventing floods in deltaic areas. The main rivers that were under continuous investigation were the Ganges and Bhagirathi-Hooghly Rivers. The dredging was most frequently executed for the Bhagirathi-Hooghly River to ensure year-round navigability and make the Kolkata (Calcutta), port workable. These interventions, however, often neglected for the other rivers like Churni due to low economic importance. Even though the Churni was important as a local commercial channel, colonial actions were rather restricted, focusing solely on maintaining basic navigability for smaller boats and preventing severe siltation that may hamper the flow. Some dredging was done to prevent river shallowing, especially in places where silt buildup reduced the water depth during the dry season. The maintenance of the Churni River was relatively optional to that of mighty rivers because, although the occasional removal of debris altered the river's natural flow patterns, substantial structural interventions like canalization or huge embankments were rare.

4.7.2 After Independence

After the independence of India, several attempts have been made for river restoration ensuring the flow and health riverine ecosystem (Table 4.13). The attempts for river conservation include continuously monitoring river discharge, water quality improvement through industrial wastewater treatment, municipal wastewater treatment, riparian vegetation management, river bank stabilization, and floodplain management. Therefore, there have been several government agencies to execute government-taken initiatives e.g. (1) action plan for pollution sources sources management, such as solid waste, biomedical waste, e-waste, municipal outfalls, sewage carrying sewers, and industrial discharges, (2) the plan emphasizing on flood plain zones (FPZ) management, collecting rainfall, preserving minimal environmental flow, and groundwater management. The severe issue of the river pollution is addressed in a variety of ways under the Ranaghat Municipality's waste management action plan. For example, initiatives toward more environmentally friendly waste management techniques are taken such as integrated waste management facilities, door-to-door waste collection, and source-level waste segregation procedures. Besides, in Nadia District, West Bengal initiatives have been taken in preventing riverbank soil erosion under the Mahatma Gandhi Rural Employment Guarantee Act (MGNREGA)

Table 4.13 Action plans and respective agencies for the Churni River restoration

Government agencies	Action plans working
SUDA	Actions designed for solid wastes and plastic wastes management in urban areas are (a) collection of solid waste at door to door level (b) establishment of facility for waste processing (c) development of information, education, and communication activities and capacity buildings of different stakeholder
WBPCB	Action plans designed for (a) management of industrial effluent (b) management of various hazardous wastes and inventorization of water polluting industries
IRD	(a) Actions designed to be implemented for flood plain zones (FPZ) protection and maintenance, for example, 0.5 km bank protection, identified by Nadia Irrigation Division, benefiting about 2000 people in 98 hectare of areas. (b) Maintaining minimum environmental flow of river
DoAg	Good agricultural practice introducing low water requiring crops; good irrigation practices; conservation of soil and water
P&RD D	Surface water preservation programs (e.g., rainwater harvesting) for areas identified—Majhdia, Batkulla, Aranghata, Bagula, Hijuli, Ranaghat, and Coopers' Camp. To coordinate with Forest Department for providing lands for tree plantation and development of biodiversity parks
SWID	Groundwater level as well as quality monitoring within the basin area
DoUD&MA	Utilization of treated wastewater

Fig. 4.30 Protection measure of the Churni River bank from erosion through vetiver grass plantation

Scheme by planting vetiver along the Churni River. The goals are to protect communities from frequent floods and river bank stabilization. Along the Churni River, vetiver plantations for river bank stabilization planning have been put into practice, using carefully selected sites. A good example of this is the brick kiln industry on the riverbank at Haskhali. The need for stabilization is pronounced due to the potential erosion caused by brick kiln industrial activities. On the left bank, a 500-m stretch has been dedicated to a vetiver plantation (Fig. 4.30).

Another noteworthy site is below the Kalinarayanpur bridge, where vetiver plantations have been established in an effort to counteract the scouring impact of the bridge. The objective of the project is to strengthen the riverbank and lessen the effects of scouring by planting vetiver here, which will increase the bridge's safety and strength. This illustrates an integrated approach to river bank management and shows a deep grasp of the interplay between human infrastructure and natural processes.

The central pollution control board (CPCB) and the West Bengal Pollution Control Board (WBPCB) have been working jointly since 1980s to monitor the water quality of the Churni River having their two significant water quality monitoring station covering the entire stretch of the Churni River. Few actions have been taken by the Indian government (GoI) to combat water pollution at national and international level. The government, for instance, has set up monitoring stations (Shibnibas and Ranaghat) to regularly check the quality of the water (Sarkar et al. 2020). Furthermore, the Carew and company (sugar and chemical producing industry) located in Bangladesh is one of the key sources of water pollution. Therefore, water pollution of the MCR has become an international issue between Bangladesh and India (Sarkar and Islam 2020). Therefore, the government has raised this issue before the government of Bangladesh. In 2019, the Bangladesh government notified first the establishment of wastewater treatment plants by the Ministry of External Affairs (GoI) resolving the bilateral problem.

4.7.3 Conclusion

The Churni River has been investigated from an integrated outlook where channel morphology, hydraulics, water quality, and habitat suitability, the issue of livelihoods all have been thoroughly examined. The analysis of channel morphology, hydraulic behavior, water quality, and riverine ecosystem all exhibit a deteriorating trend of the river. While investigating the responsible factors behind this, both natural phenomena in the forms of neotectonism and climate change, as well as human activities such as point and non-point source of pollution (discharge of industrial waste water, agrochemicals from nearby agricultural fields, urban run-off, rapid growth of brick kiln industries), and construction of road and rail crossings have appeared collectively. However, the study observed the declining discharge and high level of water pollution are the key issues intensifying the ecological stress of the river. The study of fish communities' structure has been a clear indicator of ecological stress. For example, the river has experienced the loss of fish species diversity, i.e., from 44 to 28 in the upper stretch near Majdia and 21 in the lower stretch at Ranaghat. The study has also observed that the issues have compounded on the socio-economic status of the local community, particularly the fisherman community. Therefore, all the issues have been treated from sustainable river restoration involving the wide range of stakeholders including the local community, government, and non-government agencies. Throughout this chapter and the preceding Chaps. 2 and 3, we have explored the evolutionary journey of the Bhagirathi, Jalangi, and Mathabhanga-Churni Rivers, alongside the changes in the socio-economic and cultural fabric of the region. Chapter 5 will focus deeply into integrated river basin management, emphasizing on restoring the Nadia Rivers and ensuring the sustainability of the human landscape intertwined with these waterways.

References

Action Plan for Rejuvenation of River Churni Nadia, West Bengal, *CPCB*, River Rejuvenation Committee, West Bengal (2020)

Adusumilli NC, Lacewell RD, Rister ME, Woodard JD, Sturdivant AW (2010) Effect of agricultural activity on river water quality: a case study for the Lower Colorado River Basin (No. 1371–2016–108879)

Alabaster JS, Lloyd RS (1982) Water quality criteria for freshwater fish. Butterworths

Alford WP, Weller RP, Hall L, Polenske KR, Shen Y, Zweig D (2002) The human dimensions of pollution policy implementation: air quality in rural China. J Contemp China 11(32):495–513

Allison M, Kepple E (2001) Modern sediment supply to the lower delta plain of the Ganges Brahmaputra River in Bangladesh. Geo-Mar Lett 21(2):66–74

Armantrout NB (1998) Glossary of aquatic habitat inventory terminology. Western Division of American Fisheries Society. https://doi.org/10.47886/9781888569117

Bain MB, Jia H (2012) A habitat model for fish communities in large streams and small rivers. Int J Ecol. https://doi.org/10.1155/2012/962071

Bandyopadhaya DK (2007) Banglar Nadnadi. Dey's Publishing, Kolkata. p 67

Bhatnagar A, Jana SN, Garg SK, Patra BC, Singh G, Barman UK (2004) Water quality management in aquaculture. In: Course Manual of summer school on development of sustainable aquaculture technology in fresh and saline waters, CCS Haryana Agricultural, Hisar (India), vol 3, pp 203–210

Bhattacharya P, Jacks G, Jana J, Sracek A, Gustafsson JP, Chatterjee D (2001) Geochemistry of the Holocene alluvial sediments of Bengal Delta Plain from West Bengal, India: implications on arsenic contamination in groundwater. In: Groundwater arsenic contamination in the Bengal Delta Plain of Bangladesh, vol 3084, pp 21–40

Biswas KR (2001) Rivers of Bengal. West Bengal District Gazetteers, Higher Education Department, Government of West Bengal, Kolkata

Bizzi S, Lerner DN (2015) The use of stream power as an indicator of channel sensitivity to erosion and deposition processes. River Res Appl 31(1):16–27

Brookes A, Shields Jr FD (1996) River channel restoration: guiding principles for sustainable projects

Buijs AE (2009) Public support for river restoration. A mixed-method study into local residents' support for and framing of river management and ecological restoration in the Dutch floodplains. J Environ Manage 90(8):2680–2689

Buta N, Holland SM, Kaplanidou K (2014) Local communities and protected areas: the mediating role of place attachment for pro-environmental civic engagement. J Outdoor Recreat Tour 5:1–10

Canter D (1977) The Psychology of Place. London: Architectural Press

Chatterjee K (2008) The persianization of itihasa: Performance narratives and Mughal political culture in eighteenth-century Bengal. J Asian Stud 67(2):513–543

Chatterjee M (2013) An enquiry into the evolution and impact of human interference on the Churni River of Nadia District, West Bengal. Int J Curr Res 5(5):1088–1092

Chakrabarty D, Das SK (2006) Fish community structure and ecological degradation in tropical rivers of India. Web Ecol 6(1):27–36

Chattapadhaya S (2007) Maharajendra Bahadur Krishnachandra, Dey's Publishing, Kolkata, p 52, 67

Chimwanza B, Mumba PP, Moyo BHZ, Kadewa W (2006) The impact of farming on river banks on water quality of the rivers. Int J Environ Sci Technol 2(4):353–358

Colebrooke RH (1803) Chart of the river Ganges from Colgong to Hurrisonker exhibiting the state of its islands and sand bars during the dry season of 1796–1797. https://maps-prints.com/antique-maps-india-ceylon/9379-antique-map-of-the-ganges-river-by-colebrooke-1805.html

Das BC (2014) Two indices to measure the intensity of meander. In: Mehtab Singh RB, Hassan MI (eds) Landscape ecology and water management, Proceedings of IGU International Conference, Rohtak Conference, vol 2. Springer, Japan. ISSN: 2198-3542, ISBN: 978-4-431-54870-6, pp 233–246

Das S (2022) Bartamāna, sambādapatra, In Bengali. https://bartamanpatrika.com/detailNews.php?cID=17&nID=150321&P=3

Das BC, Islam A, Biswas B (2020) Morphometry as tool to trace out the genealogy of oxbow lake. Environ Earth Sci. https://doi.org/10.1007/s12665-020-8854-3

Das G (2003) Bangla Bhashar Abhidhan. Sahitya Samsad, Kolkata, ISBN-81-85 626-08-1

Das BC (2013) Changes and deterioration of the course of river Jalangi and its impact on the people living on its banks, Nadia, West Bengal [PhD thesis, The University of Calcutta]

Das BC (2015) Modeling of most efficient channel form: a quantitative approach. Model Earth Syst Environ 1(3):1–9

Das S (2017) Evolution of drainage and morphology of upper Bhagirathi Ganga interfluve region of West Bengal with special reference to Palaeochannels. PhD Thesis. University of Calcutta, Calcutta

Das BC (2023) চূর্ণী এক চুরি করা জলের নদী: জনশ্রুতি আর সত্যের সন্ধানে. Cūrṇī ēka curi karā jalēra nadī: Janaśruti āra satyēra sandhānē. In Bengali. APRIL 2023 | Maadhukari.com. https://www.maad0hukari.com/apr-2023

Das BC, Islam A (2015) Channel asymmetry of an Ox-Bow lake: a different perspective. Int J Ecosyst 5(3A):69–74

Das SK, Chakrabarty D (2007) The use of fish community structure as a measure of ecological degradation: a case study in two tropical rivers of India. Biosystems 90(1):188–196. https://doi.org/10.1016/j.biosystems.2006.08.003

Das SK, Chakrabarty D (2007) The use of fish community structure as a measure of ecological degradation: a case study in two tropical rivers of India. Biosystems 90(1):188–196

Das BC (2024) Exploring the past to discern the present: the role of historico-geographical study in differentiating natural river channels from artificial canals. River 3(1):89–106

Das BC, Das D (2020) The Anjana: a journey from river to canal. In: Anthropogeomorphology of Bhagirathi-Hooghly River System in India. CRC Press, Boca Raton, pp 521–560

Datta DK, Subramanian V (1997) Texture and mineralogy of sediments from the Ganges–Brahmaputra–Meghna river system in the Bengal Basin, Bangladesh and their environmental implications. Environ Geol 30(3–4):181–188

Davenport MA, Anderson DH (2005) Getting from sense of place to place-based management: an interpretive investigation of place meanings and perceptions of landscape change. Soc Nat Resour 18(7):625–641

DeLoach RE (1972) Oxygen sag and stream self-purification. Journal (Water Pollution Control Federation), pp 1198–1204

Dono J, Webb J, Richardson B (2010) The relationship between environmental activism, pro-environmental behaviour and social identity. J Environ Psychol 30(2):178–186

Dunca AM (2018) Water pollution and water quality assessment of major transboundary rivers from Banat (Romania). J Chem 2018:1–8

Estrada M, Schultz PW, Silva-Send N, Boudrias MA (2017) The role of social influences on pro-environment behaviors in the San Diego region. J Urban Health 94(2):170–179

Felstiner WL, Abel RL, Sarat A (1980) The emergence and transformation of disputes: naming, blaming, claiming…. In: Theoretical and empirical studies of rights. Routledge, pp 255–306

Fielding KS, McDonald R, Louis WR (2008) Theory of planned behaviour, identity and intentions to engage in environmental activism. J Environ Psychol 28(4):318–326

Fuiman LA (1978) Descriptions and comparisons of north eastern catostomid fish larvae. M.S. Thesis, Cornell University, Ithaca, NY, 110 pp

Ganguly (2021) Cūrṇī Nadīra Tīrē….Ārṭikēla -pārṭikēla. Sāhitya. In Bengali চূর্ণী নদীর তীরে …|আর্টিকেল - পার্টিকেল |সাহিত্য| https://www.mysepik.com/on-the-banks-of-the-river-churni-sampa-ganguly-rana-ghatwb/ retrieved on 12.01.2022

Garrett JHE (1910) Bengal District Gazetteer, Nadia, Bengal Secretariat Book Depot. Reprinted in 2001, p 14, 15, 26

Garzanti E, Vezzoli G, Andò S, Limonta M, Borromeo L, France-Lanord C (2019) Provenance of Bengal shelf sediments: 2. Petrology and geochemistry of sand. Minerals 9(10):642

Ghosal U, Sikdar PK, McArthur JM (2015) Palaeosol control of arsenic pollution: the Bengal Basin in West Bengal, India. Groundwater 53(4):588–599

Ghosh TK (1991) Influence of industrial pollutants on water qualities and aquatic life of the river Churni. PhD Thesis, Fisheries Laboratory. Department of Zoology University of Kalyani, Kalyani. https://shodhganga.inflibnet.ac.in/handle/10603/211125

Ghosh TK (2015) Influence of industrial pollutants on water qualities and aquatic life of the river Churni. PhD Thesis, Fisheries Laboratory. Department of Zoology University of Kalyani, Kalyani

Gibbs L (2002) Citizen activism for environmental health: the growth of a powerful new grassroots health movement. Ann Am Acad Pol Soc Sci 584(1):97–109

Gifford R, Nilsson A (2014) Personal and social factors that influence pro-environmental concern and behaviour: a review. Int J Psychol 49(3):141–157

Guchhait SK, Islam A, Ghosh S, Das BC, Maji NK (2016) Role of hydrological regime and floodplain sediments in channel instability of the Bhagirathi River, Ganga-Brahmaputra Delta, India. Phys Geogr 37(6):476–510

Halpenny EA (2010) Pro-environmental behaviours and park visitors: the effect of place attachment. J Environ Psychol 30(4):409–421

Hawthorne TL, Toohy KR, Yang B, Graham L, Lorenzo EM, Torres H et al (2022) Mapping emotional attachment as a measure of sense of place to identify coastal restoration priority areas. Appl Geogr 138:102608

Hirst MF (1915) Report on the Nadia Rivers, reprinted in Rivers of Bengal, vol 3. Department of Higher Education-2002, West Bengal District Gazetteers, Calcutta

Hogg M, Terry DJ (2000) Social identity and self-categorization processes in organizational context. Acad Manage Rev 25(1):121

Horton RE (1932) Drainage-basin characteristics. Transact Am Geophys Union 13(1):350–361

Islam A, Guchhait SK (2018) Analysis of social and psychological terrain of bank erosion victims: a study along the Bhagirathi river, West Bengal, India. Chinese Geograph Sci 28:1009–1026

Jha VC, Bairagya HP (2011) Flood plain evaluation in the Ganga-Brahmaputra Delta: a tectonic review. Ethiop J Environ Stud Manage 4(3):12–24

Jugert P, Greenaway KH, Barth M, Büchner R, Eisentraut S, Fritsche I (2016) Collective efficacy increases pro-environmental intentions through increasing self-efficacy. J Environ Psychol 48:12–23

Kamau EC (2005) Environmental law and self-management by industries in Kenya. J Environ Law 17(2):229–244

Khakzad S, Griffith D (2016) The role of fishing material culture in communities' sense of place as an added-value in management of coastal areas. J Mar Island Cult 5(2):95–117

Klöckner CA (2013) A comprehensive model of the psychology of environmental behaviour—a meta-analysis. Glob Environ Chang 23(5):1028–1038

Knighton AD (1981) Asymmetry of river channel cross-sections: Part I. Quantitative indices. Earth Surf Process Landf 6(6):581–588

Knighton D (1998) Fluvial forms and processes: a new perspective. Hodder Arnold Publication

Koenig CE (2011) Hydrogeochemical site characterization and groundwater flow modeling of the arsenic-contaminated Gotra aquifer, West Bengal, India. Doctoral dissertation, University of British Columbia

Kuehl SA, Allison MA, Goodbred SL, Kudrass H, Giosan L, Bhattacharya JP (2005) The Ganges–Brahmaputra delta. Special Publication-SEPM 83:413

Leopold LB, Maddock T Jr (1953) The hydraulic geometry of stream channels and some physiographic implications. U. S. Geol Survey Prof Paper 252

Leopold LB, Wolman MG (1957) River channel patterns-braided, meandering and straight. U.S. Geological Survey, Prof. Paper 282B. In: Morisawa M (1968) Streams: their dynamics and morphology, McGraw Hill, New York, p 138

Leopold LB, Wolman MG (1960) River meanders. Bull Geol Soc Am 71:774. Cited from Julien PY (1985) Planform geometry of meandering alluivial channels. Civil Engineering Department, Engineering Research Center, Colorado State University, p 6

Luchs MG, Mooradian TA (2012) Sex, personality, and sustainable consumer behaviour: elucidating the gender effect. J Consum Policy 35(1):127–144

Majumder SC (1942) Rivers of Bengal Delta. Calcutta University, Kolkata

Mallik K (1910) Nadia Kahini, in Bengali. In: Ray M (1998) Pustak Bipani, Kol, 3rd ed, p 300

Miller VC (1953) A quantitative geomorphic study of drainage basin characteristics in the Clinch mountain areas. USA, Technical Report, Columbia University, Department of Geology, Issue 3, p 51

Mirza MMQ (1997) Hydrological changes in the Ganges system in Bangladesh in the post-Farakka period. Hydrol Sci J 42(5):613–631

Mondal I, Bandyopadhyay J (2022) Morphological landscape mapping of the Bhagirathi flood plains in West Bengal, India, using geospatial technology. In PK Shit, B Bera, A Islam, S Ghosh, GS Bhunia (Eds), Drainage Basin Dynamics. Geography of the Physical Environment. Springer https://doi.org/10.1007/978-3-030-79634-1_24

Mondal I, Thakur S, Bandyopadhyay J (2020) Delineating lateral channel migration and risk zones of Ichamati River, West Bengal, India. J Clean Prod 244:118740

Morisawa M (1985) Streams: their dinamacs and morphology. McGraw Hill, New York, NY, p 91

Moss B (2007) Water pollution by agriculture. Philos Trans R Soc B Biol Sci 363(1491):659–666

Mueller JE (1968) An introduction to the hydraulic and topographic sinuosity indexes. Ann Assoc Am Geogr 58(2):371–385

Mukherjee A, Fryar AE (2008) Deeper groundwater chemistry and geochemical modeling of the arsenic affected western Bengal basin, West Bengal, India. Appl Geochem 23(4):863–894

Mukherjee A, Fryar AE, Thomas WA (2009) Geologic, geomorphic and hydrologic framework and evolution of the Bengal basin, India and Bangladesh. J Asian Earth Sci 34(3):227–244

Mukhopadhaya Srijuta Rajiblochan (1805) Maharaj Krishnachandra Rayasya Charitrang, Srirampur, pp 301, 303

Mukhopadhyay B, Mukherjee PK, Bhattacharya D, Sengupta S (2006) Delineation of arsenic-contaminated zones in Bengal Delta, India: a geographic information system and fractal approach. Environ Geol 49(7):1009–1020

Mutisya DK, Tole M (2010) The impact of irrigated agriculture on water quality of rivers Kongoni and Sirimon, Ewaso Ng'iro North Basin, Kenya. Water Air Soil Pollut 213(1–4):145–149

Perkins HE (2010) Measuring love and care for nature. J Environ Psychol 30(4):455–463

Prasad PM (2004) Environmental protection: the role of liability system in India. Econ Polit Week 257–269

Rahman M, Rahaman MM (2018) Impacts of Farakka barrage on hydrological flow of Ganges river and environment in Bangladesh. Sustain Water Resour Manage 4(4):767–780

Ramkissoon H, Mavondo F (2014) Proenvironmental behaviour: the link between place attachment and place satisfaction. Tour Anal 19(6):673–688

Ramkissoon H, Smith LDG, Weiler B (2013) Testing the dimensionality of place attachment and its relationships with place satisfaction and pro-environmental behaviours: a structural equation modelling approach. Tour Manage 36:552–566

Ramkissoon H, Weiler B, Smith LDG (2013) Place attachment, place satisfaction and pro-environmental behaviour: a comparative assessment of multiple regression and structural equation modelling. J Policy Res Tour Leisure Events 5(3):215–232

Ray KC (1751) Annada Mangal, Bengali, quoted from Choudhury J (2004) Sri Chaitanyadev O Samakalin Nabadwip, in Bengali, Nabadwip Puratatwa Parisad, p 14

Ray KC (1875) Kshitish Vansabali Charita. In: Ray M (1986) Natun Sanskrita Jantra, Kolkata, pp 69–71

Ray KC (1876) Kṣitīśa banśābalī carita, natuna sanskr̥ta yantra, In Bengali (pp. 85, 107–109)

Ray KC (1986) Kshitish banshaboli Charita. In: Ray M (1986) Manjusha, Kolkata, p 57

Raymond CM, Brown G, Weber D (2010) The measurement of place attachment: personal, community, and environmental connections. J Environ Psychol 30(4):422–434

Reaks HG (1919) Report on the physical and hydraulic characteristics of the delta. Stevension-Moore S et al (1919), pp 29–132

Relph E (1976) Place and placelessness, vol 67. Pion, London

Roy S (2013) The effect of road crossing on river morphology and riverine aquatic life: a case study in Kunur River Basin, West Bengal. Ethiop J Environ Stud Manage 6(6):835–845

Roy S, Sahu AS (2017) Potential interaction between transport and stream networks over the lowland rivers in Eastern India. J Environ Manage 197:316–330

Roy S, Sahu AS (2018) Road-stream crossing an in-stream intervention to alter channel morphology of headwater streams: case study. Int J River Basin Manage 16(1):1–19

Rudra K (2010) Dynamics of the Ganga in West Bengal, India (1764–2007): Implications for science–policy interaction. Quatern Int 227(2):161–169

Rudra K (2014) Changing river courses in the western part of the Ganga-Brahmaputra delta. Geomorphology 227:87–100

Rudra K (2018) Rivers of the Ganga-Brahmaputra-Meghna Delta—a fluvial account of Bengal. Springer International Publishing, Cham. https://doi.org/10.1007/978-3-319-76544-0

Samsad S (2000) Samsad Bangla Abhidhan, Kolkata. ISBN 81-86 806-92-X

Samaddar JN (1925) The Bargi invasion of Bengal. J Royal Asiatic Soc 57(4):673–678

Sanyal T, Kaviraj A, Saha S (2015) Deposition of chromium in aquatic ecosystem from effluents of handloom textile industries in Ranaghat-Fulia region of West Bengal, India. J Adv Res 6(6):995–1002

Sarkar B, Islam A (2019) Assessing the suitability of water for irrigation using major physical parameters and ion chemistry: a study of the Churni River, India. Arab J Geosci 12(20):1–16

Sarkar B, Islam A (2020) Drivers of water pollution and evaluating its ecological stress with special reference to macrovertebrates (fish community structure): a case of Churni River, India. Environ Monitor Assess 192(1):1–31

Sarkar B, Islam A (2022) Assessing poverty and livelihood vulnerability of the fishing communities in the context of pollution of the Churni River, India. Environ Sci Pollut Res 29(18):26575–26598

Sarkar B, Islam A (2024) Assessing environmental awareness activism to river decay and recurring pollution using parallel multiple mediator model. Int J River Basin Manage 1–13

Sarkar B, Islam A, Das BC (2020) Anthropo-footprints on Churni River: a river of stolen water. In: Anthropogeomorphology of Bhagirathi-Hooghly River System in India. CRC Press, pp 433–468

Sarkar B, Islam A, Das BC (2021) Role of declining discharge and water pollution on habitat suitability of fish community in the Mathabhanga-Churni River, India. J Clean Prod 326:129426

Sarkar B, Islam A, Datta D (2023) Characterising topophilic behaviour in the wake of river decay and pollution through structural equation modelling. Environ Dev Sustain 1–32

Sarkar B, Islam A, Shit PK, Ghosh S (2022) Assessment of water pollution and aquatic toxicity of the Churni River, India. In: River Health and Ecology in South Asia. Springer, Cham, pp 303–327

Sarkar B, Das BC, Islam A, Datta D, Pawlik Ł, Quesada-Román A (2024) Temporal change in channel form and hydraulic behaviour of a tropical river due to natural forcing and anthropogenic interventions. Phys Geogr 1–35

Sassolas-Serrayet T, Cattin R, Ferry M (2018) The shape of watersheds. Nat. Commun. 9(1):3791

Scannell L, Gifford R (2013) Personally relevant climate change: the role of place attachment and local versus global message framing in engagement. Environ Behav 45(1):60–85

Seamon D, Sowers J (2008) Place and placelessness, Edward Relph. In: Key texts in human geography, pp 43–51

Schumm SA (1956) Evolution of drainage systems and slopes in badlands at Perth Amboy, New Jersey. Geol Soc Am Bull 67(5):597–646

Schumm SA (2005) River variability and complexity. Cambridge University Press, Cambridge, p 65

Sigler JW, Bjornn TC, Everest FH (1984) Effects of chronic turbidity on density and growth of steelheads and coho salmon. Trans Am Fish Soc 113(2):142–150. https://doi.org/10.1577/1548-8659(1984)113<142:EOCTOD>2.0.CO;2

Stedman RC (2002) Toward a social psychology of place: Predicting behavior from place-based cognitions, attitude, and identity. Environ Behav 34(5):561–581

Stevension-Moore et al (1919) Report on the Hooghly River and its head-waters. West Bengal District Gazetteers (pp 27, 35, 49, 50)

Sullivan D, Young IF (2020) Place attachment style as a predictor of responses to the environmental threat of water contamination. Environ Behav 52(1):3–32

Surian N, Rinaldi M (2003) Morphological response to river engineering and management in alluvial channels in Italy. Geomorphology 50(4):307–326

Thakur SK (2019) Bargi hāṅgāmā ō samakālīna rājā jamidāra, 'mudrā' mahārājā kr̥ṣṇacandra ō samakāla, naba paryāẏa, tr̥tīẏa barṣa, sampādaka- śaibāla sarakāra (pp. 173–198)

Toner PF (1986) Impact of agriculture on surface water in Ireland Part I. General. Environ Geol Water Sci 9(1):3–10

Tonge J, Ryan MM, Moore SA, Beckley LE (2015) The effect of place attachment on pro-environment behavioral intentions of visitors to coastal natural area tourist destinations. J Travel Res 54(6):730–743

Tuan Y (1974) Topophilia: a study of environmental perception, attitudes, and values. Prentice-Hall, Englewood Cliffs, NJ

Van Rooij B (2010) The people vs. pollution: understanding citizen action against pollution in China. J Contemp China 19(63):55–77

Vaske JJ, Kobrin KC (2001) Place attachment and environmentally responsible behavior. J Environ Educ 32(4):16–21

Vesely S, Klöckner CA, Brick C (2020) Pro-environmental behavior as a signal of cooperativeness: evidence from a social dilemma experiment. J Environ Psychol 67:101362

Wang X, Li J, Li Y, Shen Z, Wang X, Yang Z, Lou I (2014) Is urban development an urban river killer? A case study of Yongding Diversion Channel in Beijing, China. J Environ Sci 26(6):1232–1237

WBPCB (2020) Water quality information system. Government of West Bengal. http://emis.wbpcb.gov.in/waterquality/viewsampledatacitizen.do. Accessed 30 Dec 2020

Weinman B, Goodbred SL Jr, Zheng Y, Aziz Z, Steckler M, van Geen A et al (2008) Contributions of floodplain stratigraphy and evolution to the spatial patterns of groundwater arsenic in Araihazar, Bangladesh. Geol Soc Am Bull 120(11–12):1567–1580

Whitmarsh L, O'Neill S (2010) Green identity, green living? The role of pro-environmental self-identity in determining consistency across diverse pro-environmental behaviours. J Environ Psychol 30(3):305–314

Williams DR, Vaske JJ (2003) The measurement of place attachment: validity and generalizability of a psychometric approach. For Sci 49(6):830–840

Wohl E (2018) Rivers in the Anthropocene: the US perspective. Geomorphology

গাঙ্গুলী সম্পা (১৪.০৭.২০২১)। চূর্ণী নদীর তীরে...।আর্টিকেল-পার্টিকেল।সাহিত্য।https://www.mysepik.com/on-the-banks-of-the-river-churni-sampa-ganguly-ranaghat-wb/retrieved on 12.01.2022

Environmental Flow and Nadia Rivers Restoration: Integrated River Basin Management

5

Abstract

River decay of Nadia Rivers is quite manifested in the hydro-geomorphic and socio-ecological maps. This necessitates the resuscitation of the rivers from the standpoint of integrated river basin management. As Nadia rivers are part of the greater Ganga river system, the prescriptions for the Ganga river restoration is to be followed for the Nadia Rivers. The Ganga Action Plan (now Namami Gange) is committed to restoring the river Ganga on four pillars—a. *abiral dhara* (continuous flow), b. *nirmal dhara* (unpolluted flow), c. geological continuity, and d. ecological management. As there are various stakeholders for managing these goals, a cohesive integration is needed. Apart from these mega-scale prescriptions, some local-level initiatives for the restoration of the Nadia rivers are also important. For example, Kolkata Port Trust, IWAI and CWC are continuously monitoring the Bhagirathi-Hoghly system for better performance of the river. Similarly, Jalangi Action Plans are really crucial for the sustainable development of the Jalagi River apart from the localized movement of the various stakeholders. The topophilia of the local people residing on the riverbank of the Churni River is expressed through the spontaneous movements to save the Churni River from further deterioration. This chapter finally upholds the notion of the human-river harmony model in the context of the Nadia Rivers.

Keywords

Ganga action plan · Namami Gange · *Nirmal dhara* · *Abiral dhara* · Ecological restoration · Environmental flow · Human-river harmony model

5.1 Environmental Flow—A Concern to Save the Nadia Rivers

A significant concern regarding the "*Nadia Rivers*" is the crisis of Environmental flow discussed in the previous chapters (2, 3, and 4). Environmental flow, commonly referred to as E-flow, is the quantity, and quality of water flows required to sustain freshwater ecosystems and the human livelihood that depends upon it. The reduced water flow of the "*Nadia Rivers*" is a result of both natural and anthropogenic factors. Therefore, an integrated river basin management (IRBM) approach is essential with an emphasis on sustainable water management techniques, pollution reduction, and better water infrastructure control. Therefore the present chapters starts with the backdrop of the environmental flow crisis flowed by the E-flow management planning and the feasibility study.

B. C. Das et al., *The Nadia Rivers*, Geography of the Physical Environment,
https://doi.org/10.1007/978-3-031-85971-7_5

The current environmental flow status of the Bhagirathi-Hooghly River is influenced by a number of natural and anthropogenic factors discussed in Chap. 2. The Farakka Barrage benefits the river by lowering the siltation of the Kolkata Port, ensuring adequate water flow in the Bhagirathi-Hooghly River for navigation, and helping with irrigation for neighbouring agricultural areas. The barrage is vital to the lives of those who depend on the river for their livelihoods because it diverts water from the Ganga River, maintaining the water levels of the Bhagirathi River necessary for these economic activities, and increasing agricultural output in the region. However, the Bhagirathi-Hooghly environmental flow (E-flow) condition has become more severe due to increased human intervention. Excessive surface and groundwater extraction for urbanization and agriculture, widespread pollution from domestic sewage, and uncontrolled industrial effluent discharge have all negatively impacted water quality. These acts further reduce base flows and harm the existing ecosystem. Chapter 3 has argued that the Jalangi River has to cope with a major environmental flow problem as a result of sedimentation at the offtake point, which has gradually reduced the discharge (Das 2013) and increased the environmental flow crisis. The significant silt deposition obstructing the natural water entrance of the river has resulted in a significant reduction in discharge, especially during the dry season. The quality of the water has declined and pollution problems have become worse as a result of the river's incapacity to sufficiently dilute the increasing volumes of untreated sewage, agricultural runoff, and industrial waste. The sluggish flow is another factor that exacerbates habitat loss, particularly for aquatic species whose survival depends on steady water levels and healthy environments. Like the Jalangi River, the Mathabhanga-Churni River system (discussed in Chap. 4) is facing same problem of sedimentation at the offtake point. The natural flow is impeded by the accumulation of sediments, which causes a dramatic drop in discharge, especially during the dry season. Additionally, the accumulation of untreated home sewage, agricultural runoff, and industrial effluents has worsened water pollution leading to habitat degradation and survival challenges to the aquatic species.

Because of the Farakka Barrage, the Bhagirathi-Hooghly River faces a distinct set of challenges. While the Jalangi and Mathabhanga-Churni Rivers are naturally blocked by sedimentation at their offtake points, the flow of the Bhagirathi-Hooghly River is artificially managed by the barrage, which diverts water from the Ganges. The Farakka Barrage was built to maintain a steady flow in the Bhagirathi-Hooghly, mainly to prevent siltation in the Kolkata Port and to support navigation. This has helped sustain certain economic activities, like irrigation and shipping, even during the dry season, but human intervention has its own set of issues. In contrast, there is an absence of significant human-engineered flow management in the Mathabhanga-Churni and Jalangi Rivers indicating that they rely more on natural water flow dynamics, which are disrupted by sedimentation. As a result, these rivers face critical environmental flow, particularly during the dry season, leading to worsened water quality, pollution accumulation, and habitat degradation.

5.2 Conceptual Framework of Integrated River Basin Management

A river basin is a fundamental unit of hydrogeomorphic investigations. In classical geomorphology (Schumm 1977), a drainage basin is considered an integrated fluvial system as it contains three zones (1) production zone in the upper of the basin where water and sediment are produced (2) transfer zone where sediment load and water are transported and (3) deposition zone characterizes the accumulation of the sediment and flux of discharge to a basin outlet (Fig. 5.1). Naturally, these parts are very much integrated. Thus, any perturbations in the upper part of a basin induce drainage modification in the lower part of that system. Moreover, natural

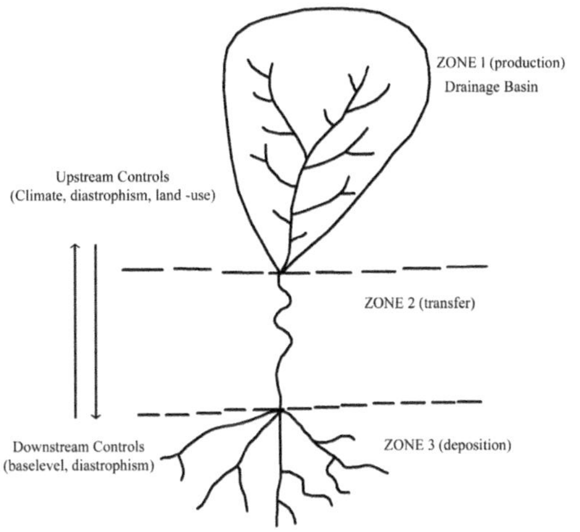

Fig. 5.1 Integrated nature of the fluvial system (Schumm, 1977)

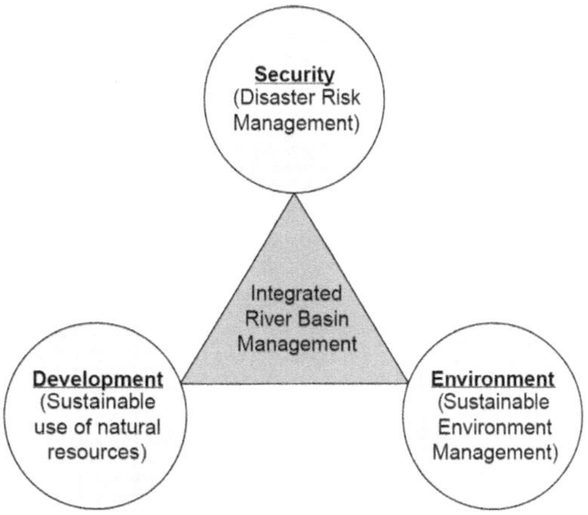

Fig. 5.2 Integrated River basin management (Phong and Shaw 2010)

resources endowment in the form of forest resources and mineral resources are quite concentrated in the upper basin area while the lower part exhibits floodplain with higher agricultural productivity and flat topography with lower relative inducing a congenial environment for the development of the infrastructure and development of the dense settlement (Shit et al. 2022).

In recent times, rampant development at the basin scale especially in the lower and middle parts of the basin has become tale-tell having detrimental effects on the fluvial system and ecological behaviour of the river. Large-scale development like dam construction, reservoir storage and other forms of human interventions like changes in land use and land cover, intensive agriculture with fertilizer-enriched agricultural drainage, and urbanization with discharge of untreated effluents into the fluvial system have degraded the rivers at the global and regional scales (Das et al. 2020). The fluvial decay is more prominent in tropical nations with excessive population pressure for these reasons. This requires initiatives for the restoration of the river for long-term ecologic and economic benefits. Thus, the concept of integrated river basin management (IRBM) has become popular recently in the wake of climate and accelerated human interventions on the geosphere.

The concept of IBRM entails three important notions—(1) long-term economic development through sustainable use of natural resources, (2) protecting environmental sustainability through some management measures, and (3) achieving security through disaster risk reduction (Fig. 5.2).

According to Clausen, (2011), the typical RBM functions include (1) Basin planning (2) Water allocation (3) Pollution control and environmental flows (4) Flood and drought management (5) Economic and financial management (6) Information management (7) Monitoring (8) Stakeholder dialogue and participation and (9) Conflict resolution. Therefore, IBRM is a holistic task in the field of water governance that includes various stakeholders. Water governance is theoretically built upon the principles of (1) effectiveness, (2) efficiency and (3) trust and engagement. These OCED principles have various aspects for sustainable river basin management (Fig. 5.3).

The global consciousness for the restoration of the river is widely observed through the implementation of the IBRM principles in various nations of the world. Tropical river basin management also has gained momentum in recent decades for the survival of the production system and humans (Pegram et al. 2013; Marques et al. 2019). The climate change

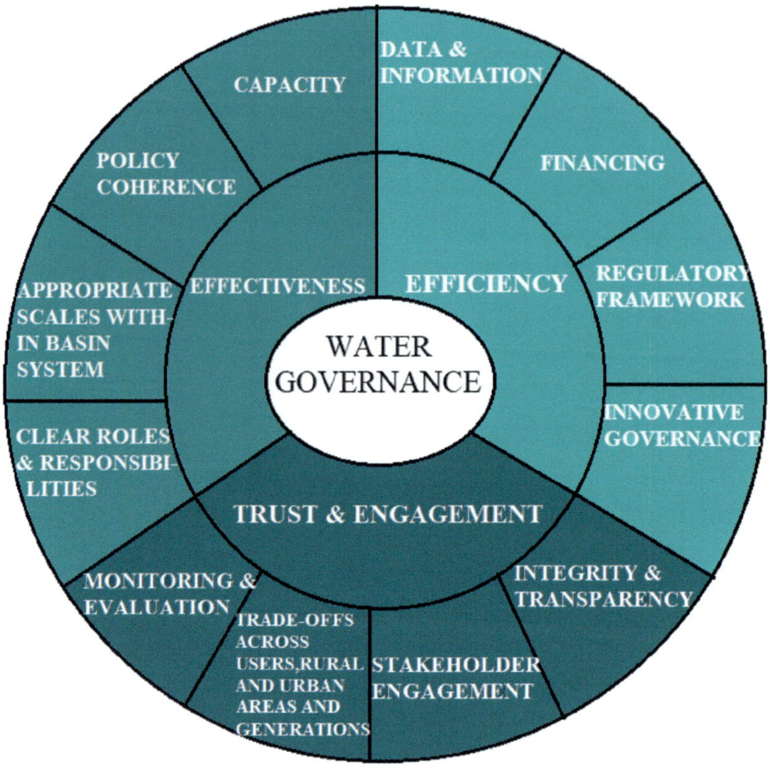

Fig. 5.3 The OECD principles on water governance, adapted from OECD (2015) (Lim et al. 2022)

scenario has induced this trend significantly in tropical developing nations (Abbass et al. 2022). The rivers of the Bengal delta are really suffering from the shortage of discharge, and loss of cubic capacity due to sedimentations (Sarker et al. 2013). In the lean months, several rivers either dried up or maintained only a feeble flow (Islam and Guchhait 2024a). Thus, flow restorations are really crucial for these rivers for the survival of the densest delta in the world.

5.3 Nadia Rivers—An Integrated Fluvial System

Nadia rivers consisting of the Bhagirathi, Jalangi, and Mathabhanga-Churni rivers are intricately linked in a system. The Jalangi River joins the Bhagirathi River at Nabadwip and the Churni River meets the Hooghly River at Shibpur near Ranaghat. The Anjana Canal (once a river) maintains a connection between the Jalangi and Churni Rivers in the Nadia district. Thus, changes in any river induce corresponding changes. However, Jalangi-Chuni connections through the Anjana have been stopped due to the conversion of the Anjana River to a canal. All these three major rivers of the Nadia district are really crucial for the development of ecology, and society. The river basin scale planning in these fluvial systems is quite problematic because none of these rivers exhibit complete basins. These basins are not like ideal basins characterized by production, transfer and deposition zones. Rather a production zones are missing in these systems or supplied from other systems. Thus, these basins are truncated in that sense. The concept of IRBM to these rivers has some limitations. However, as these rivers have cradled the densely populated civilizations, restorations are invoked for the sustainability of the society and environment.

5.4 River Basin Planning and Management

Considering the importance of the Nadia rivers, the Nadia River Division was set up by the British for the regular mapping monitoring of the rivers for better navigability of the rivers for maintaining water transport. In the Independent period, Nadia Rivers gradually decayed. In the early 1970s, river Bhagirathi maintained negligible flow during the lean months (January to May) and National Waterway No. 1 (from Banarasi to Ganga Sagar) was greatly impeded (Parua 2009). To save the Kolkata port and provide a saline-free environment for the Kolkata urban area, the Bhagirathi-Hooghly River was planned to be revived through the induction of water from the Ganga River through feeder canal from the Farakka Barrage Project constructed in 1975 (Islam and Guchhait 2017). Though the River Bhagirathi was saved through this mega-scale intervention, Jalangi and Mathabhanga-Churni were further decayed after the installations of the FBP project causing lower water availability in the downstream area of the FBP (Sarkar et al. 2021). Thus, the sources of these two rivers were almost choked for the entire year except for 1 to 2 months or even few for days during the flood seasons.

5.4.1 Ganga (Bhagirathi-Hooghly) River Restoration

5.4.1.1 Pre-British, and British Periods Initiatives

Since 1550AD, Zamindars have been constructing embankments along the east bank of the river Bhagirathi to protect the region from floods, a practice that has persisted for centuries. In 1824, the Nadia River Division extended its charge to include Bhagirathi, marking an official involvement in river management. The following year, in 1825, bandalling techniques were employed to maintain the depth on shoals below Nabadwip, indicating early efforts at river maintenance and navigation improvement (Hirst 1916).

Moving forward to the late 19th century, in 1867, Captain Forbes proposed a government-funded initiative to dig tanks for water supply, while Mr. Leonard recommended wells and khals. This year also saw small grants being sanctioned for these water-related projects, highlighting a growing recognition of the importance of water management and sanitation. By 1874, the focus had shifted to flood defence, with Mr. Wickes proposing the Jiaganj-Bhagabangola road as a secondary line of defence against floods. The Nadia River Division also proposed abandoning a portion of the Laltakuri embankment, reflecting evolving strategies in flood management and infrastructure development (Hirst 1916). In 1914, discharge observation stations were established along the river Bhagirathi, indicating advancements in hydrological monitoring. Modern dredging techniques, utilizing dredgers like Lindon Bates (1902), Alpha (1905), Rescue, and Nimotha, were employed by the Nadia River Division to maintain navigability and control sedimentation. Hirst's report in 1916 provides a comprehensive overview of the region's water management practices. He notes the historical context of embankment construction, the recent development of new embankments, and the strategic importance of certain routes in flood defence. Additionally, the report mentions planned spill measures for famine relief and ongoing efforts in river monitoring and dredging. It also highlights the preparation of malaria and sanitation-related maps, indicating a multifaceted approach to public health and environmental management in the region.

5.4.1.2 Independence Period Initiatives

(1) **Ganga Action Plan and Namami Gange projects status in West Bengal**

Observing the deteriorating conditions of the rivers in India, especially the Ganga system, the Ganga Action Plan (GAP) was launched in 1985 by the Government of India which mainly focussed on the clean Ganga mission. In 1993,

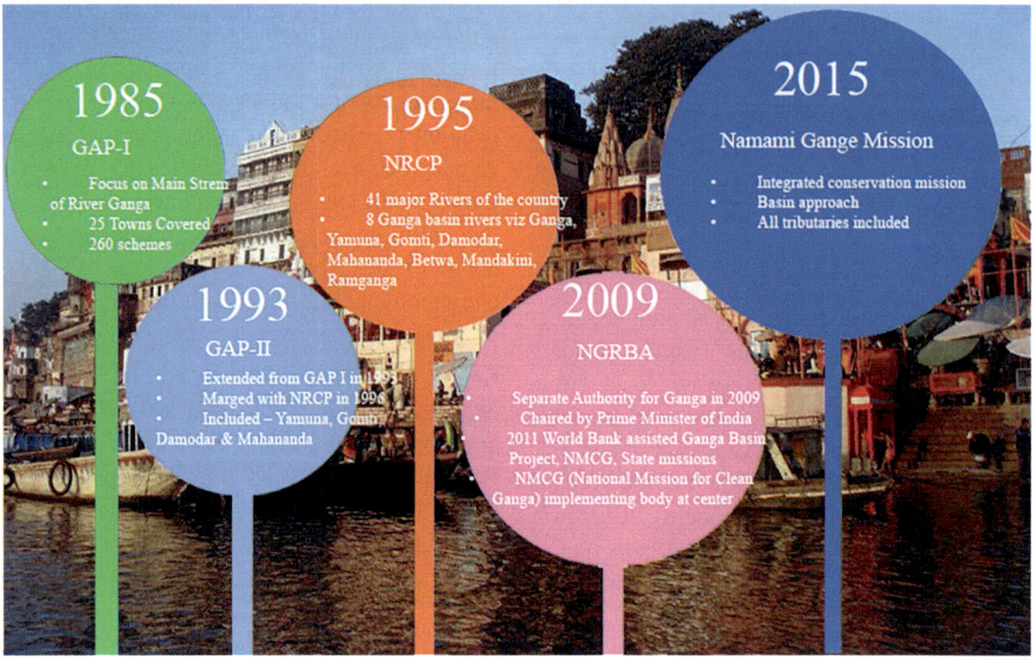

Fig. 5.4 Journey of Ganga River basin management towards IRBM approach (Note GAP stands for Ganga Action Plan, NRCP for National River Conservation Plan, NGRBA for National Ganga River Basin Authority) (Ministry of Jal Shakti 2020)

the GAP phase 2 included the Damodar and Mahananda Rivers. In 1995, a National River conservation plan was adopted considering the major tributaries of the main Ganga River. In 2009, the national Ganga basin authority was established which had a specific focus on the cleaning of the Ganga River. In 2015, a breakthrough came in the name of Namami Gange that embraced the nation of the IRBM approach where all tributaries were included (Fig. 5.4).

The government of India really strides to restore the Ganga River to its pristine conditions through the four major components (Fig. 5.5). As of now (31 July 2020), the number of projects sanctioned under the four major pillars is portrayed in Table 5.1. In West Bengal, a total of 22 projects are sanctioned of which only three are on River Bhagirathi (Jangipur, Berhampore, and Nabadwip) and only one at Nabadwip pertains to Nadia rivers (Fig. 5.6). The projects are taken for construction of sewage treatment plants, development of irrigation and waterway network, bank protection, channel diversion and sewage network development (Table 5.1). Other important insights of the projects are given in Tables 5.2, 5.3 and 5.4.

(2) **Sustainable agriculture** and afforestation for Ganga River rejuvenation

Sustainable agriculture paves way to lessening river pollution to maintain Ganga's Nirmalta and Aviralta. Namage Gange took stewardship towards developing sustainable agriculture and natural farming with the Ministry of Agriculture & Farmers' Welfare. For example, special programme was taken for organic farming on 5 km stretch along both sides of Ganga River. National Mission for Clean Ganga (NMCG) under the Namami Gange Programme is implementing Forestry Interventions in Ganga (Table 5.5). *Ganga Vriksharopan Abhiyan* for intensive plantation in collaboration with State programmes like (1) *Ganga Haritma Abhiyan* (Uttar Pradesh), (2) *Har Parisar Hara Parisar* (Bihar), (3) *Harela* (Uttarakhand), (4) *Van Mohtsav* (Jharkhand & West Bengal).

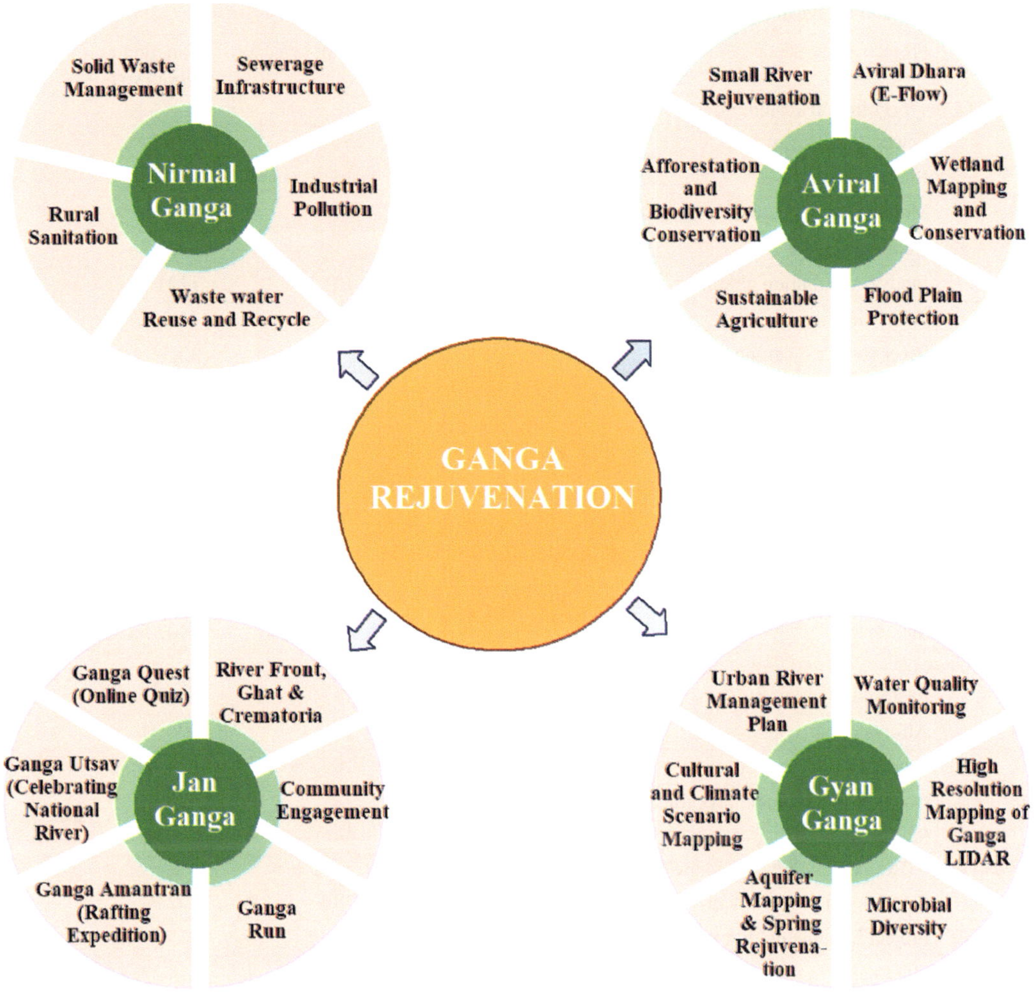

Fig. 5.5 Ganga rejuvenation through four major pillars—Aviral dhara (continuous flow), Nirmal dhara (unpolluted flow), Jan Ganga (community river interface) and Gyan Ganga (research on Ganga) (Ministry of Jal Shakti 2020)

(3) **Biodiversity conservation, wetland and cultural mapping for Ganga River Rejuvenation**

Biodiversity conservation and wetland mapping are instrumental to rejuvenating the Ganga River. There are several collaborating institute like Wildlife Institute of India (WII), Dehradun, ICAR-Central Inland Fisheries Research Institute (CIFRI), Kolkata and Centre for Environment Education (CEE) for biodiversity conservation. Moreover, Ganga Praharis are self-motivated individuals who are trained for ecological monitoring of Ganga river and its biodiversity, tree plantation techniques, awareness generation and community mobilization. Moreover, groundwater recharge through wetland and formation of State Wetland Authority are important instruments for wetland mapping and river restoration. LiDAR Mapping of ~45,000 km^2 area covering 5 states (Uttarkhand, Uttar Pradesh, Jharkhand, Bihar, and West Bengal) is another dimensions of the project. This projects also covered cultural mapping of the historical places, religious places and other intangibles heritage along the Ganga river.

Table 5.1 Namami Gange at a glance (till 31 July 2020)

Sl. No.	Projects undertaken	No of projects	Sanction cost (Rs in Cr)
Nirmal Ganga (Pollution abatement)			
1	Sewerage infrastructure	151	23,120.63
2	Bioremediation	13	242.23
3	Modular STPs decentralized treatment	1	410
4	Rural sanitation	1	1421.26
5	Industrial pollution abetment	14	1141.2
6	Ghats cleaning	3	45.54
7	River surface cleaning	1	33.53
	Sub total	184	26,414.39
Jan Ganga (People and river connect)			
8	Jan Ganga (People and River Connect)	1	336.73
9	Ghats & Crematoria (Old)	24	204.39
10	Ghats & Crematoria (New)	43	727.86
11	Ghats & Crematoria (CGF)	7	111.93
12	Ghats & Crematoria (CSR)	1	26.64
	Sub total	76	1407.55
Aviral Ganga (Ecology and flow)			
13	Restoration of kunds	1	18.48
14	Afforestation	21	306.31
15	Biodiversity conservation	8	150.54
16	Composite ecological task force Ganga Mitra	4	198.5
	Sub total	34	673.83
Gyan Ganga (Research policy and knowledge management)			
17	Institutional development	8	192.94
18	Project implementation support/research and study projects	13	165.39
	Sub total	21	358.33
	Grand total	315	28,854.11

Source Ministry of Jal Shakti (2020)

(4) **Appraisal of the Namage Gange**

The IRBM programme under Namage Gange entails the other components of river management from a holistic perspective. Sustainable agricultural practices with improved soil health and efficient use of water are recommended. Moreover, biodiversity conservation and wetland management are focussed in the scheme. Apart from the large-scale initiatives, some small-scale initiatives for the rivers are also undertaken such as bank protection works, afforestation, and embankment. In West Bengal, along the Bhagirathi River, such types of protection works (plantation of vetiver, use of geotextile for bank protection) are also taken by the other govt. and NGOs (Islam and Guchhait 2024b). Furthermore, waterbody mapping monitoring with seasonal dynamics of the rivers are measured with high-resolution LiDAR mapping using an integrated GIS system. Moreover, other tangible and intangible objects like math (temple), old trees, or heritages located along the Ganga

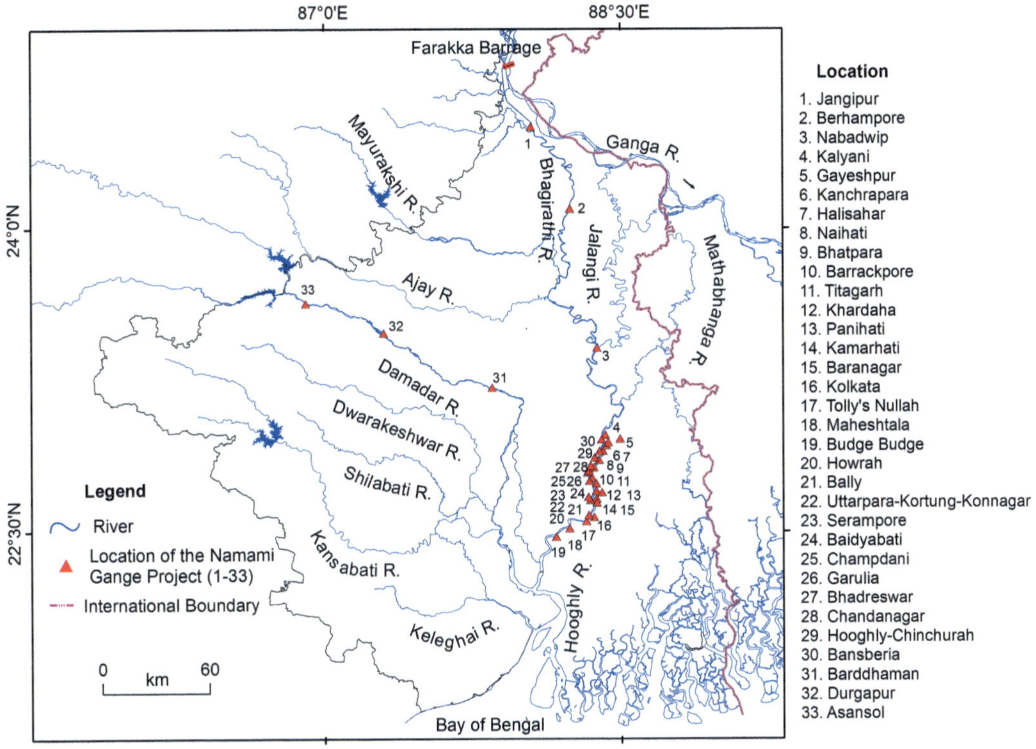

Fig. 5.6 Location of the Namami Gange projects in West Bengal (Based on Ministry of Jal Shakti 2020)

Table 5.2 Status of the Namami Gange completed projects in West Bengal

Sl. No.	Project name	Sanctioned cost (Rs. In Cr.)	STP Capacity (MLD)	Sewer network (km)
1	Sewerage scheme & STP works at Gayeshpur	168.67	8.23	61
2	Sewerage scheme & STP works at Kalyani	57.87	21	46.33
3	Sewerage scheme & STP works at Bhatpara	228.52	31	125
	Total	455.06	60.23	232.33

Source Ministry of Jal Shakti (2020)

River are aimed to be protected. This cultural mapping links the river restoration with culture and society which lie at the core of the IDBM. In West Bengal, beautification along the Ganga is done by the Govt. of West Bengal. On the one hand, it protects the river bank and creates leisure places for the city dwellers on the other. The entire programme of the Ganga River restoration is accomplished through the structured network (Fig. 5.7a) where various visions, and missions are mentioned with specific directions on how they are addressed through different implementing agencies. However, in several cases, the implementing agencies (state govt., central govt., corporate sectors, media, civil society) may differ in their viewpoints and hence implementations of the goals become challenging (Fig. 5.7b).

In the post-FBP period, the Bhagirathi-Hooghly River is monitored by the Kolkata

Table 5.3 Status of the Namami Gange projects in West Bengal in progress

Sl. No.	Project name	Sanctioned cost (Rs. In Cr.)	STP capacity (MLD)	Sewer network (km)
1	I&D works with STP at Jangipur	68.47	13	11.38
2	Interception, diversion & treatment works at Nabadwip	61.23	20	4
3	Interception, Diversion & treatment works at Behrampore	51.21	3.5	5.25
4	I&D and STP works at Kanchrapara, West Bengal	48.77	18	2.5
5	Sewerage system & STP works at Halishahr	274.76	16	226.99
6	Sewerage system with sewer network at	272.32	24	247.14
7	Rejuvenation of existing STPs & pumping station including sewerage network at North 24 Parganas Dist., West Bengal	65.54	46.24	
8	Interception, diversion & treatment works at Baranagar & Kamarahati (HAM)	172.1	60	9
9	Interception & diversion with STP at Bally (HAM)	164.93	40	
10	Interception, diversion & treatment works at Howrah (HAM)	185.22	65	
11	Integration of STPs with rehabilitation and O&M of existing assets	165.16	102	
12	Sewerage system & STP works at Budge-Budge	145.98	9.3	131.59
	Total	1675.69	417.04	637.85

Source Ministry of Jal Shakti (2020)

Port Trust. They regularly survey the Bhagirathi River (Berhampore office- Hydraulic study dept) while the Hooghly River is surveyed by the KopT hydraulic dept of the Khidurpur office. They take cross-sectional measurements, velocity, discharge, and sediment concentration, in the post-monsoon period every year and publish in the form of an annual report with hydrographic sheets. Moreover, IWAI, Govt of India also maintains a narrow (3 to 5 m) passage for the movement of the vessels safely (Islam and Guchhait 2024c). This organization along with CWC monitors the hydraulic parameters like velocity and discharge throughout the year. Thus, at present these three organizations take care of the Bhagirathi River from the perspective of the hydro-morphology of the river while the West Bengal pollution control board and central pollution control board measure the water quality parameters for river health assessment regularly.

5.4.2 Jalangi River Restoration

River Jalangi is almost a dead river with negligible flow throughout the year except for two months (August-September). Eastward titling of the Bengal Basin coupled with the construction of the FBP, LULC changes and other human interventions have led to the decaying of the river (Das et al. 2023). For the relatively small rivers like Jalangi and Churni, there is no

Table 5.4 Status of the Namami Gange projects in West Bengal under tendering

Sl. No.	Project name	Sanctioned cost (Rs. In Cr.)	STP capacity (MLD)	Sewer network (km)
1	Interception & diversion with STP at Hughly-Chinsurah (HAM)	160	29.3	20
2	Rejuvenation of exiting STPs along with lifting station & pumping station including sewerage network at Hooghly Dist., West Bengal	86.61	78.9	0
3	Interception, diversion & treatment works at Tolly's Nullah (Adi Ganga) (HAM)	307.12	26.1	114
4	Interception & diversion with STP at Maheshtala (HAM)	224.69	35	0
5	Interception and diversion (I&D) works with STP, for drains falling in river Damodar, at Durgapur, West Bengal	287.53	80	11.39
6	Detailed project report for I&D sewerage system & STP for Asansol & Kulti towns, West Bengal	384.96	95	4.5
7	I&D and STP works at Burdwan, West Bengal	234.31	50	19
	Total	1685.22	394.3	168.89

Source Ministry of Jal Shakti (2020)

Table 5.5 Afforestation in different states under the Namami Gange Programme

Sl. No.	State	Area planted so far (in ha.)	Fund sanctioned by NMCG from yr. 2016–20 (Rs. in Cr.)
1	Uttarakhand	8295	85.62
2	Uttar Pradesh	7531	102.85
3	Bihar	4836	73.86
4	Jharkhand	636	29.41
5	West Bengal	5466	45.46
	Total	26,764	337.2

Source Ministry of Jal Shakti (2020)

separate controlling body like KoPT. Rather these rivers are relatively less measured and monitored. Only the irrigation dept and a few sporadic attempts like the bridge construction authority (near Krishnagar) survey this river for their engineering works. These data indicate that the river is almost stable. For protecting this river there are two types of plans—(1) social side plan with community engagement and (2) Jalangi Action plan.

5.4.2.1 Community Initiatives and Topophilia

Save Jalangi West Bengal and other activist groups conducted various campaigns among the residents living along the banks of the Jalangi River with the aim of rekindling the bond between humans and the river. Their efforts focussed on promoting love and respect for the river, discouraging pollution by not disposing of solid waste into the water, preventing encroachment onto its banks, and ensuring that no obstacles hindered the free flow of the river (Fig. 5.8). Save Jalangi West Bengal and other activist groups also organized various events such as drawing competitions, poetry recitations, river walks, cycle rallies, and cultural programs centred around the theme of the river and its connection to humanity (Fig. 5.9a–b). Additionally,

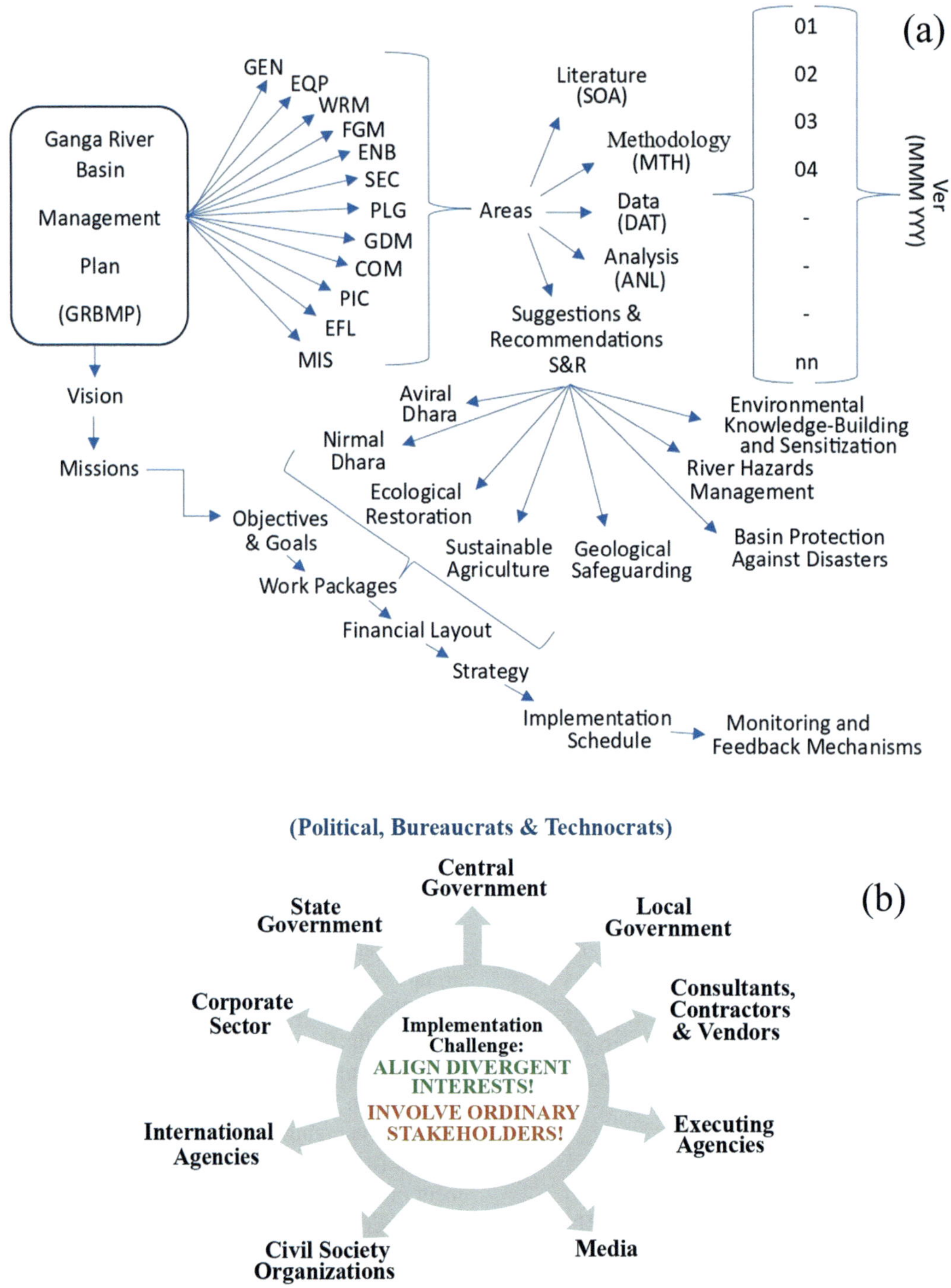

Fig. 5.7 **a** Ganga river basin management plan (GRBMP, 2015) work structure, **b** Implementing challenges in aligning the interests of key actors (Note: Abbreviation in Fig. 5.7a may be read as- NGRBA: National Ganga River Basin Authority, NMCG: National Mission for Clean Ganga, MoEF: Ministry of Environment and Forests, MHRD: Ministry of Human Resource and Development, MoWR, RD&GR: Ministry of Water Resources, River Development and Ganga Rejuvenation, GRBMP: Ganga River Basin Management Plan, IITC: IIT Consortium, PMB: Project Management Board, PICC: Project Implementation and Coordination Committee, EQP: Environmental Quality and Pollution, WRM: Water Resource and Management, ENB: Ecology and Biodiversity, FGM: Fluvial Geomorphology, EFL: Environmental Flows, SEC: Socio Economic and Cultural, PLG: Policy Law and Governance, GDM: Geospatial Database Management, and COM: Communication)

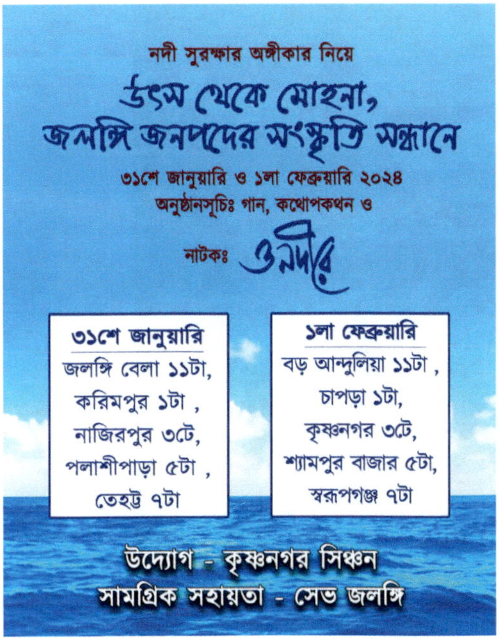

Fig. 5.8 Poster for drama shows in search of culture of Jalangi Civilization at ten different settlements all along the course of the river Jalangi. Drama 'O Nadire' (O River) was staged by 'Sinchan' group of Krishnagar in collaboration with 'Save Jalangi West Bengal' at Jalangi, Karimpur, Nazirpur, Palashipara and Tehatta on 31st January and at Bara Andulia, Chapra, Krishnagar, Shyampur Bazar, and Swarupganj on 1st February 2024

they staged river-man-related drama shows in several communities along the banks of the Jalangi River.

Save Jalangi West Bengal compelled the National Highway Authority of India to clear the accumulated earth and debris from the bridge site, which would have otherwise obstructed the natural flow of the river (Das et al. 2024). On 17 February 2024, an International seminar titled 'Revitalizing Decaying Rivers and Wetlands: Unlocking Potential Resources for Sustainable Future' was jointly organized by Krishnagar Women's College and 'Save Jalangi West Bengal'. This date holds significance as it marks the birth anniversary of poet Jibanananda Das and has been celebrated as Jalangi Diwas by Save Jalangi West Bengal for several years. The seminar, primarily orchestrated by Save Jalangi West Bengal, aimed to address the urgent need to save the Jalangi River from its accelerated deterioration caused by both natural phenomena and human activities. Distinguished speakers at the seminar were from India, the USA, and Bangladesh. The seminar underscored the natural decay process of deltaic rivers like the Jalangi while emphasizing the importance of preventing the acceleration of this deterioration. Instead, efforts should focus on implementing measures to prolong the river's lifespan for the benefit of humanity's continued existence.

5.4.2.2 Jalangi Action Plan

(a) **Action Plan 2020**

In 2020, the Nodal Agency, Municipal Engineering Directorate of the Department of Urban Development & Municipal Affairs, Government of West Bengal, submitted a proposal titled "Action Plan for Rejuvenation of River Jalangi, Krishnagar" to the Central Pollution Control Board (CPCB) in Delhi. This proposal was later approved by the West Bengal River Rejuvenation Committee (WBRRC) which was set up as per the directives of the Hon'ble National Green Tribunal. The main objective of this action plan was to rejuvenate Jalangi River through managing discharges from industrial, and municipal sources, solid e-waste and biomedical waste. This also focuses on groundwater management, environmental flow maintenance, rainwater harvesting, flood plain zone protection, riverbank plantation, irrigation practices improvement, and biodiversity park establishment.

The WBRRC forwarded the action plan to CPCB on 12 February 2020. Subsequently, the Task Team suggested revisions during its 10th Meeting on 26 February 2020, which were approved by the WBRRC in its 7th meeting on 09 June 2020. This revised plan was sent to CPCB on 09 June 2020, reviewed once again by CPCB in its 12th Task Team meeting on 11.06.2020, and further modified based on their recommendations. The finalized action plan, incorporating CPCB's suggestions, was approved by the WBRRC in its 8th meeting on 02 July 2020. The report detected a 5 km stretch of the river as "Polluted," which remains so

Fig. 5.9 **a** The environmental activist group Save Jalangi West Bengal was established with the primary aim of safeguarding the river Jalangi from pollution, degradation, and encroachment. **b** During the period of January 2nd to January 8th, 2024, the group organized a river walk to disseminate their message of conserving the Jalangi River

round the year, particularly around Krishnanagar town. The main pollutants in this stretch are primarily of a biological nature. Despite being perennial, water in this stretch is predominantly used for agricultural and fishing purposes. The proposal reported that among the industrial units listed in official reports as potential contributors to river pollution is 'The Kishan Co-operative Milk Producers' Union Ltd., located at Krishnanagar, within the river's catchment area. However, a closer geospatial and infrastructural examination reveals a significant irony in this attribution. The processing facility is situated >1.6 km away from the river, and is physically bounded by a national highway on one side and a railway line on the other. This creates a infrastructural barrier that effectively isolates the plant from the river, making any direct discharge of industrial effluent into the river highly improbable. Such spatial disconnect raises critical questions about the accuracy of attributing water pollution sources in the report. The proposed plan's estimated cost is Rs. 47.98 crore, with an additional Operation and Maintenance Expenditure of Rs. 25.16 Crore per year. The scheduled completion date for the work is set for 30 June 2021. However, except for setting some iron nets in large sewers to prevent large solid wastes (e.g. plastics) from entering the river, nothing significant was observed. Moreover, a small treatment plant at Krishnanagar Bus Stand was left incomplete and never became operational.

(b) **Action Plan 2022**

In 2022, due to repeated knocking by SAVE JALANGI West Bengal and other river activists group about the human interventions responsible for decaying of the river, Assistant Magistrate & Assistant collector, Nadia from Office of the District Magistrate, Nadia, Development & Planning Section (Vide Memo No: 297 (14)/Dev. Dated: 29.03.2022) called an Urgent Meeting on Jalangi Action Plan. Minutes of the meeting (Vide Memo No: 315 (14)/Dev. Dated: 04. 04. 2022) were as follows: District Magistrate Nadia presided over a meeting focussed on addressing various challenges concerning the Jalangi River. The Assistant Magistrate briefed attendees on the meeting's agenda, emphasizing the need to understand and tackle the river's issues through coordinated action plans.

Under the first agenda of increasing the perennial flow of the Jalangi River, stakeholders discussed the possibility of creating a pilot channel to ensure continuous water supply. Action points included conducting topographical surveys and exploring re-excavation of feeder canals. The second agenda addressed encroachments and obstructions along the river, including fishing nets, bamboo bridges, and illegal brickfields. Proposed actions involved removing encroachments and raising awareness among stakeholders, with a focus on coordination between local authorities and NGOs. Pollution and water quality were discussed under the third agenda, with concerns raised about sewage discharge and jute retting. Plans were made for sewage treatment initiatives and scientific retting methods, alongside measures to clean garbage disposal sites and promote sustainable practices. The fourth agenda centred on plantation activities along the Jalangi River bank to combat erosion and siltation. Stakeholders proposed identifying suitable areas for plantation and implementing activities through collaborative efforts. Aquatic biodiversity conservation and the proliferation of water hyacinth were addressed under the fifth and sixth agendas, respectively. Steps were proposed to share comprehensive reports on biodiversity studies and remove water hyacinth to be used for vermicompost pits. Awareness generation activities formed the seventh agenda, aiming to involve local communities in river protection efforts. The District Magistrate directed local authorities to initiate awareness programs in their respective subdivisions. In conclusion, the District Magistrate highlighted the importance of comprehensive short-term and long-term actions, emphasizing phased implementation. The meeting concluded with a vote of thanks, with plans for a review meeting in the future which was held on 23.03.2023.

(c) **Action Plan 2023**

The minutes of the meeting held are: Shri. Shashank Sethi, IAS, District Magistrate Nadia chaired the meeting, beginning with a review of the previous resolution. The discussion commenced with the first agenda focussed on increasing the flow of the Jalangi River. Bibhas Chandra Barman (Ph.D), the Director (Hydraulics) of the River Research Institute emphasized the need for a pilot channel connecting feeder river systems to ensure continuous water supply. Action points included awaiting the analysis of survey reports and coordinating with Murshidabad district. Encroachments and obstructions along the river were discussed next. Stakeholders highlighted issues such as fishing nets and illegal bamboo bridges. The Sub-Divisional Officer reported on the removal of illegal structures, with ongoing vigilance to prevent further encroachments. Pollution and water quality concerns were addressed, with a focus on sewage discharge and jute retting. Plans were made for sewage treatment initiatives and exploring alternatives to conventional jute retting methods. Additional factors contributing to pollution were discussed, with proposed solutions including promoting organic farming and exploring funding opportunities. Plantation activities along the riverbank were proposed to combat erosion and siltation. The District Magistrate instructed officials to identify suitable areas for plantation activities. The decline in aquatic biodiversity in the Jalangi River was noted, with concerns raised about the disappearance of native fish species. Action points included exploring fish ranching schemes and addressing groundwater depletion issues. Miscellaneous points were discussed, including groundwater depletion and the need for afforestation along stagnant stretches of the river. Action was directed towards investigating afforestation possibilities and fishing opportunities. (d) Action Plan 2024 This meeting, convened under the name 'District Ganga Committee' on 12th September 2024 (Memo No. 2688/NZP, Dated 06.09.2024), did not produce any discourse or resolutions significantly different from those adopted in 2022 and 2023, apart from some changes in wording. (e) Action Plan 2025 This meeting was also convened under the name 'District Ganga Committee' on 19th March 2025 (Memo: 1160/NZP Dt. 17.03.2025). Despite three earlier meetings, and observing no significant effort toward restoring Jalangi River's health, the 1st author of this book chose not to attend this 2025 meeting—unlike in the meetings held in 2022, 2023, and 2024, where he was present. The story of the Jalangi River's so-called revival efforts 'Jalangi Action Plan' illustrates a familiar pattern in river governance: annual meetings full of well-crafted resolutions, but minimal implementation on the ground. Despite repeated interventions from civil society groups such as SAVE JALANGI West Bengal, and the formal acknowledgment of the river's ecological degradation by district authorities, the institutional response has largely been procedural and symbolic. The minutes of meetings held in 2022 and 2023 provide a striking example of how administrative momentum often dissipates between planning and execution. Both meetings identified near-identical issues: diminishing perennial flow, obstruction due to encroachments, severe pollution from sewage and jute retting, erosion and siltation, and a steep decline in aquatic biodiversity. Action points were proposed, including survey work, pilot channel construction, plantation drives, removal of illegal structures, and awareness campaigns. However, little has moved beyond the drawing board. A critical finding was the failure to collect the topographical survey report from the Survey of India, which had been resolved in the 2023 meeting as a foundational step toward any hydrological intervention. This omission reflects either institutional negligence or a deeper lack of commitment, especially as the 2024 meeting proceeded without acknowledging this gap, instead suggesting a fresh survey—essentially restarting the process without reflecting on past shortcomings. This administrative loop—wherein plans are recycled, and progress is neither tracked nor enforced—has rendered the entire exercise performative. Review meetings function more as formalities than forums for accountability. Moreover, the absence of any monitoring mechanism or time-bound implementation roadmap has further weakened the potential for effective action. As such, while the language of the resolutions continues to evolve, the river's condition remains largely unchanged. The case of the Jalangi serves as a cautionary tale in river management. Unless there is a shift from periodic planning to sustained execution—backed by inter-departmental coordination, data transparency, and community involvement—such rivers will continue to deteriorate under the guise of bureaucratic concern. The repetition of annual meetings without introspection or consequence has, paradoxically, become an impediment to real progress.

5.4.3 Churni River Restoration

5.4.3.1 Artificial Channelization of the River Mathabhanga-Churni During the British Rule

Since the first quarter of the nineteenth century, Nadia rivers (Bhagirathi, Jalangi and Mathabhanga-Churni) were of main concern

Table 5.6 Chronology of anthropogenic signatures on Mathabhanga-Churni river during the nineteenth century

Year	Superintending engineer of Nadia River Division	River training works	Results	Source of information
1819	Mr. C. K. Robinson was appointed Superintendent and Collector Mathabhanga River and Nadia River Division of public works department was originated	An effort to close the offtake of River Kumar by 'bandals' (a branch of the river Mathabhanga was diverting 5/6th of the water to the south-east) was adopted	Unsuccessful attempt	Reaks (1919)
1820	Mr. May	Attention was given to clear snags	–	Reaks (1919)
1821	Mr. May	Cleared many snags including 300 timbers sunken, many wrecked boats and cut down leaning trees on banks likely fall into the river	Cleared	Garrett (1910); Reaks (1919)
1821	Mr. May	To close the offtake of the river Kumar which was still taking away 3/4th of the Mathabhanga supply	A cassion and some old boats were sunk across the offtake of the river Kumar Kumar offtake was closed	Reaks (1919)
1821	Mr. May	Straightening of the channel of the river Mathabhanga to flow water through the main channel	A cut of 1540 yards (1408.176 m) was made across a bend of the river Mathabhanga to attract the water. Mathabhanga-Churni was navigable throughout with 3feet draft	Reaks (1919)
1821	Mr. May	Closing of the offtake River Pangasi, the second branch of the river Mathabhanga takes off at 23°44′09″N and 88°52′02″E	Unsuccessful attempt	
1823	Mr. May	Dredging of Mathabhanga offtake	A dredging machine drawn by bullocks was employed but owing to deteriorated channel and severely closed offtake by sand deposit, the attempt was unsuccessful	Reaks (1919)
1824	Mr. May	Due to better condition of Jalangi and Bhagirathi, attention was given to those two rivers also and Mr. May was given additional charges	–	Reaks (1919)

(continued)

Table 5.6 (continued)

Year	Superintending engineer of Nadia River Division	River training works	Results	Source of information
1825–30	Mr. May	Two attempts to divert the water flowing away through the Pangasi	Attempts were unsuccessful	Reaks (1919)
1828–29	Mr. May	Maintaining minimum depth for navigation	A steam dredge was employed below Kumar entrance without much result due to unsuitable machinery and heavy draft (6feet) of the dredging boat	Reaks (1919)
1831–34	Mr. May	To have a satisfactory solution of Pangasi problem, a proposal was made to divert the current into a loop taking off the right bank of the Mathabhanga above the Pangasi offtake	–	Reaks (1919)
1840	Captain Smyth	–	–	Reaks (1919)
1852	Captain Smyth	Construction of a channel for 10 miles below the Pangasi by artificial banks with the use of kodalis	A passage was maintained for small boats	Reaks (1919)
	Captain Smyth			Reaks (1919)
1913	Mr. F.A.A.Cowley, Superintending Engineer, Central Circle and Deputy Secretary, Irrigation Department	Only Mathabhanga out of all the Nadia rivers is worthy to be kept artificially navigable because its offtake is just above the Sara bridge and not abnormally winding as is common for two other Nadia rivers	–	Reaks (1919)

Several cuts across the winding channel of the river to facilitate navigation through the straightened course were made during the first half of the nineteenth century. The navigation route of the river Mathabhanga-Churni was shortened by 10.46 km (12.3 km–1.84 km) by a single cut of 1.84 km just above Pangasi offtake during this period (Fig. 5.10)

Source Sarkar et al. (2020)

Fig. 5.10 Cut and straightening of tortuous path of the river Mathabhanga-Churni just above the offtake of the river Pangasi

to the British Raj and different artificial efforts were adopted to keep them navigable (Table 5.6). Profusely winding offtakes of these rivers and huge sand deposit closing offtakes made Bhagirathi and Jalangi less navigable in comparison to Mathabhanga-Churni (Sarkar et al. 2020). Efforts to keep Mathabhanga-Churni navigable are tabulated below.

5.4.3.2 People, River, and Placeness in a Riparian Context

The connection between individuals and their community with the Churni River, along with their active participation in protecting the environment, has been confirmed through empirical evidence. Their concern primarily revolves around anthropogenic pollution, particularly from Carew and Company Ltd. situated in Darshana, Bangladesh, which releases untreated wastewater into the Mathabhanga-Churni, causing severe pollution. This pollution has led to significant declines in dissolved oxygen levels and increased biochemical oxygen demand. Additionally, the industrial pollution has resulted in the disappearance of fish species in the upper reaches of the Churni River, leading to livelihood crises for local fishermen (Sarkar and Islam 2024).

As argued by Tuan (1974), the concept of a place can be understood in terms of its spatial attributes and human emotional attachment. This emotional connection, known as topophilia, is evident in various aspects such as livelihood, culture, and recreation associated with the Churni River. For generations, people have relied on the river for their livelihoods and have a deep emotional bond with it, as expressed by Ramesh Halder during interviews. Cultural activities like boating competitions and river-centric festivals further strengthen this bond, although the deteriorating condition of the river has led to fewer cultural events being organized (Sarkar et al. 2023). The community's environmental consciousness is reflected in spontaneous protests against pollution, with both local residents and environmental groups participating actively. These protests demonstrate the people's love for the river and their determination to protect it. The frequent involvement of local residents in these protests highlights the grassroots nature of the movement. Furthermore, the attitudes of the local people towards the Churni River align with Relph's theory of placeness, emphasizing the sense of security, enclosure, and comfort associated with a place (Sarkar et al. 2023). Despite facing challenges from industrial pollution, the community does not feel alienated from the river but instead exhibits a strong sense of place identity, emphasizing their depth of connection with the environment and their determination to address the issue.

5.5 Integrated Human-Harmony Model

River restoration is a complex process that entails the involvement of the various sectors of planning with different types of work. The river influences the human system through social, economic and scientific interface. Moreover, the river is intervened through various types of river training works and other basin-scale human interventions (Zhang et al. 2015). River on the one hand acts as a resource through the water system for agriculture and other uses; water disasters are also common parlance in the form of floods and droughts at various spatial and temporal scales (Fig. 5.11). Climate changes exacerbate the paces of the hydrological cycles and thereby increasing the frequency and magnitude of the

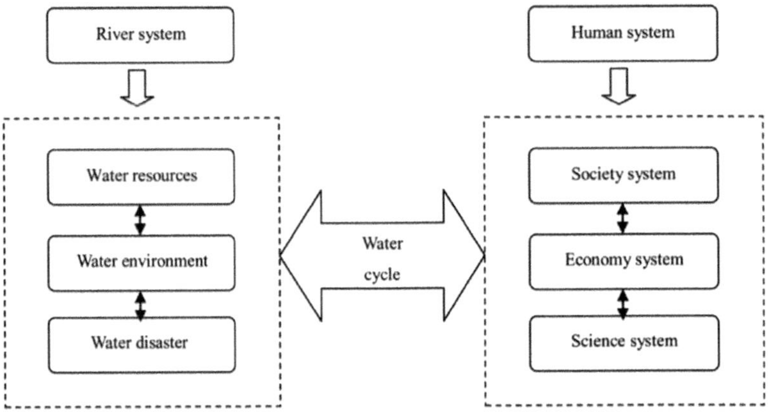

Fig. 5.11 Framework of the human-river system (Zhang et al. 2015)

hydrological extremes in recent times. Thus, developing a proper socio-hydrological system in a particular context is much required. The basic premise of this human-harmony model lies in the mutual coherence of the human and river systems. Controlling the rivers with modern tools is not desirable; rather a long-term understanding has to be developed with the fluvial system. In the context of the Bhagirathi-Hooghly River, the wider buffer zones would be created for only agriculture and not for settling purposes because of the higher chances of bank failure and settlement loss. The fluvial corridor or the living space of the river for the Jalangi and Churni would be lower and hence the buffer zones along the river for agriculture may be narrow and settlements can grow in much proximity to the river course. Thus, an extensive survey is required to adequately comprehend the river's behaviour before finally executing the plans as envisaged.

5.6 Feasibility of Integrated River Basin Management in the Context of Nadia Rivers

Due to the geopolitical issues associated with international boundaries, the feasibility problems for managing environmental flow (E-flow) across transboundary rivers are complicated. India and Bangladesh share the Ganges–Brahmaputra-Meghna basin, which includes the rivers of Bhagirathi-Hooghly, Jalangi, and Mathabhanga-Churni. International borders complicate E-flow management because they affect bilateral treaties, diplomatic connections, and competing national interests when it comes to choices about water distribution, sediment management, pollution control, and ecological sustainability. The international boundary disputes between Bangladesh and India significantly exacerbate the realistic difficulties in regulating environmental flow (E-flow) in the Jalangi and Mathabhanga-Churni rivers. Both rivers flow into Bangladesh after splitting off from the Ganges in the upstream regions, making them transboundary rivers where water management between the two countries needs to be coordinated. The natural flow capacity of these rivers is progressively declining in the absence of regular maintenance and dredging at the offtake stations. Due to the international border, India is unable to independently address the sedimentation issue since doing so could threaten downstream water supply and generate diplomatic issues. Bilateral agreements and cooperation required for the revival of the rivers.

5.7 Concluding Notes

The move towards integrated river basin management in the wake of the environmental damages and restoration of the environmental flow is quite obvious in the context of Nadia rivers. All three river basins exhibit this trend. The BRB is covered under the GAP and Namage Gange which proposes basin management from an integrated and single point treatment. No river under a basin cannot be treated separately. The BRB located in the lower Ganga basin has received attention from the perspective of maintaining continuous flow and unpolluted flow. The allocations for those goals are regularly monitored. Consequently, few improvements in the water quality and aquatic environment are observed. Other rivers like Jalangi and Churni are also getting attention from the various stakeholders including the local people's topophilia. The Jalangi action plans and initiatives from the NGOS and other groups for the restoration of these rivers are recently well known. However, more focussed and grounded efforts are required otherwise these small rivers will go into history. Moreover, the threats of climate change coupled with accelerated human interventions will also induce the processes of river decay. These contemporary issues are discussed in the next chapter with probable pathways of river restorations and way forward.

References

Abbass K, Qasim MZ, Song H, Murshed M, Mahmood H, Younis I (2022) A review of the global climate change impacts, adaptation, and sustainable mitigation measures. Environ Sci Pollut Res 29(28):42539–42559

Clausen T (2011) Scoping and planning of integrated water resources management (IWRM) and river basin management (RBM) component in Lao PDR and Thailand for the Mekong Water Dialogues, August 2011. IUCN, Gland, Switzerland

Das BC (2013) Changes and deterioration of the course of river Jalangi and its impact on the people living on its banks, Nadia, West Bengal. Ph. D. Thesis. The University of Calcutta

Das BC, Ghosh S, Islam A, Roy S (2020) An appraisal to anthropogeomorphology of the Bhagirathi-Hooghly river system: concepts, ideas and issues. In Anthropogeomorphology of Bhagirathi-Hooghly river system in India. CRC Press, pp 1–40

Das BC, Das S, Sarkar B (2023) Floods of Jalangi and Mathabhanga-Churni Rivers, Indo-Bangladesh. In Islam et al (2023). In Floods in the Ganga–Brahmaputra–Meghna Delta. Springer Nature. https://doi.org/10.1007/978-3-031-21086-0

Das BC, Islam A, Das S, Adhikary R, Sarkar R (2024) Bridge construction impacts the adjacent channel morphology and turbidity: a study on a dying distributary of River Padma. Earth Science, Systems and Society, 4, 10099. https://doi.org/10.3389/esss.2024.10099

Garrett JHE (1910) Bengal District Gazetteers Nadia 1910

GRBMP (2015) Main Plan Document. Consortium of 7 "Indian Institute of Technology"s (IITs). https://nmcg.nic.in/writereaddata/fileupload/13_GRBMP%20-%20MPD.pdf

Hirst FC (1916) Report on the Nadia Rivers, 1915. Bengal Secretariat Book Department

Islam A, Guchhait SK (2017) Search for social justice for the victims of erosion hazard along the banks of river Bhagirathi by hydraulic control: a case study of West Bengal, India. Environ Dev Sustain 19:433–459

Islam A, Guchhait SK (2024a) Context of riverbank erosion. Riverbank erosion in the Bengal Delta: an integrated perspective. Springer International Publishing, Cham, pp 23–40

Islam A, Guchhait SK (2024b) Coping strategies: towards a resilient society. Riverbank erosion in the Bengal Delta: an integrated perspective. Springer International Publishing, Cham, pp 287–316

Islam A, Guchhait SK (2024c) Riverbank erosion: a human-induced process. Riverbank erosion in the Bengal Delta: an integrated perspective. Springer International Publishing, Cham, pp 99–143

Lim CH, Wong HL, Elfithri R, Teo FY (2022) A review of stakeholder engagement in integrated river basin management. Water 14(19):2973

Marques ÉT, Gunkel G, Sobral MC (2019) Management of tropical river basins and reservoirs under water stress: experiences from Northeast Brazil. Environ MDPI 6

Ministry of Jal Shakti (2020) Namami Gange programme—at a Glance. Director general national mission for clean Ganga, Department of water resources, river development and Ganga Rejuvenation 1st Floor, Major Dhyan Chand National Stadium, India Gate, New Delhi 110002, India. Accessed 20, 25 April 2024. https://nmcg.nic.in/pdf/NGP-At%20a%20Glance%20(Final%20Version%20Printed).pdf

Parua PK (2009) Only a barrage can save!. The Ganga: water use in the Indian Subcontinent, pp 79–106

Pegram G, Yuanyuan L, Le Quesne T, Speed R, Jianqiang L, Fuxin S (2013) River basin planning principles: procedures and approaches for strategic basin planning. Asian Development Bank, GIWP, UNESCO, and WWF-UK

Phong T, Shaw R (2010) Chapter 12 river basin management for effective disaster risk reduction in the face of changing climate. In Climate change adaptation and disaster risk reduction: issues and challenges. Emerald Group Publishing Limited, pp 265–289

Reaks HG (1919) Report on the Physical and Hydraulic Characteristics of the Rivers of The Delta. In (Moor et al. 1919). Report on the Hooghly River and Its Head-Waters. Vol-I. The Bengal Secretariat Book Depot, Calcutta, 1919, In Biswas (2001), Rivers Of Bengal, Vol-II, Govt Of West Bengal, P- 87,107

Sarkar B, Islam A (2024) Assessing environmental awareness activism to river decay and recurring pollution using parallel multiple mediator model. Int J River Basin Manage 1–13

Sarkar B, Islam A, Datta D (2023) Characterising topophilic behaviour in the wake of river decay and pollution through structural equation modelling. Environ Dev Sustain 25(12):15043–15074

Sarkar B, Islam A, Das BC (2021) Role of declining discharge and water pollution on habitat suitability of fish community in the Mathabhanga-Churni River, India. J Cleaner Prod 326:129426

Sarkar B, Islam A, Das BC (2020) Anthropo-footprints on Churni river: a river of stolen water. In Anthropogeomorphology of Bhagirathi-Hooghly river system in India. CRC Press, pp 433–468

Sarker MH, Akter J, Rahman MM (2013) Century-scale dynamics of the Bengal delta and future development. In Proceedings of the international conference on water and flood management, pp 91–104

Schumm SA (1977) The fluvial system. Wiley

Shit PK, Bera B, Islam A, Ghosh S, Bhunia GS (2022) Introduction to drainage basin dynamics: morphology, landscape and modelling. Drainage basin dynamics: an introduction to morphology, landscape and modelling, pp 1–9

Tuan YF (1974) Topophilia: a study of environmental perception, attitudes, and values. Columbia University Press

Zhang J, Tang D, Ahmad I, Wang M (2015) River–human harmony model to evaluate the relationship between humans and water in river basin. Curr Sci 1130–1139

Future Challenges and Way Forward

6

Abstract

The transformation of the Nadia rivers under the natural processes coupled with human intentions is well demonstrated in earlier chapters, particularly in Chaps. 2, 3, 4, and 5. This chapter focuses on future challenges such as climate change, sea-level rise, and socio-cultural changes. Under the increasing threats of climate change and sea-level rise, river degradations tend to escalate which requires special attention from the various stakeholders. Thus, future challenges and future directions of work are indicated in this chapter.

Keywords

Climate change · Sea-level rise · Social vulnerability · Population explosion · Compound risk · Nadia River division

6.1 Future Challenges

Climate change-induced channel modifications are quite common in the literature. The projections indicate more warming of the local climatic conditions. The warming of the temperature intends to accelerate river water temperature causing enhanced decomposition and poor water quality. On the other hand, the extremities in climate tend to exacerbate the frequency and magnitude of floods and drought at the basin scales. Intergovernmental panel on climate change (IPCC) also warns that by 2050 most rivers may lose their environmental flow thereby restricting static life and endangering the existence of human civilizations. Already the effects of global warming are being felt. Rivers are getting dried up. The Ministry of Earth Sciences released its inaugural climate change assessment for the Indian region in 2020, highlighting a significant 6% decline in monsoon rainfall from 1951 to 2015. Notable reductions were observed over the Indo-Gangetic plains and the Western Ghats. Concurrently, data on the discharge and sediment load of the river Ganga during monsoon months indicated declining trends at most gauging sites (Zakwan and Ahmad 2021). Further analysis revealed a decreasing gradient in water availability across the Gangetic plain, from west to east (Deltares 2018). Currently, the river Jalangi relies predominantly on rainfall, with rare contributions from the river Padma (Das et al. 2024). However, with a diminishing flow in the Ganga, it is projected that Jalangi may cease to receive supply from Padma shortly. The escalating trend of declining rainfall exacerbates the water scarcity issues faced by the river, potentially preventing it from maintaining its minimum environmental flow. Consequently, sections of the river, particularly the upper 51 km, face the risk of desiccation or

B. C. Das et al., *The Nadia Rivers*, Geography of the Physical Environment,
https://doi.org/10.1007/978-3-031-85971-7_6

severe degradation. The proliferation of algae, hyacinth, and other aquatic plants is expected to extensively cover the river, even downstream to Krishnanagar, forming intermittent pools. This ecological imbalance could lead to a depletion of fish stocks, endangering the livelihoods of fishermen and ultimately resulting in the abandonment of their occupation. Sea-level rise compounds the climate-induced risk (Siegert and Pearson 2021). The lower part of the basins especially Bhagirathi-Hooghly will feel the impact of sea-level rise on the morphological hydraulics and ecosystem services. Moreover, in West Bengal, a greater sea-level rise is anticipated due to a greater rate of land subsidence by 1.5 mm/year (Becker et al. 2020). Salinity intrusion will be rising causing threats to the freshwater ecosystems and human civilizations.

It is also found that population projections indicate the explosion of the population especially in developing nations. The Nadia rivers' basins are densely populated. For example, Nadia District has a higher population density (1316 person km^2 as per the 2011 census). The escalations in the populations have corresponding effects on the landscape alternations in the form of LULC changes water diversion, elevated pollution levels, etc. Leveraging the field data coupled with the secondary reports indicates a nuanced control of the basin topography and channel hydrology and morphology. The Jalangi River is almost dried up in the upper and middle parts and a similar condition for the Churni. This trend is expected to exacerbate the killing of the rivers and their associated ecosystems. As the Bhagirathi is under the Indo-Bangladesh water-sharing treaty, the complexity of water sharing might increase in the face of the lower availability of flow in the Farakka Barrage Project (FBP) reservoir (Islam and Guchhait 2017). Thus, poor water quality due to escalated human interventions at the Bhagirathi River Basin may be another future challenge (Islam et al. 2020). The issues of riverbank erosion in Murshidabad and Nadia districts and channel bar formation in the lower part of the river near Nabadwip and further below due to the transport limited condition of the river might further aggravate (Guchhait et al. 2016; Islam and Guchhait 2020). Moreover, the Churni River receives industrial wastewater from the Darshana Sugar industry in Bangladesh (Sarkar et al. 2024). As this river is transboundary, the problem of wastewater discharge may prevail escalating the pollution level of the river (Sarkar et al. 2020). Jalangi River is already disconnected from the source and the upper part of the riverbed is used for agricultural practice (Das and Bhattacharya 2020). The intensive use of land including the use of river beds for agriculture or settlement will increase because people are distancing themselves from the river and relying on the built environment.

6.2 Way Forward

Given this context, future research endeavors should prioritize integrated basin management strategies to optimize the utilization of available water resources. At the channel or reach scale, action-research initiatives are recommended, fostering community engagement in activities such as fish farming, agriculture, aquaculture, duck farming, water tourism, and water entertainment. These proactive measures hold the potential to lengthen the life of dying rivers, ensuring the sustainability of water resources intertwined with the socio-economic and cultural fabric along its banks.

The study includes future research directions and identifies critical areas that require more investigation to completely address the Churni River's degradation. Considering that agriculture is a significant source of revenue along the river, it is important to comprehend how water pollution impacts crop quality and productivity. To ensure food security for Churni River-dependent communities and preserve sustainable agricultural production, it is critical to comprehend the effects of water pollution on crops and water resource management. Effective river management also requires a thorough examination of environmental flow and sustainable river behavior. Studying sediment transport dynamics, flow patterns, and ecological reactions to flow changes are potential subjects for this research. These kinds of realizations are essential to preserving the natural order and enabling river-dependent human endeavors like irrigation and fishing. Studying the effects of river deterioration on groundwater quality is essential due to the relationship between surface water and groundwater. This study may involve tracking the quality of groundwater along the river and figuring out how contaminants in surface water impact groundwater. It is crucial to comprehend these relationships to manage water resources sustainably and to safeguard human populations and ecosystems that depend on groundwater. Finally, investigating integrated river basin management techniques and river restoration is essential to restoring the Churni River's health. In addition to physical restoration, this entails tackling the socio-economic issues that contribute to river deterioration. It is possible to design more effective and sustainable solutions for restoration projects that will ensure the long-term health of the Churni River and the well-being of the people who depend on it by embracing physical-social drivers, such as land use practices and community participation.

The Mathabhanga-Churni River is a transboundary river system crossing in the India-Bangladesh border and offering substantial socio-economic and environmental opportunities for both nations. However, the river is facing some challenges declining discharge coupled with sedimentation and water pollution that hinder the functions of the aquatic ecosystem. The joint venture to safeguard and restore the Mathabhanga-Churni River will be beneficial for both countries. Therefore, the prospects may be addressed in the following ways. Collaboration regarding data collection and monitoring the entire river system is crucial which means this transboundary river could immensely benefit from designing a unified monitoring framework for capturing river water quality and health of the ecosystem. There is an urgent need for future collaborations to ensure the restoration of natural river functions along the Jalangi and Mathabhanga-Churni rivers, the adoption and implementation of different measures like native vegetation plantation, riverbanks stabilization with vetiver grass, and the creation of buffer in the riparian zones for permitting selective anthropogenic intervention within the defined threshold. Additionally, the rivers can be effectively restored by implementing large-scale restoration projects with financial support from international collaboration. Finally, for long-term success, the participation of the local community is essentially required.

The present piece of work may grow interest among future stakeholders and researchers for exploring river degradation modeling based on the consideration of river basin dynamics from the perspectives of process approach, form approach, and management and futuristic approach involving climate change, sea-level rise, and socio-economic changes.

References

Becker M, Papa F, Karpytchev M, Delebecque C, Krien Y, Khan JU et al (2020) Water level changes, subsidence, and sea level rise in the Ganges–Brahmaputra–Meghna delta. Proceedings of the National Academy of Sciences

Das BC, Bhattacharya S (2020) The Jalangi: a story of killing of a dying river. In: Anthropogeomorphology of Bhagirathi-Hooghly river system in India. CRC Press, pp 381–431

Das BC, Islam A, Das S, Adhikary R, Sarkar R (2024) Bridge construction impacts the adjacent channel morphology and turbidity: a study on a dying distributary of River Padma. Earth Sci Syst Soc 4(1), 10099.

Deltares (2018) Strategic basin planning for Ganga river basin in India. Ganga River Basin model and WIS report and documentation. https://cwc.gov.in/sites/default/files/ganga-river-basin-model-and-wis-report-and-documentation.pdf. Accessed 30 April 2024

Guchhait SK, Islam A, Ghosh S, Das BC, Maji NK (2016) Role of hydrological regime and floodplain sediments in channel instability of the Bhagirathi river, Ganga-Brahmaputra Delta, India. Phys Geogr 37(6):476–510

Islam A, Guchhait SK (2017) Search for social justice for the victims of erosion hazard along the banks of river Bhagirathi by hydraulic control: a case study of West Bengal, India. Environ Dev Sustain 19:433–459

Islam A, Guchhait SK (2020) Characterizing cross-sectional morphology and channel inefficiency of lower Bhagirathi River, India, in post-Farakka barrage condition. Nat Hazards 103(3):3803–3836

Islam A, Guchhait SK, Hoque MM (2020) Anthropogeomorphology of the Bhagirathi river. In: Anthropogeomorphology of Bhagirathi-Hooghly river system in India. CRC Press, pp 41–82

Sarkar B, Islam A, Das BC (2020) Anthropo-footprints on Churni river: a river of stolen water. In: Anthropogeomorphology of Bhagirathi-Hooghly river system in India. CRC Press, pp 433–468

Sarkar B, Das BC, Islam A, Datta D, Pawlik Ł, Quesada-Román A (2024) Temporal change in channel form and hydraulic behaviour of a tropical river due to natural forcing and anthropogenic interventions. Phys Geogr 45(5), 483–517

Siegert M, Pearson P (2021) Reducing uncertainty in 21st century sea-level predictions and beyond. Front Environ Sci 9:751978

Zakwan M, Ahmad Z (2021) Trend analysis of hydrological parameters of Ganga River. Arab J Geosci 14:163

Index

B. Das et al., *The Nadia Rivers*, Geography of the Physical Environment, https://doi.org/10.1007/978-3-031-85971-7

W

GPSR Compliance

The European Union's (EU) General Product Safety Regulation (GPSR) is a set of rules that requires consumer products to be safe and our obligations to ensure this.

If you have any concerns about our products, you can contact us on ProductSafety@springernature.com

In case Publisher is established outside the EU, the EU authorized representative is:

Springer Nature Customer Service Center GmbH
Europaplatz 3
69115 Heidelberg, Germany

Batch number: 10755764

Printed by Printforce, the Netherlands